Lecture Notes in Artificial Intelligence   2298

Subseries of Lecture Notes in Computer Science
Edited by J. G. Carbonell, and J. Siekmann

Lecture Notes in Computer Science
Edited by G. Goos, J. Hartmanis, and J. van Leeuwen

Springer
Berlin
Heidelberg
New York
Barcelona
Hong Kong
London
Milan
Paris
Tokyo

Ipke Wachsmuth   Timo Sowa (Eds.)

Gesture and Sign Language in Human-Computer Interaction

International Gesture Workshop, GW 2001
London, UK, April 18-20, 2001
Revised Papers

 Springer

Series Editors

Jaime G. Carbonell,Carnegie Mellon University, Pittsburgh, PA, USA
Jörg Siekmann, University of Saarland, Saarbrücken, Germany

Volume Editors

Ipke Wachsmuth
Timo Sowa
Universität Bielefeld, Technische Fakultät
33594 Bielefeld, Germany
E-mail: {ipke,tsowa}@techfak.uni-Bielefeld.de

Cataloging-in-Publication Data applied for

Die Deutsche Bibliothek - CIP-Einheitsaufnahme

Gesture and sign language in human computer interaction : revised papers /
International Gesture Workshop, GW 2001, London, UK, April 18 - 20, 2001.
Ipke Wachsmuth ; Timo Sowa (ed.). - Berlin ; Heidelberg ; New York ;
Barcelona ; Hong Kong ; London ; Milan ; Paris ; Tokyo : Springer, 2002
   (Lecture notes in computer science ; Vol. 2298 : Lectures notes in
   artificial intelligence)
   ISBN 3-540-43678-2

CR Subject Classification (1998): I.2, I.3.7, I.5, I.4, H.5.2
ISSN 0302-9743
ISBN 3-540-43678-2 Springer-Verlag Berlin Heidelberg New York

Springer-Verlag Berlin Heidelberg New York
a member of BertelsmannSpringer Science+Business Media GmbH

http://www.springer.de

© Springer-Verlag Berlin Heidelberg 2002
Printed in Germany

Typesetting: Camera-ready by author, data conversion by DA-TeX Gerd Blumenstein
Printed on acid-free paper   SPIN 10846440   06/3142   5 4 3 2 1 0

Preface

The international gesture workshops are interdisciplinary events for those researching gesture-based communication and wishing to meet and exchange ideas across disciplines. A focus of these events is a shared interest in using gesture and sign language in human-computer interaction. Since 1996 international gesture workshops have been held roughly every second year, with fully reviewed post-proceedings typically published by Springer-Verlag.

Held at City University, London, April 18 - 20, 2001, and attended by more than 70 participants from all over the world, Gesture Workshop 2001 was the fourth in a highly successful workshop series. It was organized in association with the British HCI Group, a specialist group of the British Computer Society. As its predecessors in 1999 (Gif-sur-Yvette), 1997 (Bielefeld), and 1996 (York), GW 2001 aimed to encourage multidisciplinary collaboration by providing a platform for participants to share, discuss, and criticize both research in progress and more complete research with a multidisciplinary audience.

Consistent with the steady growth of research activity in this area, a large number of high-quality submissions were received, which made GW 2001 an exciting and important event for anyone interested in gesture-related and technological research relevant to human-computer interaction. In line with the practice of previous gesture workshops, presenters were invited to submit their papers for publication in a subsequent peer-reviewed publication of high quality. The present book is the outcome of this effort. Representing the research work from 10 countries, it contains a selection of 25 articles, and 8 short papers reflecting work in progress.

An invited contribution by Annelies Braffort (LIMSI/CNRS France) is concerned with ethical aspects of research on computer science and sign language and posits that linguists specialized in sign language as well as deaf people should be involved when developing tools that are meant to be helpful to the deaf. Six sections of reviewed papers pertain to the following themes:

- *Gesture Recognition*
- *Recognition of Sign Language*
- *Gesture and Sign Language Synthesis*
- *Nature and Notation of Sign Language*
- *Gestural Action and Interaction*
- *Applications Based on Gesture Control*

The work presented in these papers encompasses a multitude of research areas from among: automatic recognition, interpretation, and synthesis of gestures and sign language; computational representation of gestures and sign language; user issues and application paradigms; gesture and human-movement tracking techniques; gesture and sign language perception and production; modeling, analysis, and coding techniques; coverbal gestures; temporal relations between gestures and other modalities; system architectures; applications and demonstrator systems.

The present volume is edited by Ipke Wachsmuth and Timo Sowa, University of Bielefeld, who agreed to take over editorship when the organizers and co-chairs of GW 2001, Marilyn Panayi and David Roy (Zeric Ltd, London), could not continue with their editorial roles due to unforeseen circumstances. On behalf of all the participants of GW 2001, we would like to express our deepest thanks to Marilyn and David who worked tirelessly to make Gesture Workshop 2001 a successful event with a program of a very high standard.

Thanks also to the local committee, Bencie Woll, Chair of Sign Language and Deaf Studies at City University, and her colleagues Martha Tyrone and Klimis Antzakas, for hosting the event and contributing to a well-attended and lively meeting.

Last but not least, the editors are grateful to the authors of the papers, as well as to the international reviewers. As a result of their hard work this volume will serve as a timely and comprehensive reference for researchers and practitioners in all the related disciplines.

A final note. In connection with the preparation of this volume, a Gesture Workshops website was created at `http://www.TechFak.Uni-Bielefeld.DE/ags/wbski/GW/` as a permanent reference and portal for those interested in the workshop series and their proceedings. It contains links to all previous (and prospectively future) gesture workshops and related publications, and it will hopefully prove helpful to those wishing to join, or further interact with, an interdisciplinary and international gesture community.

February 2002   Ipke Wachsmuth, Timo Sowa

Reviewers

Table of Contents

Gesture and Sign Language Synthesis

Nature and Notation of Sign Language

Gestural Action & Interaction

Applications Based on Gesture Control

Research on Computer Science and Sign Language: Ethical Aspects

Annelies Braffort

LIMSI/CNRS, Bat 508, BP 133, F-91 403 Orsay cedex, France
Annelies.Braffort@limsi.fr

Abstract. The aim of this paper is to raise the ethical problems which appear when hearing computer scientists work on the Sign Languages (SL) used by the deaf communities, specially in the field of Sign Language recognition. On one hand, the problematic history of institutionalised SL must be known. On the other hand, the linguistic properties of SL must be learned by computer scientists before trying to design systems with the aim to automatically translate SL into oral or written language or *vice-versa*. The way oral language and SL function is so different that it seems impossible to work on that topic without a close collaboration with linguists specialised in SL and deaf people.

1 Introduction

Sign languages (SL) are the primary mode of communication for many deaf people in many countries. SL have a history which induces some ethical problems when hearing researchers want to carry out studies related to these languages.

In particular, when computer scientists want to design recognition or synthesis systems, they must be aware of what SL is and what part of SL they are dealing with. Actually, SL and oral languages are completely different and classical systems used for natural oral (or written) language processing cannot be used for SL processing.

First, SL are essentially based on spatial properties, combined with properties of iconicity. Most of the systems dedicated to automatic translation of SL to spoken language, and *vice-versa*, do not take into account the spatial component and this iconicity. Yet, linguistic studies on iconicity show that this kind of modelling is essential when dealing with "real" (natural) SL [1,2,3].

2 Historical and Present Context

A summary of the historical context of French Sign language can be found in [4]. Before the 17th century, deaf people were not considered to be able to testify by themselves. "Someone who doesn't speak can't reason". During the 17th century, the Church began to consider that deaf education was possible. During the 18th century, a priest, l'Abbé de l'Épée, began to study the French Sign language (FSL). He said that

I. Wachsmuth and T. Sowa (Eds.): GW 2001, LNAI 2298, pp. 1-8, 2002.
© Springer-Verlag Berlin Heidelberg 2002

"gestures express human thought, like an oral language." He invented methodical signs, in order to teach written French to deaf children. This was a mixture of natural signs used by the children and artificial signs created by him, in order to copy literally an artificial gestural grammar from the French grammar (articles, gender, tense, etc.).

During the 18th century, the French National Assembly promoted a law allowing deaf people to obtain the Human Rights. During this period, a real bilingual education was proposed to deaf children. Former deaf students became teachers for deaf children, and schools for the deaf were growing in number.

In the middle of the 19th century, there was a significant growth in Deaf culture, numerous associations were created, and the first studies in linguistics appeared. But there were constant quarrels between those that adhered to an *oralist* education (no SL at all) and those that adhered to a *bilinguist* education. These arguments culminated in 1880, at a conference in Milan (Italia) where FSL was completely forbidden in the education of deaf children.

During the 20th century, some slow evolution appeared: in 1968, the access to expression was given to linguistic minorities. In 1991, the ban was suppressed. In 1995, an ethical committee pronounced in favour of the teaching of FSL when cochlear implants are implanted on young children.

Nowadays, the education of deaf children remains mainly oralist, and the level of educational success is low. FSL is not officially considered as a language, but only as a method of re-education: In France, many educational institutes for deaf children are under the Ministry of Health rather than the Ministry of Education !

Thus, in France, there are often problems of mutual acceptance between the hearing and deaf communities. For example, there are often deaf people who do not agree with the fact that research on SL is carried out by hearing researchers.

In this context, hearing researchers should ask themselves these questions:

- Is there really an *interest* in my research for the deaf community ?
- Is what I call Sign Language in my papers *really* Sign Language ?

3   French Sign Language Functioning: Main Principles

The principle characteristic of FSL, and of SL in general, is the use of the gestural-visual channel. Thus, there is a compromise between articulation, visual interpretation and comprehensibility. Signs are articulated with a constant purpose of a economising in articulation, while retaining a meaningful message which can be easily interpreted by another person.

Thus the meaning is conveyed by means of hand parameters (generally named hand-shape, movement, orientation and location) but also gaze (direction, closing), facial expression, mouth movements, and chest movements. These parameters occur simultaneously and are articulated in space. Each of these parameters possesses a syntactical function and a semantic value, allowing whole parts of a sentence in oral language to be expressed with a single sign.

An interesting point relates to the structured use of space to express semantic relations. The space located in front of the signer is called *signing space*. The entities of the discourse (persons, objects, events…) are most of the time located in this signing space. Then spatio-temporal syntactic rules and semantic representations are used to build the sentence. For example, spatial relations of localization between entities are generally established without using specific signs. It is the sign order and the representation in space which allows interpretation.

Fig. 1. Example of FSL sentence: "The cat is inside the car"

Fig. 1 shows the FSL sentence "The cat is inside the car". In this example, the utterance is expressed with four signs [5]:

- the standard sign [CAR] ;
- the ["C" classifier], performed by the dominant hand. It allows the car to be located in the signing space. The hand shape represents the frontier of the car. During this time, the non-dominant hand still represents the "car" ;
- the standard sign [CAT], performed by the non-dominant hand, while the car location is maintained by the dominant hand ;
- the ["C" + "X" classifiers]: the cat location is given by the non-dominant hand ("X" classifier, representing the cat's legs), related to the car location given by the dominant hand ("C" classifier).

The sign order is less important than their arrangement in space. But we can draw out some general principles. The larger and the more static objects are signed before the little or moving ones. The general rule is that one signs from the more general, the context, to the more detailed, the action.

The more important property of SL, not often studied by researchers, is the *level of iconicity* [1,2]. It can be observed in "real" natural SL that the standard signs, which are listed in the dictionaries, are not the only signs used. Because the language is highly contextual, the sentence structure is dictated by the semantic level, and a lot of signs are built dynamically, depending on the structure of the context, which is represented by means of the signing space.

In conclusion, the production of the sentence and of the signs is highly driven by the semantic level. The SL functioning is essentially a "top-down" one.

4   Computer Science Application Dedicated to Sign Languages

The declared objectives of studies on processing systems dedicated to sign language are to design automatic recognition, or synthesis, systems of SL sentences. This supposes that the aim is to translate an oral (or written) sentence into a sign language sentence, or *vice-versa*.

Some projects presented during the London Gesture Workshop, like the European project VISICAST [6], show that within delimited sets of vocabulary and sentence structures, some realistic animation in SL can now be achieved, by using motion capture and sentence generation. Some indispensable parameters must be added for a better understanding by deaf people, like gaze, and perhaps a higher level of iconicity. But the current results are quite promising. The VISICAST project brings together partners from various fields, including computer science and linguistics, and also deaf users.

In the field of automatic gesture recognition, unfortunately, projects involving linguists, deaf people, and computer scientists altogether are quite rare. Thus, most of the studies that claim to be dedicated to SL are actually dedicated to what computer scientists think SL is: often only gestures with the hands, performing standard signs, with a linear grammar such as in oral languages.

The following sections propose an overview of the approaches generally encountered and the limitations they imply. For automatic recognition systems, the different steps generally performed are the *capture* of gestures, their *modelling*, their *recognition*, and finally the *translation*.

4.1   Capture

The capture of gestures constitutes an important area of research. At present, the numerous systems developed in this area use one or two colour video cameras to track the signer's movements. The problem consists of choosing relevant visual data to represent gestural data. Because gestures are articulated in 3D space, it is not easy to design an accurate representation from 2D images. Cameras allow us to track the whole body, including the face, but generally, only hand movements are studied.

The use of data gloves can help to manage 3D information on hand-shape, and the use of 3D trackers can help to deal with 3D information on location and orientation. But this information is limited to the hand.

Most of the time, only the hand parameters are considered. But in SL, gaze, facial expressions, and chest movements contribute to the sign definition and the sentence meaning. Only capturing hand gestures cannot be hoped to be sufficient if the aim is really to design a translation system.

4.2   Modelling

Modelling of gestures is designed in order to build recognition systems. The work consists in choosing a set of primitives, which are extracted from the data collected during the capture, and to use them as input during the recognition process.

At present, the modelling of gestures is most of the time limited to hand gestures, and all the signs are represented by the same primitives.

But in SL, from a modelling point of view, two kinds of signs, at least, must be differentiated [7]:

- The signs from the standard lexicon, for which only inter- and intra-signer variability can occur;
- All the other signs, whose production depends on the context, and for which at least one parameter can differ from one production to another. These signs can be viewed as parameterised gestures.

The standard signs can be represented by primitives dedicated to the four hand parameters (hand-shape, movement, orientation and location). But the non-standard signs can be represented only by primitives present in all the possible productions.

For example, a particular kind of sign, defined as "classifier", is used to refer to an entity previously signed in the sentence. The classifier can be used to locate the entity in the signing space (like the second sign in Fig.1), to show a displacement of this entity, to express a spatial relation in regard to another entity (like the fourth sign in Fig.1), etc. Thus, the movement, the orientation and the location of a classifier depend entirely on the context. The normal hand parameters cannot be used to represent a classifier. Only the hand-shape can be used to represent this kind of sign.

When dealing with "real" sentences of SL, an important part of the work consists of evaluating, for each type of non standard sign, the primitives which could be used in a recognition system. This work cannot be done without the participation of linguists specialised in SL, and even more, in the iconic side of these languages.

4.3   Recognition

Recognition consists in transforming continuous numerical data into a sequence of discrete symbols. These tools must classify and segment the data representing the SL sentence.

Most of the systems are based on Hidden Markov Models (HMM) [7,8,9,10,11]. HMM are well-suited to deal with temporal data, but they are not able to discriminate learned data from others. All the vocabulary expected must be previously learned by the system.

These systems can be evaluated, in terms of recognition rates, for a given set of learned vocabulary, and their performance measured on a test vocabulary.

These techniques are generally well controlled and provide satisfactory results in oral or written language. However, they are not designed to take into consideration the spatial data that are present in natural SL. So, only a very few parts of the sentence can be processed by the recognition step, which is reduced to the recognition of standard signs. Most of the signs, the non standard ones, cannot be processed by a recognition tool, because their production depends on the spatial context.

4.4   Translation

As explained before, the way gestural and spoken languages function is very different. The recognition step provides a sequence of symbols which are not comprehensible enough to convey the whole meaning of the sentence. Because of the constant use of non-standard signs in SL sentences, the spatial context must be represented in the system, in order to model spatio-temporal syntactic rules allowing at least a few non standard signs to be interpreted, like classifiers for example, or directional verbs [7,12].

There is a lack of research on how to manage this kind of interpretation. Thus, the so-called translation systems are actually limited to sentences composed only of standard signs. This results in a low level of translation. These kinds of sentences are of course absolutely not "real" (natural) SL.

And there is also a lack of studies on semantic level [5], where meaning would be represented completely independently of the language, allowing the translation of the utterance meaning rather than the conversion of a syntactic structure to another.

We are actually very far away from the automatic translation system envisioned by some computer scientists !

5   Conclusion: The Necessity of an Ethic

The considerations above show that the classical approach of computer scientists is greatly unsatisfactory. This is generally caused by a strict "bottom-up" approach, where no information coming from the higher processing levels (context, semantic levels) is used at the lower processing levels (capture, modelling, recognition). Moreover, the proposed approaches are often limited to the application of a specific technique from inside a specific discipline. But studies on SL should always be multidisciplinary by definition.

Unfortunately, the real aim of the studies is often to validate a given technique, which is supposed to be able to give a better recognition rate, a better precision in image processing, etc. It appears inappropriate to use the terms "Sign Language recognition" in this context.

If the real aim is to deal with Sign Language, all the different varied and complex elements of language must be taken into account, that is to say that the whole of SL must be studied, including the iconic elements. In this context, computer scientists cannot work on SL without the help of deaf people and linguists, even if linguists still have a lot of work to do in order to formalise the functioning of the iconic part of SL.

While waiting for their research results, computer scientists can work on a reduced part of SL, limited to standard signs. This is at least a first step toward a very long term objective. But in this case, computer scientists must stop pretending to deal with SL any more. They must be more precise and specify that the aim is not to deal with the whole of SL, but to tackle the recognition of only the standard part of SL.

In conclusion, computer scientists who want to work on SL must take into account the problematic history of SL, which implies to take care of the real needs of the Deaf communities. Moreover, they must learn at least their main linguistic properties of SL. The way oral language and SL function is so different that it seems impossible to work on automatic "translation" systems without a close collaboration with linguists specialised in SL. This is such a fascinating research subject that it deserves to be treated with respect.

References

1. Cuxac C.: La langue des Signes Française (LSF), les voies de l'iconicité. Faits de Langues #3, OPHRYS (2000). In French.
2. Cuxac C.: French Sign language: Proposition of a structural Explanation by Iconicity. In Braffort A. et al. (eds.): Gesture-Based Communication in Human-Computer Interaction, LNAI 1739, Springer p.165-184 (1999).
3. Sallandre M.-A. and Cuxac C.: Iconicity in Sign Language: A theoretical and methodological point of view. In Wachsmuth I. and Sowa T. (eds): gesture and Sign language in Human-Computer Interaction, LNAI 2298, Springer (2002).
4. Moody B. : La Langue des Signes. Histoire et Grammaire. Vol 1. IVT, Paris (1983). In French.
5. Lejeune F., Braffort A. and Desclés J.-P.: Study on Semantic Representation of French Sign Language Sentences. Wachsmuth I. and Sowa T. (eds): gesture and Sign language in Human-Computer Interaction, LNAI 2298, Springer (2002).
6. Verlinden M., Tijsseling C. and Frowein H.: A Signing Avatar on the WWW. Wachsmuth I. and Sowa T. (eds): gesture and Sign language in Human-Computer Interaction, LNAI 2298, Springer (2002).
7. Braffort A.: ARGo: An Architecture for Sign Language Recognition and Interpretation. In Harling P. et al. (eds.): Progress in Gestural Interaction, Springer p.17-30 (1996).
8. Starner T., Weaver J. and Pentland A.: Real-Time American Sign Language Recognition using a Desk- and Wearable Computer-Based Video. In IEEE Transaction on Pattern Analysis and Machine Intelligence, 20(12):1371-1375 (1998).
9. Vogler C. and Metaxas D.: Parallel Hidden Markov Models for American Sign Language Recognition. ICCV, Kerkyra, Greece (1999).
10. Vogler C. and Metaxas D.: Toward Scalability in ASL Recognition: Breaking down Signs into phonemes. In Braffort A. et al. (eds.): Gesture-Based Communication in Human-Computer Interaction, LNAI 1739, Springer p. 211-224 (1999).

11. Heinz H., Bauer B. and Kraiss K.-F.: HMM-Based Continuous Sign Language
     Recognition using Stochastic Grammars. In Braffort A. et al. (eds.): Gesture-
     Based Communication in Human-Computer Interaction, LNAI 1739, Springer
     p. 185-196 (1999).
12. Sagawa H. and Takeuchi M.: A Method for Analysing Spatial Relationships
     Between Words in Sign Language Recognition. In Braffort A. et al. (eds.):
     Gesture-Based Communication in Human-Computer Interaction, LNAI 1739,
     Springer p.197-209 (1999).

An Inertial Measurement Framework for Gesture Recognition and Applications

Ari Y. Benbasat and Joseph A. Paradiso

MIT Media Laboratory
77 Massachusetts Ave., NE18-5FL Cambridge, MA 02139, USA
{ayb,joep}@media.mit.edu

Abstract. We describe an inertial gesture recognition framework composed of three parts. The first is a compact, wireless six-axis inertial measurement unit to fully capture three-dimensional motion. The second, a gesture recognition algorithm, analyzes the data and categorizes it on an axis-by-axis basis as simple motions (straight line, twist, etc.) with magnitude and duration. The third allows an application designer to combine recognized gestures both concurrently and consecutively to create specific composite gestures can then be set to trigger output routines. This framework was created to enable application designers to use inertial sensors with a minimum of knowledge and effort. Sample implementations and future directions are discussed.

1 Introduction

Inertial measurement components, which sense either acceleration or angular rate, are being embedded into common user interface devices more frequently as their cost continues to drop. These devices hold a number of advantages over other sensing technologies: they directly measure important parameters for human interaction and they can easily be embedded into mobile platforms. However, in most cases, inertial systems are put together in a very *ad hoc* fashion, where a small number of sensors are placed on known fixed axes, and the data analysis relies heavily on *a priori* information or fixed constraints. This requires a large amount of custom hardware and software engineering for each application, with little possibility of reuse. We present a solution in the form of a generalized framework for inertial gesture recognition. The system consists of a compact inertial measurement unit (IMU), a light-weight gesture recognition algorithm, and scripting functionality to allow specified combinations of such gestures to be linked to output routines of a designer's choice. We envision that the IMU could easily be incorporated into almost any device, and a designer would be able to quickly and easily specify the gestures to be detected and their desired response. Sample implementations of the framework are discussed, as well as future possibilities. A more detailed discussion can be found in [5].

I. Wachsmuth and T. Sowa (Eds.): GW 2001, LNAI 2298, pp. 9–20, 2002.
© Springer-Verlag Berlin Heidelberg 2002

2   Related Works

There are two major threads in the academic literature on the uses of inertial components in gestural interfaces. The first is in the area of musical input. Sawada[18] presented an accelerometer-only system that can recognize ten gestures based on a simple feature set. These gestures are then used to control a MIDI (Musical Instrument Digital Interface) instrument. The Brain Opera, a large-scale interactive musical exhibit produced by the Media Laboratory, included an inertially-instrumented baton as an interface[14], again to a MIDI instrument. In both of these cases, the gesture recognition techniques were quite specific to the application at hand and are therefore difficult to generalize.

The second area is the use of inertial sensors as a stand-alone interface for palmtop computers. The Itsy system from Compaq[4] uses accelerometers both as a static input, with the user tilting the device to scroll images, and a dynamic input, where fanning the device zooms the image in or out. Notable design ideas and concepts for such devices are presented by Small and Ishii[19] and Fitzmaurice[10]. While interesting, the input spaces of devices such as the Itsy tend to be limited as they consider only orientation (either static or dynamic), completely ignoring the kinetic information from the accelerometers.

The work most closely related to ours is that of Hoffman et. al.[13], who used a glove-based accelerometer system for the recognition of a subset of German sign language. The key differences lie in the complexity: our system (as described later) is designed around simpler algorithms (direct data stream analysis versus hidden Markov models) and less densely sampled data (66 Hz versus 500 Hz). Further, our inclusion of gyroscopes increases the range of gestures which can be recognized and the wireless nature of our sensing package allows for its use in more diverse situations.

3   Sensor Hardware

This project's direct lineage can be traced to the Expressive Footwear[15] project, where a printed circuit card instrumented with dynamic sensors (gyroscopes, accelerometers, magnetic compass), among a number of others, was mounted on the side of a dance shoe to allow the capture of multi-modal information describing a dancer's movements. This data was then filtered for a number of specific features, such as toe-taps and spins, which were used to generate music on the fly. While very successful, this system could not be generalized because the chosen set of sensors measured only along specific axes (as suggested by the constraints of the shoe and a dancer's movement) and the circuit card was too large for many applications. Therefore, we decided to create a new system that would contain the sensors necessary for full six degree-of-freedom (DOF) inertial measurement. The system had to be compact and wireless, to allow the greatest range of possibilities. While there are currently a number of 6 DOF systems commercially available, they were all unsuitable for a number of reasons. Crossbow Technologies offers the DMU-6X inertial measurement unit[9] which

Fig. 1. Current IMU hardware and reference frame

has excellent accuracy, but is quite large ($> 30\,\text{in}^3$). The Ascension Technology miniBird 500 [8] magnetic tracker is the smallest available at $10mm \times 5mm \times 5mm$ making it particularly easy to use. However, the closed loop nature of the sensor requires that it be wired, and the base unit is somewhat cumbersome. Also, both of these systems are fairly expensive and neither matches our specification in terms of ease of use (small, wireless).

The physical design of our wireless inertial measurement unit is a cube 1.25 inches on a side (volume $< 2\,\text{in}^3$) and is shown in figure 1. Two sides of the cube contain the inertial sensors. Rotation is detected with three single axis Murata ENC03J piezoelectric gyroscopes[1]. Acceleration is measured with two two-axis Analog Devices ADXL202 MEMS accelerometers[2]. The sensor data are input to an Analog Devices ADuC812 microcontroller (on the remaining side) using a 12-bit analog-to-digital converter. The raw sensor values are then transmitted wirelessly at an update rate of 66 Hz using a small RF Monolithics transmitter module[3] to a separate basestation, which connects to a data analysis machine via a serial link. The complete system operates at 3 V, draws 26 mA while powered, and runs for about 50 hours on two batteries placed in parallel. These batteries, together with the planar transmit antenna, are also small enough to fit inside of the cube formed by the hardware (see above, at left). The total cost of the system, in prototype quantities, is approximately US $300.

In the two years since the design of our hardware, low-cost inertial packages have become much smaller. Analog Devices currently markets the ADXL202E with dimensions of 5 mm × 5 mm × 2 mm, and is in preproduction of the ADXRS150 gyroscope, which will be available in a similar package. It is now readily possible to make a (almost) flat IMU with a footprint of less than $1\,\text{in}^2$, and this is an opportunity that we are currently pursuing.

Finally, there is no absolute need to use the wireless link to a remote computer. We argue that some of the most interesting devices will be those which in-

---

[1] Max. angular velocity $300°/\sec$. Sensitivity $0.67\,\text{mV}/°/\sec$.
[2] Max. acceleration $\pm 2\,\text{g}$. Pulse width output, sensitivity $12.5\%/\,\text{g}$
[3] Frequencies: $315, 916\,\text{MHz}$. Max transmission speed 19.2kbps.

tegrate enough processing power to perform the software functions of the framework (recognition and matching) on board. Such devices would then have a sense of their own motion and could respond to it *in situ*.

4   Problem Statement

Gesture recognition is a wide-ranging research topic. Therefore, it is important at this point to discuss the specifics being addressed in this paper. Our goal was to design generalized algorithms to utilize the hardware described above as fully as possible. Since inertial sensors measure their own motion, the system will either have to be worn by the subjects or embedded in an object manipulated by them. Therefore, we chose to examine (as a first problem) human arm movement, with a gesture defined as any non-trivial motion thereof. It is assumed that these movements will convey information, though their interpretation is considered outside the scope of this paper[4]. Further, to take advantage of the compact, wireless nature of the system, we would like to be able to create stand-alone applications. Therefore, the algorithms should be as efficient as possible, with incremental increases in accuracy being readily traded for increases in speed or algorithmic simplicity.

5   Atomic Gesture Recognition Algorithm

5.1   Atomic Gestures

We define atomic gestures as those that cannot be further decomposed, and which can be combined to create larger composite gestures. The value of this definition is that it should only be necessary to recognize a small set of atoms which span the space of human gestures; thereafter, any gesture of interest can be synthesized from its atoms. The fundamentally important choice of atomic gestures to recognize was made through an examination of both the raw sensor data and the human kinetics literature.

Figure 2 shows the parsed (by the full algorithm) accelerometer traces from a number of simple arm movements and figure 3 shows the curves created by a straight-line and by a there-and-back motion on one axis. Note that this multi-peaked structure is representative of all human arm motion[11]. A straight-line motion will create a two-peaked trace, while the there-and-back motion creates a three-peaked trace[5]. Therefore, we define the atomic gestures simply by the number of contained peaks. While this may not provide a complete basis for the space of human gesture, this decomposition exploits the structure in the data to greatly simplify the gestures recognition with few drawbacks (see section 7.1).

---

[4] In fact, the point of the framework is that meaning is to be added in the scripting stage by an application designer.

[5] A single-peaked trace would imply a net change in velocity. The arm's limited range makes constant velocity motion impossible for more than short periods.

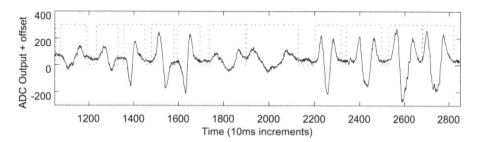

Fig. 2. Parsed acceleration data from simple human arm motion

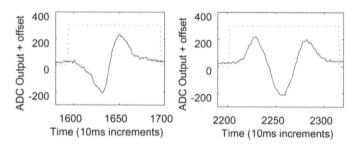

Fig. 3. From figure 2, a straight-line and a there-and-back gesture

Further, we parameterized these gestures in terms of magnitude (length) and duration, which is again fundamental to human motion[2].

We simplify the recognition by examining the data streams on an axis by axis basis (where the axes are those defined by the body frame of the sensor). This allows us to exploit the symmetry in the problem and run the same algorithm repeatedly in a one-dimensional space, which is more efficient than doing so once in a high-dimensional space. This also means that the recognition algorithm can be used with any number of sensors, allowing applications with incomplete, as well as multiple, IMUs.

The literature[6] suggests the definitions given above are not unduly limiting. In the case of atomicity, the study of French Sign Language by Gibet et al.[12] showed that the most common gestures were either straight-line or twist (the simplest gyroscope gesture in our scheme) and further that they most often lay on one of the three main axes (or planes). As for the parameterization, the widely used Hamburg Notation System[16] includes modifiers for both speed and size of gestures.

---

[6] While the cited references are both from sign language research, it is not our intent to suggest that this system is designed for use in that application, only that their descriptions of arm motion are both interesting and generally applicable.

5.2   Activity Detection

In the activity detection stage, sections of increased energy in the data are flagged on an axis by axis basis for further analysis. To do so, the variance of the data over a fixed window is calculated. Since the variance is proportional to $\Sigma(x^2) - (\Sigma x)^2$, this can be done very efficiently by keeping a running sum and running sum of squares of the data. Ranges where the variance is greater than a set threshold are considered to be periods of activity. The window size and threshold value can be found analytically, based on the sensor noise floor and the minimum attack speed considered to represent a deliberate motion. The threshold should err on the side of false positives, as these will later be rejected by the gesture recognition algorithms.

While the activity detection algorithm is designed primarily to filter the data before it is passed on to the recognition system, other purposes can be served in annotating the data stream. For example, consider a medical system that not only does real time recognition but also caches data for further analysis. While storing detailed data for an entire days' worth of motion is achievable, storing only those data segments where activity was noted would be much more efficient.

5.3   Gesture Recognition

Given a flagged area, the next step is to determine whether a gesture is present therein, and if so which atomic gesture it is, and what its parameters are. Because efficiency is a key goal for the gesture recognition algorithm, very general, powerful, and cycle-hungry systems such as hidden Markov models[17] (HMM) are not applicable. Instead, an algorithm was designed that took advantage of *a priori* knowledge of the structure of human arm muscle motion described above.

Since the velocity of the arm is zero at the ends of the gesture, the integral of the acceleration across it must be zero as well (after subtracting any baseline change due to change in orientation). Therefore, recognition is accomplished simply by tracking across an area of activity, and recording the number of peaks and their integral. A minimum peak size is assumed (to reject noise and dithering) and smaller peaks are subsumed into the previous peak (if possible). If the ratio of the net area under the peaks to the sum of the absolute value of their areas is below a fixed threshold (random walk noise must be accounted for even in the ideal case), a valid gesture is said to be present. Its duration is simply the duration of the area of activity, and its magnitude is proportional to the absolute sum divided by the duration. The parameters are determined after the recognition stage, which allows it to be done in a single pass, rather than via a multi-step search using expectation-maximization (or similar techniques), as is often the case with HMMs[20].

For gyroscope gestures, the integration criterion no longer holds (as the data is one less derivative away from absolute motion). Still, in testing, a simple threshold on peak sum proved effective in distinguishing gesture from noise. Duration found as above and magnitude is now proportional to the absolute sum itself.

To improve recognition rates, gestures are further analyzed to account for two special cases. Broken gestures – those made up of two or more areas of activity – are tested for by combining areas of activity (less than a fixed time apart) with no valid gesture and reevaluating the combination. Composite gestures – where two or more gestures are within a single area of activity – are found by splitting and reexamining areas of activity at the point where two adjoining peaks whose masses have the same polarity. In figure 2, the data around 18000 ms was originally parsed as a broken gesture, and the data between 25000 ms and 28000 ms was originally parsed as a composite gesture.

It is interesting to note that even the current modest data sampling rate and accuracy may be more than is needed. In the case of a recorded sample stream of atomic gestures, the same recognition rate was achieved with both the full data stream and one downsampled to 30 Hz and 8 bits[5]. This suggests that the redundancy in human gesture could be even further exploited than we have here, and more lightweight algorithms, and therefore devices, could be created.

6   Scripting

Until this stage in the framework, an application designer needs not have concerned themselves with the details. Their only (required) role in the framework is in the final scripting phase. In this phase, the designer can combine gestures both consecutively and concurrently to create composite gestures of interest. Matches on individual atoms can be restricted to those with certain parameters, and OR and AND logical combinations are allowed between atoms (though the user can add new matching functions if desired). Such logical combinations can then be placed in temporal order to create full composite gestures. These gestures are then connected to output routines.

Given a composite gesture they wish to recognize, a designer only has to perform it a few times, note the atomic gestures recognized and their order, and then write a simple script to recognize that combination. It can then be tied to any output functionality they desire (auditory, graphical, etc.).

It is also possible at this stage to analyze additional sensor data of interest (e.g. bend sensors in a data glove), though the designer would need to provide recognition algorithms for that data. The output of those algorithms could then be added to the atomic gestures found by the inertial gesture recognition and matched and composed in the same fashion.

Note that there is no necessity to use the scripting system provided. The atomic gestures can be used as inputs to an HMM or other recognition algorithm, thereby fulfilling a role similar to that of phonemes.

7   Applications and Implementations

While the purpose of this paper is to present a framework for easily creating inertial sensing based applications, that potential cannot be evaluated without considering applications created using it. This section will discuss the limitations

that must be considered when creating applications, and then describes both a
large-scale application and a stand-alone implementation of the system.

7.1   System Limitations

While we attempted to create as general a wireless system as possible, various
features of both inertial sensing generally and this framework specifically impose
certain restrictions on the gestural applications that can be implemented. The
most important constraint to consider is the lack of an absolute reference frame.
Given the class of sensors chosen (based on their price and size), it is not possible
to track orientation relative to a fixed frame for longer than approximately five
seconds[5]. Therefore, it is necessary for the body reference frame itself to have
some meaning associated to it, such that the application designer will be able to
construct output functions that are appropriate regardless of the instrumented
object's orientation in the world frame.

The second constraint, imposed by the set of atomic gestures chosen, is that
the system cannot track multi-dimensional gestures, except for those which are
separable in space and time. An arbitrary straight line can always be decomposed, and therefore can be recognized, while a rotating object tracing the same
linear path would not be recognizable because the local axis of acceleration
changes with time, making the decomposition impossible. However, the physical
constraints of the gestural system will often prevent such movements, especially
in the case of human movement.

The final set of constraints are those imposed by the algorithms used for
analysis and gesture recognition. Gestures must come to a complete stop to be
found by the activity detection algorithm, making fluid motions and transitions
hard to pick up. Note that as long as movement on one axis comes to a stop, the
recognition of at least part of the gesture can be attempted. For example, while
drawing a square shape can be done repeatedly with a single motion, each line
segment will contain a start and a stop on the appropriate axis.

7.2   Sample Application

This system was first used in (void*) [6], an interactive exhibit created by the
Synthetic Characters group at the MIT Media Lab. In this exhibit, a user could
control one of three semi-autonomous virtual characters, causing them to dance.
Drawing our inspiration from Charlie Chaplin's famous 'buns and forks' scene in
The Gold Rush[7], we created an input device whose outer casing was two bread
rolls, each with a fork stuck into the top, thereby mimicking a pair of legs(see
figure 4). An IMU was placed inside each of the buns, with the data sent to a
separate computer for recognition. A variety of gestures (kicks, twirls, etc) with
both one- and two-handed versions were recognized and used as commands to
the virtual characters.

An HMM-based gesture recognition system was used in this project with
reasonable success both at SIGGRAPH '99 and in demonstrations at the Media
Laboratory. Its main source of error was overconstraint of the HMM parameters,

Fig. 4. Buns and Forks Interface to (`void*`)

leading to a lack of consistency not only among users, but also among slight variations in the IMU position over time. Adjusting the parameters had limited success in improving this situation.

The gesture recognition for (`void*`) was recently redone using the framework described above, and achieved a similar level of accuracy as the original implementation[5]. The superiority of the framework lies, by design, in its speed and ease of use. It occupies few processor cycles, freeing them up for more complex output tasks. Also, the script took less than two hours to write, with most gestures being trivially defined (e.g. a kick is composed of concurrent twist and straight-line atoms). In contrast, the HMM-based system was so complex that it was not able to process data at even half the sensor update rate and each training cycle for each gesture took numerous hours.

7.3   Stand-Alone Implementation

To demonstrate the ease of implementation of these algorithms, we built a gesture recognition system into a Palm III personal digital assistant (PDA). While a PDA is not the best platform for a gesture-based interface, since it is difficult for the user to visually track the screen as it is moved, this platform was chosen because of its ubiquity and ease of use.

Our implementation used a reduced sensor set (only 2 accelerometers) because of space restrictions within the Palm III case. It was otherwise complete, providing not only the recognition algorithms, but also simple scripting capabilities, allowing atoms to be interactively defined and combined (figure 5). Individual recognized atoms and their parameters were displayed in a separate output mode. Therefore, a designer can simply perform a gesture, see the atoms it creates on the output screen, and then use the GUI shown above to encode the composite gesture it represents and specify the output when detected (currently only a text message or beep sound are available).

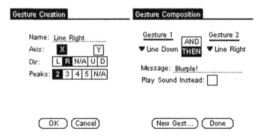

Fig. 5. Gesture creation and composition

The Palm III uses a Motorola DragonBall processor (on par with the original Macintosh) which runs at 16 MHz and has no floating point unit. Further, both program and data memory are limited (and shared with other applications). Nonetheless, we were able to implement our framework, running at 50 Hz, in approximately 200 lines of C code. Several features made this possible: the algorithms use only integer operations, few calculations are done at each time step, and the same analysis is run on each axis. This experience suggests that even faster implementations could be achieved by optimizing the algorithms for this platform.

8   Future Directions

8.1   Short Term

The next step for this project, as alluded to in Section 3, is to examine the creation of much smaller (particularly thinner) versions of the inertial hardware and the application possibilities which they would open. These sensor packages would be designed around a stacked configurable circuit board architecture, with a high-bandwidth transceiver on the bottom, a general microprocessor and ADC in the middle, and sensor board on top. This work, being done in collaboration with the National Microelectronics Research Centre in Cork, Ireland, will exploit not only the smaller inertial sensors now on the market, but also advanced fabrication techniques such as bare die bonding and stacking, to make as compact a system as possible.

One goal in building such systems (of the many they would enable) is to extend the work done in the Expressive Footwear project. Instead of measuring only the feet of a single dancer, the vision is of whole ensembles of dancers with both wrist and ankles instrumented. This would present challenges not only in terms of infrastructure, but also in the need for quick and efficient gesture recognition such that the large volume of data from the performers can be analyzed and responded to in real time.

8.2   Long Term

Given the ability to create simple portable computers (wearables, PDAs, etc.), the question is how this portability can be exploited. The most intriguing is the concept of a generalized inertial gesture system, which can switch from application to application simply by attaching it to a new object and downloading (using little bandwidth) the scripts for a new set of gestures. These objects would then have a sense of their own movement and the ability to respond thereto.

The value of such a system can be seen in the example of a shoe-based physical therapy system which cues the wearer to damaging or awkward movements. Often, a person who is walking incorrectly will take notice only because their feet (ankles, shins, etc.) begin to hurt. However, they will likely be incapable of correcting the situation, because the feedback received has only a weak temporal coupling to the action, which makes learning more difficult[1]. Further, this feedback is based on the outcome of an action rather than its quality, which would be far more valuable[3]. If the shoe-based system mentioned above emitted different tones (e.g.) for different types of incorrect motion, it would be possible to greatly shorten the feedback latency, and therefore correct problems far more effectively.

9   Conclusions

We built an inertial gesture recognition framework comprising three parts. A compact inertial measurement unit is used to collect data from the object of interest and wirelessly transmit it to a personal computer. This data is then analyzed with a windowed variance algorithm to find periods of activity, and a generalized gesture recognition algorithm is applied to those periods. Gestures are recognized in an atomic form on an axis-by-axis basis using a number of physically based constraints, and those atoms can be combined into more complicated gestures using an output scripting system. This system was designed for use by application designers and allows output functions to be linked to specific gesture inputs. The overall framework was light-weight enough to be implemented on Palm III and this work points to a future where interface designers can use easily configured inertial sensors in a wide variety of settings.

Acknowledgements

We would like to express our thanks to our collaborators in the Synthetic Characters Group, who provided both the impetus and wonderful output functionality for the original version of the IMU. We appreciate the support of the Things That Think Consortium and the sponsors of the MIT Media Laboratory. Mr. Benbasat also acknowledges the support of the Natural Sciences and Engineering Research Council of Canada and the Toshiba Corporation.

References

1. J. Anderson. Skill acquisition: Compilation of weak-method problem solutions. *Psychological Review*, 94(2):192–210, 1987. 19
2. C. Atkeson and J. Hollerbach. Kinematic features of unrestrained vertical arm movements. *Journal of Neuroscience*, 5(9):2318–2330, 1985. 13
3. W. Balzer, M. Doherty, and R. O'Connor, Jr. Effects of cognitive feedback on performance. *Psychological Bulletin*, 106(3):410–433, 1989. 19
4. J. F. Bartlett. Rock 'n' scroll is here to stay. *IEEE Computer Graphics and Applications*, 20(3):40–45, May/June 2000. 10
5. A. Y. Benbasat. An inertial measurement unit for user interfaces. Master's thesis, Program in Media Arts and Sciences, Massachusetts Institute of Technology, September 2000. 9, 15, 16, 17
6. B. Blumberg et al. (void*): A cast of characters. In *Conference Abstracts and Applications, SIGGRAPH '99*, pages 169–170. ACM Press, 1999. 16
7. C. Chaplin, director. The Gold Rush, 1925. 100 mins. 16
8. Ascension Technology Corp. http://www.ascension-tech.com/. 11
9. Crossbow Technology, Inc. http://www.xbow.com/html/gyros/dmu6x.htm. 10
10. G. W. Fitzmaurice. Situated information spaces and spatially aware palmtop computers. *Communications of the ACM*, 36(7):38–49, July 1993. 10
11. T. Flash and N. Hogan. The coordination of arm movements: An experimentally confirmed mathematical model. *Journal of Neuroscience*, 5:1688–1703, 1985. 12
12. S. Gibet et al. Corpus of 3D natural movements and sign language primitives of movement. In *Proceedings of Gesture Workshop '97*, pages 111–121. Springer Verlag, 1997. 13
13. F. Hoffman, P. Heyer, and G. Hommel. Velocity profile based recognition of dynamic gestures with discrete hidden Markov models. In *Proceedings of Gesture Workshop '97*, page unknown. Springer Verlag, 1997. 10
14. T. Marrin and J. Paradiso. The digital baton: a versatile performance instrument. In *Proceedings of the International Computer Music Conference*, pages 313–316. Computer Music Association, 1997. 10
15. J. A. Paradiso, K. Hsiao, A. Y. Benbasat, and Z. Teegarden. Design and implementation of expressive footwear. *IBM Systems Journal*, 39(3&4):511–529, 2000. 10
16. S. Prillwitz et al. Hamnosys. Version 2.0; Hamburg Notation System for sign languages. An introductory guide. In *International Studies on Sign Language and Communication of the Deaf Vol. 5*, page unknown, 1989. 13
17. L. Rabiner. A tutorial on Hidden Markov Models and selected applications in speech recognition. *Proceedings of the IEEE*, 77:257–86, 1989. 14
18. H. Sawada and S. Hashimoto. Gesture recognition using an accelerometer sensor and its application to musical performance control. *Electronics and Communications in Japan, Part 3*, 80(5):9–17, 1997. 10
19. D. Small and H. Ishii. Design of spatially aware graspable displays. In *Proceedings of CHI '97*, pages 367–368. ACM Press, 1997. 10
20. A. Wilson. *Adaptive Models for the Recognition of Human Gesture*. PhD thesis, Program in Media Arts and Sciences, Massachusetts Institute of Technology, September 2000. 14

Interpretation of Shape-Related Iconic Gestures in Virtual Environments

Timo Sowa and Ipke Wachsmuth

Artificial Intelligence Group, Faculty of Technology, University of Bielefeld
D-33594 Bielefeld, Germany
{tsowa,ipke}@techfak.uni-bielefeld.de

Abstract. So far, approaches towards gesture recognition focused mainly on deictic and emblematic gestures. Iconics, viewed as iconic signs in
the sense of Peirce, are different from deictics and emblems, for their
relation to the referent is based on similarity. In the work reported here,
the breakdown of the complex notion of similarity provides the key idea
towards a computational model of gesture semantics for iconic gestures.
Based on an empirical study, we describe first steps towards a recognition
model for shape-related iconic gestures and its implementation in a prototype gesture recognition system. Observations are focused on spatial
concepts and their relation to features of iconic gestural expressions. The
recognition model is based on a graph-matching method which compares
the decomposed geometrical structures of gesture and object.

1 Introduction

Gesture as a communicative modality that supplements speech has widely attracted attention in the field of human-computer interaction during the past
years. Applications benefit in various ways from the multimodal approach that
combines different kinds of communication. The synergistic use of gesture and
speech increases the bandwidth of communication and supports a more efficient style of interaction. It potentially increases naturalness and simplicity of
use, since oral and gestural competences are present by nature, while WIMP[1]
competence has to be learned. In virtual environments, where users may move
around, it can be even impossible to use WIMP-style interfaces. Gesture and
speech complement each other, so that some concepts we wish to communicate
are more easily expressed in one modality than in the other. Speech, the linear
and structured modality, is advantageous for abstract concepts, while gesture,
as an inherently space-related modality, supports the communication of concrete and spatial content. This property of gesture makes its use in an interface
particularly helpful for space-related applications.

The interpretation of iconic gestures in spatial domains is a promising idea
to improve the communicative capabilities of human-computer interfaces. This
paper presents an approach to utilize co-verbal iconic gestures as information

---

[1] Windows, Icons, Menus, and Pointing interfaces

I. Wachsmuth and T. Sowa (Eds.): GW 2001, LNAI 2298, pp. 21–33, 2002.

carriers about shape properties of objects. The application background of our research is the area of virtual design and construction, in which spatial concepts play a significant role. In particular, we focus on an intuitive interaction with the machine that (ideally) requires no learning. To approach this aim, we base the system design on the results from an empirical study about object descriptions. Before we describe this in detail, we give a brief analysis of the nature of iconic gesture and review related research.

1.1 Iconic Gestures and Shape

In the gesture typology by McNeill [11], shape-related or shape-describing hand movements are a subset of *iconic* gestures. This gesture type is used simultaneously with speech to depict the referent. Semiotically, the term *iconic* is derived from the notion of icons according to the trichotomy of signs suggested by Peirce [17]. An icon obtains meaning by *iconicity*, i.e. similarity between itself and its referent. Meaning thus becomes an inherent part of the sign.[2] The second and third type of the sign typology are index and symbol. Both require additional knowledge to be intelligible. For indices this is a shared situation and for symbols a shared social background.[3] Iconicity in gestures is multifaceted, it may occur with respect to geometric properties of referents, spatial configurations, or actions.[4] Our modelling approach is confined to the first facet, the representation of spatial properties.

The problem of gesture recognition from the viewpoint of engineering is quite often reduced to a pattern classification task. Following this paradigm, a given data stream (images in the case of video-based recognition or data from other sensor types) is classified according to a pre-defined set of categories. The class then determines the meaning of the gesture resulting in a direct mapping of gestural expression onto meaning. This engineering viewpoint is put forward, for example, by Benoit et al. [1]:

> "[The recognition of] 2D or 3D gestures is a typical pattern recognition problem. Standard pattern recognition techniques, such as template matching and feature-based recognition, are sufficient for gesture recognition." (p. 32)

The (pure) pattern recognition approach works well for emblems and application-dependent gestures which are defined in a stylized dictionary. However, the iconic gestures we consider in our approach are not part of an inventory or a standardized lexicon. Iconicity may refer to any spatial property or configuration one can think of in a given scenario. The number is potentially infinite, likewise

---

[2] This claim presupposes that sender and receiver share similar cognitive capabilities to perceive the sign and to mentally transfer it to its referent.

[3] The gestural correlates of indices and symbols are, respectively, deictic gestures and emblems.

[4] This subdivision is used in [15] to define pictomimic, spatiographic, and kinemimic gestures.

are the possibilities of a gestural realization. Even identical content may be expressed differently due to personal preferences. This fundamental property of iconic gestures, and co-verbal gestures in general, is expressed by McNeill [11]:

> "The gestures I mean are the movements of the hands and arms that we see when people talk ... These gestures are the spontaneous creations of individual speakers, unique and personal. They follow general principles ... but in no sense are they elements of a fixed repertoire." (p. 1)

With McNeill, we assume that for a computerized interpretation of iconic gestures to be successful and scalable, a paradigm different from pattern classification is necessary. The detection of similarity between gesture and referent requires an analysis of their components and their components' relationships.

1.2   Related Research

There is already a multitude of recognition systems for emblems and manipulative gestures used for tasks like robot control, navigation in virtual reality, sign language recognition, or computer games. Overviews of the research in this field were compiled, for example, in [19,16]. We found only very few approaches that take up iconic gestures in human-machine communication.

One attempt to recognize and interpret co-verbal iconic hand gestures was realized in the ICONIC system [7,21]. The prototype allows a user to interact with objects in a virtual environment, for example, to move an object with a speech command "move the teapot like this" and an accompanying dynamic gesture which indicates the direction. The interpretation of a gesture is speech-driven. Whenever speech suggested the possibility of a gesture ("like this" in the example), the system looked for an appropriate gesture segment that complements the spoken utterance. Static gestures are used to indicate places of objects, whereas dynamic gestures indicated movements. Iconicity with respect to shape properties was evaluated for the hand posture. It was correlated to the shape of the object referenced verbally so that hand rotation can be applied. The authors call the process of correlating object-shape and hand-shape *Iconic Mapping*. In the following sections, we will use the same term in a more generalized sense for the mapping between gestural expression and a reference object.

Though the ICONIC approach takes iconic gestures explicitly into account, its interpretative strategy is "caught" in language-like structures. A gesture, as a unit of meaning, augments and specifies just the semantics of the spoken utterance. Thereby, the approach does not consider the very nature of the imaginal aspects of gestures. This makes it impossible to model compositional properties where subsequent gestures use space in the same way to produce a "larger" gestural image.

On the empirical side, there is a lot of research on gestural behavior, but most of it is focused on different conversational situations, for example narrative discourse, specialized on certain gesture types, like deictics, or it considers isolated, non co-verbal gestures.

In her Ph.D. thesis [4], Hummels describes a study on the use of autonomous gestures for object design. It was found that different strategies like cross-section, extrusion, contour, usage mimics, and surface design were employed to visualize a given object. The author further emphasizes that there "is no one-to-one mapping between the meaning of the gesture and structural aspects of the hands" ([4], Section 3.53). There are yet several inter-individual consistencies with respect to prevalent combinations of posture and function. These include indication of height with two flat hands or between thumb and index finger, or the cutting of material using the flat hand.

Hummels' analysis is detailed and insightful, however it remains open in how far the results from autonomous gestures used there can be transferred to co-verbal gestures we are interested in.

The lack of empirical research on the use of co-verbal gestures in our target scenario caused us to conduct a study about the properties of shape-related gestures. A short summary of the main insights is given in the following section. More details can be found in [20]. In Section 3 we present an interpretation model for iconic gestures that considers iconicity and imaginal aspects. A prototype implementation for shape-related gestures is decribed in Section 4.

2   Properties of Shape-Related Gestures: A Study

As a basis for our modeling and implementation approach we conducted an empirical study. Its aim was to analyze the gestural behavior people naturally exhibit facing the task of object description in an experimental setting that closely resembles the targeted conversational situation. A total of 37 subjects were asked to describe parts from a virtual construction application displayed on a projection screen (Fig. 1). Subjects were equipped with electromagnetic 6DOF-trackers, mounted at the wrists and the neck. For hand-shape evaluation they wore data-gloves on both hands; and speech was recorded with a microphone headset (Fig. 2, left). The descriptions were video-taped from a frontal view (Fig. 2, right shows a snapshot from the video corpus). Although the use of the hands was mentioned in the instructions, gestural explanations were not explicitly enforced. Subjects were told to explain the objects' appearance in a

Fig. 1. The stimuli objects used in the empirical study

way that other subjects, who would watch the video afterwards, would be able to imagine the object.

As an outcome, our corpus data consists mainly of co-verbal gestures that were recorded with accompanying speech on a common time-line. The evaluation was focused on form features of the gestural expression and their relation to geometrical properties of the stimuli objects. In sum, our observations suggest that a limited set of gestural features exists to which spatial entities of the objects can be assigned. The majority of spatial entities are object extents in different dimensions, often, but not in each case, oriented like the stimulus object. The quantitative reproduction of extent is not always given, but it was comparable in relation to other extents of the object described. The significant features were linear and circular movements, the distance "vector" between the palms in two-handed gestures, the aperture of the hand in "precision-grip"-like gestures, the orientation of the palm, and a curved hand-shape. These gesture features can be combined to form a gesture like the one shown in Fig. 7 (cf. Section 4). The movement describes the dominant object's extent while the hand aperture indicates the subordinate extent. In this case, an elongated thin, or flat object could be meant.

Besides the parallel use of features in a single gesture, features may also occur sequentially in a series of gesture phrases. The cubical object (cf. Fig. 1; right), for example, was quite often described by three subsequent gestures indicating the extent in each spatial dimension. All three gesture phrases belong to a common gestural image, they form *one idea unit* [11,5]. We found that most people structurally decomposed the objects and described the parts one after another. The round-headed screw, for instance, was decomposed into the head, the slot on top, and the shaft. All parts were described independently, but their spatial relations were often retained, even beyond the scope of an idea unit. This cohe-

Fig. 2. Gesture and speech recording devices (left): Data gloves, motion trackers, and microphone headset. Example from the corpus (right): Showing the vertical extent of an object

sive use of space establishes a persistent gestural discourse context (also called *virtual environment* by Kita [6]), to which subsequent utterances may relate.

3   A Model of Gesture and Speech Recognition

3.1   General Framework

The prevailing psycholinguistic theories on gesture and speech production assume at least two sources of multimodal utterances. Their origin is the speakers working memory which contains, according to the nomenclature by Levelt, propositional and spatial representations [10]. Hadar and Butterworth call the corresponding mechanisms conceptual processing and visual imagery [3]. Both mechanisms influence each other in that they mutually evoke and even modify representations. It is assumed that conceptual representation is the predominant driving force of speech while visual imagery promotes gesture production (even though there is disagreement about the details of the coordination). On the surface, the production process results in co-expressive gestural and spoken utterances. The production model by McNeill assumes basically the same two origins of an utterance which he calls analytic processes and imagery [13,12]. McNeill yet refuses an information processing approach in favor of a dialectic process in which gesture and speech, viewed as materialized thinking, develop from a common source *(growth point)*.

If we assume that gestures are communicative in the sense of conveying information to a recipient, then they should contribute to visual imagery and conceptual representations of the receiver. Current approaches towards speech and gesture integration just consider the conceptual representation which is built up from the fusion of oral and gestural information. However, a provision for the visuo-spatial properties of utterances, including effects like spatial cohesion, requires visual imagery. These considerations provide the basis for an abstract model of the recognition process which is shown in Fig. 3. The recipient perceives speech and gesture and analyzes them according to their structure. The

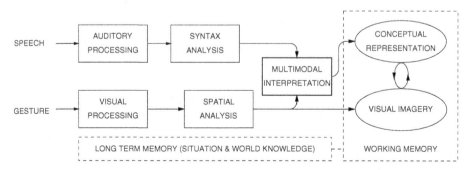

Fig. 3. An abstract processing model of gesture and speech recognition

propositional part of meaning is determined by integrating both modalities to a common conceptual representation. The multimodal interpretation step takes the categorical, language-like properties into account, in which emblems, deictics, iconic gestures, and – of course – speech, can be analyzed. At the same time a persistent image of the gestural part is built up. The property of persistence is essential for the creation of a visual discourse context that serves as a reference frame for subsequent utterances. Analogously to the production models, both representations do not exist independently from each other since conceptual processing may evoke imagery and vice versa. With this abstract recognition model in mind, we will now focus on the technical realization of the processing steps from spatial analysis to the construction of visuo-spatial imagery.

3.2   Computational Model

The application scenario of our approach towards a computational treatment of imagery is a reference identification task. The idea is to have a system that can identify a given object from among a set of objects based on a gestural description. A computational model of this process, which substantiates the spatial analysis, is shown in Fig. 4. Instead of visual information from gesture, the input consists of movement and posture data provided by data gloves and motion trackers, so that visual processing is not necessary. The spatial analysis is subdivided into three processing steps. At first, gestural features like handshape, movement, or holds are recognized. A segmentation module divides the meaningful gesture phases from subsidiary movements and meaningless postures. Following [2], we assume that it is generally possible to determine the meaningful part on the basis of movement and posture features without considering speech. The features of active phases are abstracted to basic spatial entities. Spatial entities are the building blocks of imagery, they are stored in the visual imagery component where spatial relations between different entities are determined. That way, a structured spatial representation of the gestural input is constructed.

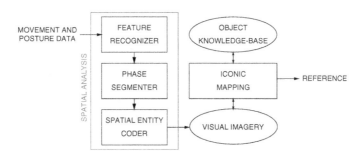

Fig. 4. A model of a recognition system for reference object recognition with shape-related gestures

A mapping mechanism compares internal object representations with the image and selects the object whose representation matches best.

4 Implementation Approach

Based on the computational model, we implemented a first prototype. It allows the user to select an object from a set of objects displayed on the screen by iconic gestures. The setting and instrumentation is comparable to the environment that we used for the empirical study. Fig. 5 shows a user and the wall display with some differently shaped objects. The recognition system is embedded in a virtual reality framework called AVANGO, a scene-graph based software environment that allows easy access to high-level VR programming [22].

Fig. 5. The prototype system: Selecting reference objects by gestural shape description

4.1 Feature Recognizer

Hand Features The recognition system provides a module that transforms (device-dependent) posture data from the CyberGloves into a general description of the hand-shape. The module is based on a kinematic model of the human hand which includes forward-kinematics to compute hand segment positions from given joint angles [8]. Put into the scene graph, the model can be visualized, thus providing a control mechanism for the correctness of the data mapping onto the model during development and testing (Fig.6). The input data for the hand module consists of a stream of 18-dimensional vectors that represent the measurements of the bimetal bending sensors in the CyberGloves. The vector components are numbers that correspond to the joint angles in a (more or less) linear fashion. However, since the CyberGloves measurement points do not perfectly match to the joints of the model and since the hand model is underspecified

Fig. 6. Visualization of hand model (left) and real hand with CyberGlove (right)

by the glove data, some transformations and dependencies are applied to compute the missing model data. The resulting model represents the hand-shape and hand-internal movements of the user quite accurately, provided that the hand does not have contact with real objects or with the body.[5]

The hand module provides a set of fuzzy variables that express the bending of the fingers in a symbolic way, derived from HamNoSys notation [18]. The module differentiates between five basic finger shapes: angular, rounded, bent, rolled, and stretched. The set of the five corresponding variables is computed for each finger (except the thumb, which is handled separately) and for the whole hand. By this, the module can provide information like "the left hand is rounded" or "the index of the right hand is stretched" to a subsequent module. Further fuzzy information concerns the position of the thumb which may be aligned or in opposition to the other fingers. This facilitates the detection of "precision grip"-like gestures used to indicate, for example, short distances.

Movement Features Data on the location and orientation of the wrists is provided by the tracking devices. This data can be combined with the output of the hand module to compute, for example, the absolute position of the fingertips or the center of the palm. From these reference points, different movement features are detected. Currently we evaluate holds and linear movement segments. Both features may indicate meaningful gesture phases for static or dynamic gestures [20]. The detection itself is based on the PrOSA-framework, a development from our group that facilitates rapid prototyping of evaluation systems for gestural movements in VR-scenarios [9].

4.2 Phase Segmenter

Our prototype system currently supports a simple gesture segmentation based on the typology of movement phases suggested by Kendon [5]. The typology defines

---

[5] In this case the heuristics may fail, but iconic gestures are normally non-contact movements.

gesture units as movements that occur between two resting positions. Any unit may consist of one or more *gesture phrases*, whereas each phrase contains a gesture stroke and an optional preparation and retraction phase. The resting space in our implementation is the area below chest height. Movement segments that begin or end in the resting space are counted as preparation or retraction phases and are not considered for further analysis. A retraction to resting space also signals the end of a gesture and initializes the iconic mapping process. There are three types of gesture strokes our prototype can recognize: holds in one hand, synchronized holds in both hands, and linear movements of one hand. Transitional movements between sequential strokes can be filtered out.

4.3 Spatial Entity Coder

According to our observations, we regard the most prominent meaning categories, spatial *extent* and *diameter* (a special case of extent), as basic spatial entities. The spatial entity coder abstracts from the concrete realization of the gesture through the transformation to a set of spatial entities. This processing step thus takes the variability of gestural expressions to convey a concept into account. Spatial entities are stored in the visual imagery component and for each new entry, spatial relations to existing entries are determined. The system checks for *orthogonality, dominance* (of one extent over another), and *equality* of two extents. Thus, the representation of the imaginary content may be visualized as a graph with spatial entities as nodes and spatial relations as links (Fig. 7).

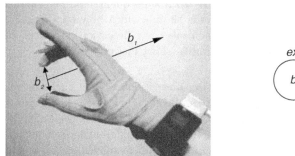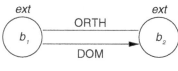

Fig. 7. Spatial representation of a dynamic gesture (hand moves right along b_1 direction): Spatial extents b_1 and b_2 are recognized and checked for spatial relations. b_1 is orthogonal and dominant to b_2

4.4 Iconic Mapping

The iconic mapping module determines the most appropriate object from the object knowledge base that matches the content of visual imagery. The output of the iconic mapping may be used as a reference object in an application command. The mapping is based on a comparison of the gesture representation graph with the pre-defined object graphs by means of subgraph matching [14]. The mapping process evaluates for each object graph how much cost is expended to transform the gesture graph into a true subgraph of the model. The cost functions for insertion, deletion, and modification of nodes and links are defined individually. Our prototype computes the cost function according to some simple heuristics: Generally, it is assumed that the object model is complete, meaning that the gesture graph usually does not contain additional nodes and links. Therefore, a deletion of nodes and links is more expensive than an insertion. Node types can be `extent` and `diameter`, where extent is a superclass of diameter. The transformation from extent nodes to diameter nodes is therefore cheaper than vice versa. The transformation cost for each node and link total to a sum that represents the mismatch between gesture and object. Fig. 8 exemplifies the recognition process for a cubical object.

From the three-phase gesture unit indicating the cube's extent in all dimensions a gesture model graph with three nodes for the spatial entities is built. Entity relations `ORTH` for orthogonality and `EQ` for equality are determined. The graph is then matched against the object database. Note that a complete match is not necessarily needed to recognize the object. Even if only two of the three object extents are indicated by gesture, the cube interpretation would match the gesture model better than, for example, the bar.

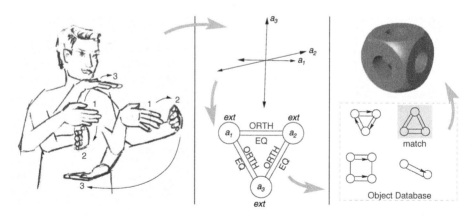

Fig. 8. The recognition process: A three-phase gesture unit (left) is mapped on a static model of imagery features (middle) and compared to object models (right)

5 Conclusion and Further Work

We presented an approach towards the interpretation of shape-describing gestures that is applicable in virtual design and virtual construction tasks. The model goes beyond a simple gesture-to-meaning mapping by way of decomposing the gestural utterance into spatial entities and their relations. That way, a *static* spatial representation of the *dynamic* gesture is built up. Our approach emphasizes the importance of imagery in addition to conceptual structures to construct a visuo-spatial discourse context. AI techniques from image understanding, i.e. graph-based representations and image recognition by means of subgraph isomorphism, are used for realization of a running system.

The inclusion of speech in our prototype system remains an open issue for further research. To achieve it, a cross-link between imagistic and conceptual memory is necessary. Another point for improvement is gesture segmentation. A detailed analysis of gestural features that indicate gesture strokes is still in progress. Likewise an evaluation of the overall system performance is subject to future work.

Acknowledgment

This research is partially supported by the Deutsche Forschungsgemeinschaft (DFG) in the Collaborative Research Center "Situated Artificial Communicators" (SFB 360).

References

1. Christian Benoit, Jean-Claude Martin, Catherine Pelachaud, Lambert Schomaker, and Bernhard Suhm. Audio-visual and multimodal speech systems. In D. Gibbon, editor, *Handbook of Standards and Resources for Spoken Language Systems - Supplement Volume D*. to appear. 22
2. P. Feyereisen, M. Van de Wiele, and F. Dubois. The meaning of gestures: What can be understood without speech? *European Bulletin of Cognitive Psychology*, 8:3–25, 1988. 27
3. U. Hadar and B. Butterworth. Iconic gestures, imagery, and word retrieval in speech. *Semiotica*, 115(1/2):147–172, 1997. 26
4. Caroline Hummels. *Gestural design tools: prototypes, experiments and scenarios*. PhD thesis, Technische Universiteit Delft, 2000. 24
5. A. Kendon. Gesticulation and speech: Two aspects of the process of utterance. In M. R. Key, editor, *The Relationship of Verbal and Nonverbal Communication*, pages 207–227. Mouton, The Hague, 1980. 25, 29
6. Sotaro Kita. How representational gestures help speaking. In McNeill [13], chapter 8, pages 162–185. 26
7. David B. Koons, Sparrell Carlton J., and Thorisson Kristinn R. *Intelligent Multimedia Interfaces*, chapter 11. MIT Press, Cambridge, Mass., USA, 1993. 23
8. Stefan Kopp and Ipke Wachsmuth. A knowledge-based approach for lifelike gesture animation. In W. Horn, editor, *ECAI 2000 - Proceedings of the 14th European Conference on Artificial Intelligence*, pages 663–667, Amsterdam, 2000. IOS Press. 28

9. Marc Erich Latoschik. A general framework for multimodal interaction in virtual reality systems: PrOSA. In *VR2001 workshop proceedings: The Future of VR and AR Interfaces: Multi-modal, Humanoid, Adaptive and Intelligent*, 2001. in press. 29

10. W. J. Levelt. *Speaking*. MIT press, Cambridge, Massachusetts, 1989. 26

11. D. McNeill. *Hand and Mind: What Gestures Reveal about Thought*. University of Chicago Press, Chicago, 1992. 22, 23, 25

12. David McNeill. Catchments and contexts: Non-modular factors in speech and gesture production. In McNeill [13], chapter 15, pages 312–328. 26

13. David McNeill, editor. *Language and Gesture*. Language, Culture and Cognition. Cambridge University Press, Cambridge, 2000. 26, 32, 33

14. Bruno T. Messmer. *Efficient Graph Matching Algorithms for Preprocessed Model Graphs*. PhD thesis, University of Bern, Switzerland, 1996. 31

15. Jean-Luc Nespoulous and Andre Roch Lecours. Gestures: Nature and function. In Nespoulous, Perron, and Lecours, editors, *The Biological Foundations of Gestures: Motor and Semiotic Aspects*. Lawrence Erlbaum Associates, Hillsday N. J., 1986. 22

16. Vladimir I. Pavlovic, Rajeev Sharma, and Thomas S. Huang. Visual interpretation of hand gestures for human-computer interaction: A review. *IEEE Transactions on Pattern Analysis and Machine Intelligence*, 19(7):677–695, July 1997. 23

17. Charles Sanders Peirce. *Collected Papers of Charles Sanders Peirce*. The Belknap Press of Harvard University Press, Cambridge, 1965. 22

18. Siegmund Prillwitz, Regina Leven, Heiko Zienert, Thomas Hanke, and Jan Henning. *HamNoSys Version 2.0: Hamburg Notation System for Sign Languages: An Introductory Guide*, volume 5 of *International Studies on Sign Language and Communication of the Deaf*. Signum Press, Hamburg, 1989. 29

19. Francis Quek, David McNeill, Robert Bryll, Susan Duncan, Xin-Feng Ma, Cemil Kirbas, Karl E. McCullough, and Rashid Ansari. Gesture and speech multimodal conversational interaction in monocular video. Course Notes of the Interdisciplinary College "Cognitive and Neurosciences", Günne, Germany, March 2001. 23

20. Timo Sowa and Ipke Wachsmuth. Coverbal iconic gestures for object descriptions in virtual environments: An empirical study. Technical Report 2001/03, Collaborative Research Center "Situated Artificial Communicators" (SFB 360), University of Bielefeld, 2001. 24, 29

21. Carlton J. Sparrell and David B. Koons. Interpretation of coverbal depictive gestures. In *AAAI Spring Symposium Series: Intelligent Multi-Media Multi-Modal Systems*, pages 8–12. Stanford University, March 1994. 23

22. Henrik Tramberend. Avocado: A distributed virtual reality framework. In *Proceedings of the IEEE Virtual Reality*, pages 14–21, 1999. 28

Real-Time Gesture Recognition by Means of Hybrid Recognizers

Andrea Corradini[1,2]

[1] Oregon Graduate Institute of Science and Technology
Center for Human-Computer Communication
20000 N.W. Walker Road, Beaverton, OR 97006, USA
andrea@cse.ogi.edu
http://www.cse.ogi.edu/CHCC
[2] Department of Neuroinformatics, Technical University of Ilmenau
D-98684 Ilmenau, Federal Republic of Germany
andrea.corradini@informatik.tu-ilmenau.de
http://cortex.informatik.tu-ilmenau.de

Abstract. In recent times, there have been significant efforts to develop intelligent and natural interfaces for interaction between human users and computer systems by means of a variety of modes of information (visual, audio, pen, etc.). These modes can be used either individually or in combination with other modes. One of the most promising interaction modes for these interfaces is the human user's natural gesture.

In this work, we apply computer vision techniques to analyze real-time video streams of a user's freehand gestures from a predefined vocabulary. We propose the use of a set of hybrid recognizers where each of them accounts for one single gesture and consists of one hidden Markov model (HMM) whose state emission probabilities are computed by partially recurrent artificial neural networks (ANN).

The underlying idea is to take advantage of the strengths of ANNs to capture the nonlinear local dependencies of a gesture, while handling its temporal structure within the HMM formalism. The recognition engine's accuracy outperforms that of HMM- and ANN-based recognizers used individually.

1   Introduction

The keyboard has been the main input device for many years. Thereafter, the widespread introduction of mouse in the early 1980's changed the way people interacted with computers. In the last decade, a large number of input devices, such as those based on pen, 3D mouse, joystick, etc., have appeared. The main impetus driving the development of new input technologies has been the demand for more natural interaction systems. Natural user interfaces can be obtained by integrating various interaction modes [9], by using natural tangible objects [23], or by using natural human gestures [22,34].

Even in the field of robotics, where most efforts have been hitherto concentrated on navigation issues, there has been a growing interest in body-centered

I. Wachsmuth and T. Sowa (Eds.): GW 2001, LNAI 2298, pp. 34–47, 2002.
© Springer-Verlag Berlin Heidelberg 2002

input and the idea of using gestures and/or speech to interact with a robot has begun to emerge. A new generation of service robots operating in commercial or service surroundings both to give support and interact with people has been employed [20,33,38] and is currently gaining a great deal of attention.

The longterm objective of this research work is to develop a perceptual interface permitting an autonomous mobile robot to act as shopping assistant in a supermarket environment by exploiting visual and acoustic information from gestures and speech, respectively. The goal is the recognition and interpretation of customer demands, requests and needs. Therefore, the robot should be capable of getting into contact with, following, lead, supporting and interacting with any human customer. In this work, we exclusively concentrate on the visual part of that interface.

2 Gestures and HCI: Definitions

Every human language has several expressions indicating hand, arm, body movements and gestures. As from any dictionary we can obtain a set of meanings for the word gesture used as a noun in a common sense, within the scientific community gestures have been defined in several different ways. Some researchers refer to them as three–dimensional movements of body and/or limbs, others as motion and orientations of both hand and fingers [28,40]. Fewer researchers regard gestures as hand markings entered with a mouse, fingertip, joystick or electronic pen [24,31], some others refer to them as human facial expression or lip and speech reading [16,35].

Whereas in common use, gestures have meanings easily understood within the culture and the country in which they are expressed [14,19,25,26], in the context of human-computer interaction (HCI) they must first be exactly defined and made meaningful in their own right. Nevertheless the deterministic nature of computer input constrains one to fix the meaning of the gestures to be recognized.

In accordance with the tendency to see gestures as dynamic and posture as static throughout this paper we make use of the following definitions.

Definition 1 (Posture/Pose) *A posture or pose is defined solely by the (static) hand locations with respect to the head position. The spatial relation of face and hands determines the behavioral meaning of each posture.*

Definition 2 (Gesture) *A gesture is a series of postures over a time span connected by motions.*

In this context, we propose a feature-based bottom-up strategy to track hands and head throughout the actual gesture. Thus, motion information is encoded into the trajectory described by the movement of both hands with respect to the head. Assuming that same gestures originate similar trajectories, gesture recognition amounts to trajectory discrimination within the input feature space.

3   From Postures to Gestures

3.1   Posture Analysis and Feature Extraction

In conformity with the definitions we give in section 2, the localization of head and hands is a mandatory task for any further processing task. In [7,10] we report a reliable method which successfully localizes and segments hands and head from color images. That procedure represents the starting point for the work described in this paper.

Fig. 1. From left to right: input image, head and hands modelled by blob regions, segmentation's outcome as binary image characterizing the spatial relation between head and hands with chosen frame of reference as overlay

Figure 1 shows a typical segmentation's outcome. The resulting binary image consists of up to three blob regions, each modelled about the locations of the head and the hands. Since the segmented image is a binary description of the current posture in the image, we extract from it one 15-dimensional feature vector to characterize that posture in a more compact way. We carefully choose the vector's elements in order for them to capture both shape of and spatial relations between the blobs and incorporate prior knowledge by means of invariants under translation and scaling transformations.

As the first invariant elements, we take seven out of the ten normalized central moments up to the third order. We drop out both zeroth and first order moments as they do not contain any useful information. The computation of the remaining feature vector elements is carried out with the goal of compensating for shift variations of the person gesticulating in front of the camera. We use for each image a suitable coordinate system by fixing its origin point at the current head's center of mass [10]. Then, we calculate the last feature components relating to the head position and regardless to the position of the user within the image. Table 1 summarizes and describes in detail the invariants.

The individual vector components have values which differ one another in a significant manner even by several orders of magnitude. Because that size dissimilarity does not reflect their relative importance we further rescale them by performing a linear transformation known as whitening. It allows for correlation among the variables considered and removes certain dominant signals to get a more Gaussian distribution of the data [5].

Table 1. Description of the feature vector components

Feature Nr.	Symbol	Description
$1 \ldots 7$	$\nu_{pq} \mid p, q = 1 \ldots 3$	Normalized central moments of second and third order.
8,9	ϑ_1, ϑ_2	Values in the range $[-180, 180]$ indicating the angles in degree between the x axes and the segment connecting the origin with the centroid of right and left hand, respectively (see figure 1).
10,11	ς_1, ς_2	Let ρ_1 and ρ_2 indicate the length of the segments connecting the origin with the centroid of the right and the left hand, respectively (see figure 1), these value are defined as $\varsigma_1 = \frac{\rho_1}{\max\{\rho_1, \rho_2\}}$ and $\varsigma_2 = \frac{\rho_2}{\max\{\rho_1, \rho_2\}}$.
$12 \ldots 15$	$v_{Rx}, v_{Ry}, v_{Lx}, v_{Ly}$	The first (last) two values represent the normalized velocities of the right (left) hand centroid along the Cartesian coordinates. We use $\max\{\rho_1, \rho_2\}$ as normalization factor.

Blob tracking [39] is employed to keep track of hand and head locations from one image to the next over an input video stream. This offers a partial solution to both the occlusion problem and the case of head-hand/hand-hand overlapping. In such cases, the segmentation typically results in fewer than three blobs (see figure 2), yet tracking helps recover hand and head positions.

3.2 Gesture as Postures' Temporal Sequence

Such a view-based spatial description of postures strongly limits the number of postures which can be detected. The static information embodied in each individual feature vector is a context-independent snapshot of a wider dynamic process associated with the evolution of the gesture. The sequence of time-ordered multidimensional feature vectors extracted from the input video stream originates a trajectory within the feature space. This trajectory encodes the gesture which underlies the movements of both hands and head in the space over the time. Different gestures originate different trajectories while performances of the same gesture are expected to yield close spatial paths over a common temporal scale. In this framework, gesture recognition amounts to time-scale invariant sequence recognition.

Fig. 2. From left to right: input image with hands and head aligned on the camera axis, the segmentation results in one single region consisting of three overlapping blobs, and finally one possible lateral view of the input image. By using only one monocular camera in front of the user, we cannot determine how far the hands are from the camera. Thus, any posture characterized by a given head-hand spatial relation originates the same segmented image when moving hands or head along the camera axis

For their capabilities to deal with temporal variability we use Hidden Markov models (HMM) as main component of our gesture recognizer. HMMs embody a form of dynamic time warping for model evaluation and allow for a statistical representation of multiple exemplar gestures in a single model. Thus, we utilize as many HMMs as there are gestures to be recognized.

4   The Gesture Recognition Engine

4.1   HMM/ANN Hybrid Architecture: Motivations

Hidden Markov models are a kind of parameterized statistical model of time-sequential data which have been used with particular success in many applications like e.g. speech recognition [2,6,37], and handwriting recognition [21,32]. HMMs make it possible to deal with piecewise stationary data sequences by providing time-scale invariability during recognition. Basically, they consist of a finite number of states with associated state transitions and state output probabilities to jointly model a set of input sequences.

Though powerful and flexible, such parametric models typically suffer from some weaknesses which limit their capabilities and generality. Notable is their lack of contextual information since the probability that a certain feature vector is emitted by a state exclusively depends on this current state. Furthermore, standard HMMs assume predefined, though trainable, parametric distribution functions for the probability density functions associated with their states.

Artificial neural networks (ANNs) are another connectionist tool that has been utilized for temporal sequence processing [3,13,15,18,27,36]. ANNs are able to approximate almost any nonlinear function without needing any assumption about the statistical distribution of the patterns in the input space. Context information can be incorporated by choosing both an adequate architecture and a suitable training algorithm. However, connectionist models do not completely address the problem of time alignment and segmentation limiting their ability to deal with the time sequential nature of the gestures.

Thus, an attractive and promising approach to gesture recognition can be obtained by combining the capabilities of both HMMs and ANNs into a hybrid system [4,6]. Such a strategy can take advantage of the strengths of artificial neural networks to capture the temporally nonlinear local dependencies, while remaining in the HMM formalism to handle the temporal structure and the elastic time alignment between movements and states. That can be accomplished by forcing the connectionist network outputs to represent probabilities to use as estimates for HMM state distribution functions.

4.2   HMM Topology

For the choice of the model topology there is no established standard to rely on. However, left-to-right models seem to be more suitable for our purposes because they permit to associate time with model states in a straightforward way (figure 3). Concerning the number of states within each model, it varies from four (stop gesture) to nine (waving gestures), according to the average time duration of the corresponding gesture in the training data.

We employ the Baum-Welch algorithm [29] to train each HMM so as to model the feature vector sequences from the associated gesture in the learning data set. This algorithm is an iterative procedure based on the maximum likelihood criterion and aims to maximize the probability of the samples by performing a dynamic time warping between input patterns and states.

Recognition consists in comparing an unknown sequence with the HMM. That gesture associated with the model which best matches the input sequence is chosen as the recognized movement.

4.3   Probability Estimation by ANNs

The demand for the states to model the input distribution without assumptions on the shape of the probability density functions, leads us to the use of artificial neural networks. ANN-derived probabilities are more robust than parametric densities traditionally used because they are obtained from discriminant training. Moreover, some ANN paradigms take the context into account [15,18,36].

The most direct way to take the context into account by static ANNs is to turn the largest possible part of the temporal pattern sequence into a buffer on the network input layer. Hence, each time, an input subsequence is presented to the network by feeding the signal into the input buffer and next shifting it at successive time intervals. This buffer must be chosen in advance and has to be large enough to contain the longest possible sequence and thus maintain as much context information as possible. However, a large buffer means a large number of model parameters and implicitly a large quantity of training data. To incorporate the context into such static feed-forward networks already during the design phase, we provide the network with an additional carefully chosen

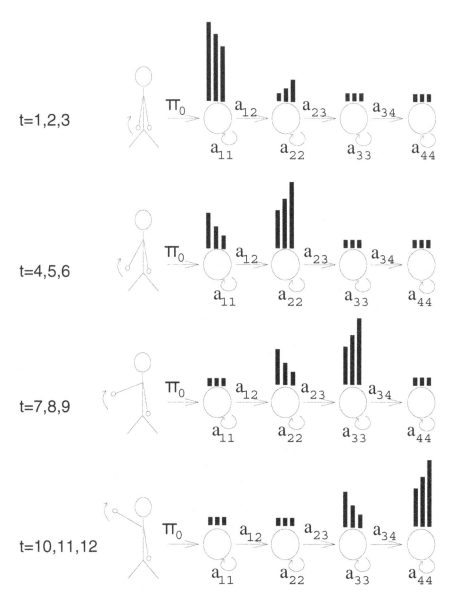

Fig. 3. Left-to-right HMM modeling a pointing gesture by aligning states with gesture segments. The coefficient a_{ij} denotes the transition probability between state i and state j, while π_0 is the initial probability. At time instant t, any state emits one single probability (here three sequential time instants are depicted) which increases from left to right as the arm moves up

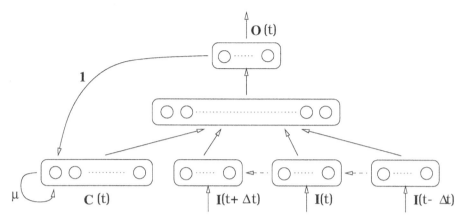

Fig. 4. Taking context into account is quite easy with a buffered partially recurrent ANN. The network input is split into more fields: a buffered input (here with $\Delta t = 1$) and the context unit. Concerning the buffered input, the central unit is the current feature vector while left and right fields contain its time-neighboring vectors. The layers are fully connected forwards

set of fixed, and hence not trainable, recurrent connections. Feedback and self-connections furnish some additional memory since they encode the information from the most recent past and do not complicate the training phase at all. The architecture of this kind of partially recurrent network is depicted in figure 4.

The input layer is divided into two parts: the context unit and the buffered input units. The context unit holds a copy of the activations of the output layer from the previous time steps. The buffered input field consists of several groups of units, each group representing one feature vector of the input sequence. Indicating with $(2\Delta t + 1)$ the buffer width we arbitrary choose $\Delta t = 2$. The current output is always associated with the center of that input buffer while the feature vectors constituting the input sequence are stepped through the window time step by time step. There are as many output neurons as the number of HMM states[1]. Providing 120 units in the hidden layer the resulting network topology is $(75 + \#HMM states)$-120-$\#HMM states$. The feedback connections between output and context unit are kept constant to the value 1. After iterating over the time, the updating rule for the context unit $\mathbf{C(t)}$ at time instant t after iterating can be expressed as

$$\mathbf{C(t)} = \mu\mathbf{C(t-1)} + \mathbf{O(t-1)} = \sum_{i=0}^{t-1} \mu^{t-i}\mathbf{O(i)} \tag{1}$$

---

[1] Equivalently, we can use as many connectionist models as the HMM states. In such a case, each ANN must have both one individual output unit and at least as many input units as the dimension of the feature vectors. Further, the dimension of the context unit is allowed to be different from state to state.

where $\mathbf{O(t)}$ is the output vector at time t. Thus, the context unit accumulates the past network history in a manner depending on the choice of the parameter μ which has to be chosen in the range $[0, 1]$ to avoid divergence. The more we choose it close to 1 the more the memory extends further back into the past. There is a trade off between sensitivity to detail and memory extension in the past so that we choose the value $\mu = 0.5$.

ANNs for directly computing the emission probabilities associated to HMM states can be used only under the condition for the network outputs to represent probabilities. In the one-of-more classification case the outputs of any network trained by minimizing the mean square error function can be interpreted as posterior probability [30]. Moreover, to be interpreted as probabilities the network outputs must sum up to one and lie in the range between zero and one. Due to these constraints, we train the ANN in classification mode in conformity with the one-of-more coding, i.e. one output for each class with all target vector elements equal to zero except for the correct class where it is one. We use the least mean square function as error function. The output layer units perform the soft-max function [8] to avoid further normalization. As usual, the hidden neurons compute the sigmoid activation function. The training data set consists of vector sequences, together with their binary target vectors. Given the training vector, only one element of its target vector is set to one according to the HMM state which is expected to output the higher probability value when presenting that training vector.

In figure 5 the whole hybrid HMM/ANN system is depicted. We do not jointly train the Markov model and the connectionist networks. First we train the recurrent networks and then we apply the Baum-Welch reestimation algorithm to determine the HMM parameters.

5   Preliminary Results

The data collected is extracted from video sequences of five subjects performing 90 repetitions of each individual gesture. Half data is used for training, the remaining for testing. The categories to be recognized are five. The subjects are led into a room with constant background and instructed how the meaningful gestures look. They are further instructed to look at the camera and execute the movements. The video is digitized at 120×90 spatial, 12Hz temporal and 24-bit (three color channel each 8 bit) photometric resolution.

Table 2 summarizes the achieved performance concerning the recognition task. There is an acceptance threshold associated with each Markov model. An input is not classified if after feeding it into each HMM, either the difference between the highest and the second highest output is not over that heuristically determined threshold, or all the outputs are under its value.

The recognition occurs in real-time though off-line over sequences whose first and last frame are assumed to correspond respectively to the start and the end of the gesture performance. In that way, we do not need to worry about gesture segmentation.

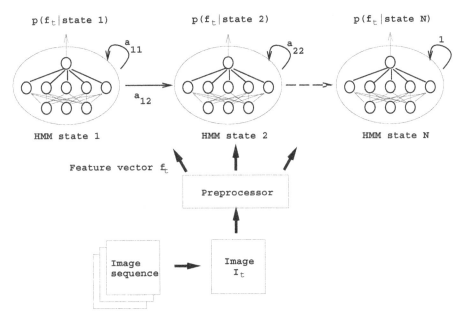

Fig. 5. Structure of the ANN/HMM hybrid system. ANNs are utilized for the classification of arbitrarily distributed feature vectors and HMMs for modeling of sequences of them. At each time instant t one feature vector $\mathbf{f_t}$ is extracted from the current image frame I_t and further presented to each HMM state. The state emission probability is then computed by ANNs, furnishing the local likelihood of the input vector

6   Conclusion and Open Questions

In this on going work we deal with hands-free gesture input from a sequence of monocular images. To bound the scope of the problem we created a database that contains a set of well defined primitives for the gestural language we use. We define a *gestobulary* (gesture vocabulary) which is easily recognizable and does not require special training for the users. For recognition we made use of a hybrid ANN/HMM architecture where connectionist models are used to overcome some intrinsic weaknesses of the Markov formalism.

The recognition rate achieved with this system is higher than that achieved by other hybrid ANN/HMM architectures and conventional non-hybrid HMM-based systems we proposed in the past [12,11]. The complexity (i.e. the number of parameters) of this architecture is higher as well. Of course, we do not believe that an increased number of parameters will necessarily guarantee better recognition rates. However, despite the current results might also strongly depend on the implementation and the limited data at hand, we claim that they are justified and supported by some theoretical investigations. We plan to closely investigate how the increased complexity affects the recognition rate, and to test

Table 2. Recognition results using hybrid ANN/HMM

Gesture	% of not classified patterns	% of false classified patterns	Recognition rate in %
stop	4.5	3.3	92.2
waving right	6.0	3.7	90.3
waving left	5.7	4.5	89.8
go right	4.8	3.9	91.3
go left	4.8	4.1	91.1

the system on a larger gestobulary and with an extended set of training data. The limited availability of labeled data suggests to further adopt the procedure of cross-validation [17] for ANN training. The use of a clusterer over the training data set associated with each gesture could be exploited as well, to avoid the hand-labeling procedure otherwise necessary for the determination of the target vectors. Such a clustering approach will speed up the learning phase, make the model fully automatic, and create vector training pairs that fit the real distribution of the data inside the input feature space.

Since the HMMs are trained with the non-discriminative Baum-Welch algorithm, poor discriminative capability among different models is to be expected [29]. Indeed, by maximizing the maximum likelihood instead of the maximum a posteriori, the HMMs only generate high probabilities for its own class and do not discriminate against models. A jointly ANN and HMM discriminative learning is expected to improve the robustness of the system and has to be investigated.

The overall system has to be understood as work in progress, undergoing continuous changes. Up to now, the major constraints are that the gesture recognition process cannot be used for on-line recognition since we do not segment the gestures with respect to start and end point. The posture segmentation uses only skin color which causes problems when other skin-colored objects are in the scene. Eventually, an additional problem arises from the lack of three dimensional information when creating the posture image. A motion of the hand along the camera plan cannot be perceived. Such movements always amount for a not-moving hand held at the same x-y location (see figure 2). The use of an additional camera/stereo image copes with this issue and permits the system to extract depth information.

Acknowledgments

This work is supported in part by the European Commission through the TMR Marie Curie Grant # ERBFMBICT972613; also in part by the Office Naval Research grants: N00014-99-1-0377, and N00014-99-1-0380.

References

1. Amari S.-I., "Dynamics of pattern formation in lateral-inhibition type neural fields", Biological Cybernetics, 27:77–87, 1977.
2. Backer J., "Stochastic Modeling for Automatic Speech Understanding", Speech Recognition, Reddy D. eds, pp. 521–542, Academic Press, New York, 1975. 38
3. Bengio Y., "A Connectionist Approach to Speech Recognition", International Journal of Pattern Recognition and Artificial Intelligence, 7(4):3–22, 1993. 38
4. Bengio Y., "Markovian Model for Sequential Data", Neural Computing Surveys, 2:129–162, 1999. 39
5. Bishop C. M., "Neural Networks for Pattern Recognition", Clarendon Press, 1995. 36
6. Bourlard H., and Morgan N., "Hybrid Connectionist Models for continuous Speech Recognition", Automatic Speech and Speaker Recognition: Advanced Topics, Lee, Soong and Paliwal eds., pp. 259–283, Kluwert Academic, 1997. 38, 39
7. Braumann U.-D., "Multi-Cue-Ansatz für ein Dynamisches Auffälligkeitssytem zur Visuellen Personenlokalisation", PhD thesis, TU Ilmenau (Germany), 2001. 36
8. Bridle J. S., "Probabilistic Interpretation of Feedforward Classification Network Outputs with Relationship to Statistical Pattern Recognition", Neurocomputing: Algorithms, Architectures and Applications, Soulie' F. and Herault J. eds., NATO ASI Series, pp. 227–236, 1990. 42
9. Cohen P. R., Johnston M., McGee D. R., Oviatt S., Pittman J., Smith I., Chen L., and Clow J., "QuickSet: Multimodal interaction for distributed applications", Proceedings of the 5th International Multimedia Conference, pp. 31–40, 1997. 34
10. Böhme H.-J., Braumann U.-D., Corradini A., and Groß H.-M., "Person Localization and Posture Recognition for Human-robot Interaction", Gesture-Based Communication in Human-Computer Interaction: International Gesture Workshop, Lecture Notes in Artificial Intelligence 1739, pp. 105–116, 1999. 36
11. Corradini A., and Groß H.-M., "Implementation and Comparison of Three Architectures for Gesture Recognition", Proceedings of the IEEE International Conference on Acoustics, Speech, and Signal Processing, 2000. 43
12. Corradini A., Böhme H.-J., and Groß H.-M., "A Hybrid Stochastic-Connectionist Architecture for Gesture Recognition", special issue of the International Journal on Artificial Intelligence Tools, 9(2):177–204, 2000. 43
13. Dorffner G., "Neural Networks for Time Series Processing", Neural Network World, 6(4):447–468, 1996. 38
14. Efron D., "Gesture, Race and Culture", Mouton & Co. (The Hague), 1972. 35
15. Elman J. L., "Finding Structure in Time", Cognitive Science, 14:179–211, 1990. 38, 39
16. Essa I. A., and Pentland A.,"Facial Expression Recognition using Dynamic Model and Motion Energy", MIT Media Laboratory, Technical Report 307, 1995. 35
17. Hjorth J. S. U., "Computer Intensive Statistical Methods Validation, Model Selection, and Bootstrap", Chapman & Hall, 1994. 44
18. Jordan M. I., "Serial Order: A Parallel Distributed Processing Approach", Advances in Connectionist Theory, Elman L. and Rumelhart E. eds., Lawrence Erlbaum, 1989. 38, 39
19. Kendon A., "Current Issues in the Study of Gesture", Current Issues in the Study of Gesture, Nespoulous J.-L., Perron P., and Lecours A. R. eds pp. 200–241, 1986. 35

20. King S., and Weiman C., "Helpmate Autonomous Mobile Robot Navigation System", Proc. of the SPIE Conf. on Mobile Robots, pp. 190–198, Vol. 2352, 1990. 35
21. Kundu A., and Bahl L., "Recognition of Handwritten Script: a Hidden Markov Model based Approach", Proceedings of the IEEE International Conference on Acoustic, Speech, and Signal Processing, pp. 928–931, 1988. 38
22. LaViola J. J., "A Multimodal Interface Framework for Using Hand Gestures and Speech in Virtual Environments Applications", Gesture-Based Communication in Human-Computer Interaction: International Gesture Workshop, Lecture Notes in Artificial Intelligence 1739, pp. 303–314, 1999. 34
23. McGee D. R., and Cohen P. R., "Creating tangible interfaces by augmenting physical objects with multimodal language", Proceedings of the International Conference on Intelligent User Interfaces, ACM Press, pp. 113-119, 2001. 34
24. McKenzie Mills K., and Alty J. L., "Investigating the Role of Redundancy in Multimodal Input Systems", Gesture and Sign-Language in Human-Computer Interaction: International Gesture Workshop, Lecture Notes in Artificial Intelligence 1371, pp. 159–171, 1997. 35
25. McNeill D., "Hand and Mind: what gestures reveal about thought", The University of Chicago Press (Chicago, IL), 1992. 35
26. Morris D., Collett P., Marsh P., and O'Shaughnessy M., "Gestures: their origin and distribution", Stein and Day, 1979. 35
27. Mozer M. C., "Neural Net Architectures for Temporal Sequence Processing", Weigend A. and Gerschenfeld N. eds., Time Series Prediction: Forecasting the Future and Understanding the Past, Addison-Wesley, pp. 243–264, 1993. 38
28. Pavlovic V. I., Sharma R., and Huang T. S., "Visual Interpretation of Hand Gestures for Human Computer Interaction: A Review", IEEE Transactions on Pattern Analysis and Machine Intelligence, 19(7):677–695 , July 1997. 35
29. Rabiner L. R., and Juang B. H., "An Introduction to Hidden Markov Models", IEEE ASSP Magazine, pp. 4–16, 1986. 39, 44
30. Richard M. D., and Lippmann R. P., "Neural Networks Classifiers Estimate Bayesian a posteriori Probabilities", Neural Computation, 3:461–483, 1992. 42
31. Rubine D, "Specifying Gestures by Example", Computer Graphics, 25(4), pp. 329–337, 1991. 35
32. Schenkel M. K., "Handwriting Recognition using Neural Networks and Hidden Markov Models", Series in Microelectronics, Vol. 45, Hartung-Gorre Verlag, 1995. 38
33. Schraft R. D., Schmierer G., "Serviceroboter", Springer Verlag, 1998. 35
34. Sowa T., Fröhlich M., and Latoschik M. E., "Temporal Symbolic Intgration Applied to a Multimodal System Using Gestures and Speech", Gesture-Based Communication in Human-Computer Interaction: International Gesture Workshop, Lecture Notes in Artificial Intelligence 1739, pp. 291–302, 1999. 34
35. Stove A., "Non-Emblematic Gestures for Estimating Mood", Gesture and Sign-Language in Human-Computer Interaction: International Gesture Workshop, Lecture Notes in Artificial Intelligence 1371, pp. 165–171, 1997. 35
36. Waibel A., Hanazawa T., Hinton G. E., Shikano K., and Lang K. J., "Phoneme Recognition Using Time-Delay Neural Networks", IEEE Transactions on Acoustic, Speech, and Signal Processing, 37(12):1888-1898, 1989. 38, 39
37. Waibel A., and Lee K., "Readings in Speech Recognition", Morgan Kaufmann, 1990. 38
38. Waldherr S., Thrun S., and Romero R., "A Gesture-based Interface for Human-robot Interaction", to appear in: Autonomous Robots, 2000. 35

39. Wren C. R., Azarbayejani A., Darrell T., and Pentland A. P., "Pfinder: Real-Time Tracking of the Human Body", IEEE Transactions on PAMI, 19(7):780–785, 1997. 37
40. Wu Y., and Huang T. S., "Vision-Based Gesture Recognition: A Review", Gesture-Based Communication in Human-Computer Interaction: International Gesture Workshop, Lecture Notes in Artificial Intelligence 1739, pp. 103–116, 1999. 35

Development of a Gesture Plug-In for Natural Dialogue Interfaces

Karin Husballe Munk

Computer Vision and Media Technology Laboratory
Aalborg University, Denmark
khm@cvmt.auc.dk

Abstract. This paper describes work in progress on the development of a plug-in automatic gesture recogniser intended for a human-computer natural speech-and-gesture dialogue interface module. A model-based approach is suggested for the gesture recognition. Gesture models are built from models of the trajectories selected finger tips traverse through physical 3D space of the human gesturer when he performs the different gestures of interest. Gesture capture employs in the initial version a data glove, later computer vision is intended. The paper outlines at a general level the gesture model design and argues for its choice, as well as the rationale behind the entire work is laid forth. As the recogniser is not yet fully implemented no test results can be presented so far. Keywords: bi-modal dialogue interface, human natural co-verbal gestures, gesture recognition, model-based, spatial gesture model, finger tip trajectory.

1 Introduction

When humans communicate face-to-face in a dialogue, their utterances serve two purposes: exchange of meaning, and regulation of the exchange acts [2,4,11]. For both purposes varying modalities are used. The regulating utterances in themselves and the shifts in modality used for expression of an utterance are made in order to achieve a robust, efficient, and smoothly running communicative process. The modalities used include for the both utterance functions, content delivery and process control, auditory and visual behaviours. Included are speech acts, non-verbal utterances, intonation variation, facial expression, gaze direction, and hand and body gestures.

A commonly adopted metaphor for an ideal human-computer interface is that of a human-human-like dialogue [4,5]. The dialoguing dyads are the human user and some anthropomorphic "something" on equal footing with the human. Thus to take the metaphor seriously and intend its realisation must imply provision of the above-mentioned communicative capabilities and skills for the computer part.

As one step-on stone toward this goal, a hand gesture detection and recognition plug-in module is under development at our laboratory. This paper describes the current work in progress on this module. The gesture recogniser is intended for

I. Wachsmuth and T. Sowa (Eds.): GW 2001, LNAI 2298, pp. 47–58, 2002.
© Springer-Verlag Berlin Heidelberg 2002

integration with a discourse manager module, which takes input from that and from a speech recogniser [15,17] in parallel. Together the modules comprise a bi-modal input channel aimed towards a virtual computer-generated character, a humanoid agent. The input channel is employed as part of a full bi-modal interface in a pre-designed sample application. Here virtual characters confront the user immersed in a virtual reality environment [10,16] for the experience of certain dramaturgical events [8]. Fig. 1 sketches the intended context of operation for the gesture recogniser. The bi-modal input channel is meant for continuous operation, with no external signalling to the system of the start and the end of the single utterances from the user.

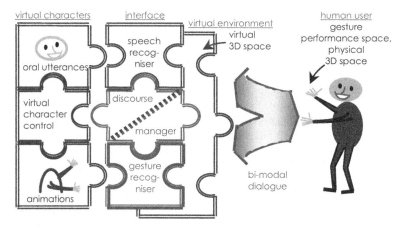

Fig. 1. The intended context of operation for the plug-in gesture recogniser: in a speech-and-gesture dialogue interface towards virtual character(s) in a virtual environment set-up

2   Design Considerations

An overall design consideration for the gesture recogniser concerns a thorough decision on the task to be fulfilled. To make the recogniser as purposeful as can be, this issue is attended to in two regards. The first is a determination of the definition set for the recogniser, also denoted the set of items of interest. That is, which items the recogniser is to detect in the midst of everything, and recognise, perhaps as an instance from a more general class. The second is a discussion at a principal level of which processes the recogniser is to execute. In specific focus is distribution of responsibility between the gesture recogniser and the discourse manager to which the recogniser is to be plugged. Finally processing time requirements are commented.

2.1   Choice of Gestures of Interest

The set of items of interest for a gesture recogniser obviously is a set of gestures. There may here be a distinction between the term *'gesture'* understood as *'gesture meaning'*, that is, the semantics carried in some gesture expressed, and *'gesture'* understood as *'gesture form'*, i.e. the physical shape of the expression. From a pragmatic point of view all gesture recognition commence with recognition of gesture form, and mapping to semantic content is a sort of add-on processing. As laid forth in the next section, for the present plug-in recogniser this is even charged to a different module, the discourse manager. Thus for any gesture recogniser basically, and in particular for the one, we develop, entirely, the set of items of interest is a set of gesture forms. Each gesture form is understood to be manifest in space and time in a delimited segment of hand movements and or changes in hand or finger configuration. That is, a gesture form occupies a finite volume in space, and endures for a finite time. For operation in continuous scenarios as sketched above, a particular aspect of the recognition process is to pick out gesture candidate segments from the continuous input of user hand movement observations. This is discussed further in Sect. 2.2.

In a first version the set of gesture forms of interest is desired to establish a vocabulary of forms that is manageable by being limited, yet plausible and worthy by being used frequently. When deciding on which gestures to include it then seems natural to base on empirical studies of humans in face-to-face dialogues. But the wide and intricate variation of forms humans use in conversational gesturing [2,11], makes this a tedious, and perhaps not passable path to tread. To catch all forms used even for expressing one single meaning may turn out almost impossible. The other way around, to list all possible different gesture forms is perhaps quite as laborious and unachievable. E.g. let a gesture form be characterised by the trajectories in space the finger tips traverse. Then the listing requires for each finger tip a quantization of the set of all traversable trajectories in the 3D physical space of the gesturer. The quantization must divide the set into ensembles of trajectories denoted to be of the same form, that is, having shapes that appear alike inside the ensemble but separably different from shapes of trajectories of adjacent ensembles. Though constrained by human physiology, still traversable trajectory shapes span a vast continuum and the number tends to infinite.

In conclusion, the all-exhaustive approach for decision on the gestures of interest seems in no form feasible. Still the suggested gesture recogniser may realise a valuable input channel while concentrating on a few distinct gesture forms. Ideally:

- The gestures of interest shall be a comprehensive yet tractable set of gesture forms considered to be natural choices of form for expressing meanings considered to be natural to express via gestures at the incidents when turning to the gesture modality seems a natural choice. The 'gesture-prompting' incidents considered, shall be those expected to be typical to occur during execution of the expected typical application(s) to be served by the dialogue interface. Most frequent gesture forms shall be given priority.
- The gestures of interest shall pose a well-structured and well-bounded obvious set, to ease the human user in comprehending supported forms from unsupported.

Previous studies have shown that human natural communicative gestures posses an innate variation even within the same single form [2,3,11]. Gestures of the perceptually same form, as judged by human observers, exhibit quite a range in the precise curves followed in 3D space by hand and fingers. The variation is ubiquitous both from person to person, and when the same person repeats the gesture form. Hence when designing the set of gesture forms of interest to include a limited number of distinct gesture forms, this is understood to mean perceptually different forms.

In the remainder of the paper the designating *'perceptually'* is left out for short but to be understood when speaking of a specific gesture form, unless otherwise stated. Also, *'gesture'* is used as short for *'gesture form'*, leaving out the designating *'form'*, where it is thought not to lead to misunderstanding.

For the plug-in gesture recogniser we develop, a set of gestures of interest for a prototype version has been selected in cooperation with a group of linguists at the Centre for Language Technology (Copenhagen) who works with the interaction between gestures and speech [15,16]. Fig. 2 tabulates the gesture meanings for which a chosen subset of intuitively possible forms (1-3 per meaning) is included in the set of interest, and some of their properties of relevance for this work. As noted from the table the set spans the domain of natural, co-verbal gestures in different senses.

Whether this set fulfills the above prescriptions for the choice of gestures of interest cannot be finally evaluated theoretically. A set can only be validated through implementation and subjection of the resulting recogniser to a group of arbitrary users of different familiarity with similar interfaces, gesturing styles, ages, etc.

meaning	denotation	function level & function	form	form characteristics
pointing gestures	hand signal, emblem, deictic (concrete)	semantic, sentence level & content provision	standard	static, end-posture, non-quantifiable
size-indicating gestures	conversational topic / hand signal, emblem (by pre-agreed form)		non-standard (limited variety) / fix (pre-agreed)	static, end-posture, quantified
waving gestures	conversational topic (form non-standard) / hand signal, emblem (fix form)		non-standard (culture specific standard elements	dynamic, course of hand movements, non-quantifiable
turn gestures	conversational - interactive, beat, meta-gesture	pragmatic, discourse, meta-level & process regulation	non-standard (tends to degree of common elements)	

Fig. 2. The set of gestures of interest includes subsets of forms for the meanings listed. Legend: *(meaning)*: the conveyed meaning, *(denotation)*: the super-ordinate gesture type, *(function level & function)*: the linguistic level of operation and function for *(meaning)*, *(form)*: the physical gesture form, *(form characteristics)*: properties and essence of *(form)* [2,11]

2.2   Domain of Responsibility

As stated above, from a pragmatic point of view, all gesture recognition begins by recognition of gesture form. And for the presented plug-in recogniser, here it also ends; interpretation of gesture semantics is postponed for the subsequent discourse manager module to handle. This is due to the following:

In general the mapping between communicative meaning expressed and gesture form is a multiple-to-multiple mapping in human natural face-to-face dialogues [2,3,11]. The actual meaning conveyed by a gesture of a certain form changes largely e.g. with the instantaneous processing context for the gesture issue, that is, the state of the discourse process. On the other hand, the gesture form chosen for signalling a certain meaning depends on situation, recent and simultaneous other utterances whether by gesture or other modality, style and age of the gesturer, and adopted conventions of the dialogue dyads. Hence when the semantics of gesture inputs in an integrated multi-modal interface are to be interpreted, one must take into consideration both current and previous inputs by the other modalities, and other contextual factors. In such interfaces typically one function exists which surveys user input from all sources and modalities, and keeps a log of human-computer dialogue history. For the bi-modal input channel of which the present plug-in gesture recogniser is intended to be part, such function lies in the discourse manager [16,17]. This module is therefore the one capable of and also in charge of interpreting the semantics of the gesture inputs. To summarise, in multi-modal interfaces ideally the output from the gesture recogniser is not some fixed meaning interpretation tied to the recognised gesture form, but rather an agreed label for the perceived shape of the form. E.g. for a gesture waving towards the gesturer himself, ideally the output is not an annotated semantic <Come>, but rather the tag "Waveto", using a somewhat descriptive label, or "shape7", as labels are in principle arbitrary. And annotation of the communicated meaning is left for the discourse manager or a similar module.

For any recogniser the canonical sequence of processes to consider to engage in may be expressed: i) input registration or data capture, ii) description or representation, iii) classification or identification, and iv) interpretation. For a recogniser in continuous operation as intended for the present, the single processes are repeated continuously and concurrently, providing a continuous update of the internal representation each produces. For a gesture recogniser the time span covered by an entity of an internal representation is expected to increase throughout the process sequence from i) to iv). Hence the processes are here expected to repeat themselves in slower and slower loops, with i) data capture repeating fastest.

The actual recognition of an item of interest may be perceived as taking place in process iii) or iv). For a gesture recogniser, it lies close to translate the former to be detection and identification of the occurrence of certain delimited patterns of hand movements and or changes in hand formation. The latter then is identification of the semantic content embodied in the movements or changes. In other words, process iii) identifies gesture physical form, process iv) identifies gesture conversational meaning. Hence in this translation of the canonical processes, i)-iii) all concern exclusively gesture outline in time and space. And the conclusion above on the ideal distribution of tasks in a multi-modal interface then gives, for the plug-in gesture recogniser at hand responsibility is limited to processes i)-iii).

In brief, process i) translates to acquisition of input hand movements and hand formation changes. Process ii) translates to description of these in such a way and at such a level of abstraction, that it is possible and computer-tractable in process iii) to evaluate whether or not they contain a gesture of interest, and which. For cases of continuous input as intended for the present gesture recogniser, the question of where in the input the gesture is found, corresponding to when was the gesture performed, is often of equally high importance. To do the evaluation at all, in principle the continuous input must in a pre-step to the description process be divided into suitable pieces, each to be described on its own. Each piece is to cover quite exactly a time interval of either a single gesture performance or no gestures of interest. When the gesture start and end are unknown as in the intended context of operation, this temporal segmentation is a critical issue. In the gesture recogniser here presented, the problem is circumvented like this: At first all possible starting points for a gesture are hypothesised also to be one, and then in a later phase of processing the false hypotheses are discarded. The selection of candidate starting points is closely linked to the formalism used for the descriptions eventually produced. In Sect. 3 this is explained to further depth.

2.3   Timing Requirements

For the plug-in gesture recogniser in the intended scenario, timing requests may be put up as follows. Human conversational gestures precede or occur simultaneously with the speech phrase they accompany [2,5,11] if such exist; gestures may occur having no correspondence in speech. Thus one request could be that recognised, speech-accompanying gestures are delivered at least as fast as the speech recogniser deliver recognition of the accompanied phrase. But this need not be. As long as the detected gestures and speech phrases are all properly time stamped, they may subsequently be correctly paired in the discourse manager. Hence for both un- and accompanying gestures the timing requirement is set by the capabilities of that module, and by the total action-reaction loop from user to virtual character addressed, or system as such. The acceptable delay from user utterance to character or system answer or otherwise reaction, varies with the complexity of the utterance, and the expected complexity of the response. For simple utterances requesting simple responses no more than a few seconds is assumed tolerable before start of the response. This delay available for processing time is to be shared amongst all modules in the system coming into play in sequence in order to present the response to the user. With reference to Fig. 1, for a gesture prompting e.g. an oral or gesture response, these include the gesture recogniser, the discourse manager, the virtual character control module, and the module handling either character animation or character oral utterances. With equal shares, the timing requirement on the gesture recogniser is then to produce a valid output within about 0.5-0.75 seconds from the end of a gesture of interest.

3   Method and Techniques

3.1   Approach

Approaches undertaken in previous work by others on the recognition of hand gestures may roughly be divided into two classes: what may be largely categorised the connectionists' approach applying trained networks like Hidden Markov Models or neural networks, and the model-based approach implying use of distinct, explicit, apparent models [6,7,18]. Typically the former runs directly on the input measurement signal without any foregoing conversion or interpretation of data. The latter on the contrary most often depends on exactly this; from the input measurement signal, features thought to embed exclusive properties of each of the gestures to recognise are derived, and further operation is based solely on analysis of these features. It has been claimed that although the connectionists' approach has shown promising with implemented systems realising fair recognition rates, and therefore have been widely pursued in recent years, in order to achieve the last gap to 100 % the most prosperous and perhaps necessary path to follow leads through model-based approaches [6]. The work here presented takes on a model-based approach.

Specifically two types of models are developed and used in the gesture recogniser:

- a kinematic model of the hand defining hand geometry and articulation structure,
- a 3D spatial model for each gesture of interest, residing in physical performance space of the gesture itself and of the human gesturer.

Following a note on input data registration, the next sections discuss these models, their choice, use, and establishment to further detail.

3.2   Gesture Capture

In general gestures or hand movements may be registered for computer analysis either visually by means of one or more video cameras capturing images of hand and fingers which through analysis may render their varying positions in space [6,7,13,18,19], or more directly a data glove and or a magnetic tracker mounted on the hand may provide measurements on the changing finger formation and or overall hand position respectively [1,12,20]. This work is pre-planned to use both manners, visual and direct, as substitutes for each other in two different successive settings.

The current first setting uses a 16-sensor data glove for finger configuration by measurement of finger joint flexions [21], in conjunction with a magnetic tracker mounted at the centre of the back of hand for overall hand position and orientation [22]. The next setting will use vision equipment of one or two video cameras instead.

3.3   Hand Model

As explained in the next section, the developed model for a gesture is based on the trajectories in 3D physical performance space that some selected points on the hand traverse during performance of the gesture. In the current setting in order to calculate such trajectories from the provided measurements of finger joint flexion and overall hand position and orientation, a kinematic hand model is requested and used.

The hand model specifies hand articulation structure through location of finger joints and pertaining motion degrees of freedom, and hand geometry through spatial dimensions of finger segments and palm of hand. The model is inspired by the human hand skeletal structure, and by the model suggested in [9], but differs from this in location of thumb-part and other minor details.

To obtain ground true trajectories the hand model must be individually adjusted to fit each gesturing person and reflect the gesturer's exact lengths and widths of hand segments. In contrast, even for the same person gestures of the same form exhibit an innate, unavoidable variation in the precise true curves followed in 3D space, as noted earlier. This natural variation may well subsume any calculated trajectory inaccuracy or distortion induced by hand model misfit. The ubiquitous presence of the variation also argues for basing the gesture recognition on gesturer-independent gesture models that are designed to accommodate occurring differences in the outline observed for same gesture. Hence such method is undertaken in the present gesture recogniser, as discussed below. And, for these reasons a fixed hand model is used, reflecting hand and finger dimensions of an arbitrary 'average' person.

3.4   Gesture Model

As stated above, the gesture models designed must cater for unavoidable variation in the performances of the same gesture form in a proper manner. The models must be wide for all instances of the same gesture to be classified together, and still narrow for different gestures to be separable. In this work the following course is adopted:

For each gesture of interest a model is defined based on the trajectories that some selected points on the hand traverse in 3D physical space. In focus are specifically centre points of finger tips. Several facts promote this choice: it constitutes a model basis simple to comprehend, and tip trajectories are straight forward calculable from the current input measurements. Further it supports replacement of the current gesture capture devices of data glove and tracker by vision equipment in a subsequent setting in that finger tip trajectories are readily obtainable from image sequences [14,19], which time progression of parameters directly measurable by the data glove and tracker are not.

The gesture model not only is based on finger tip trajectories traversed in 3D physical performance space of the gesture, it also resides in this space itself.[1] Each gesture model consists of a trajectory model for the trajectory of the tip of each of one or more fingers on one or both hands as considered necessary and sufficient for safe distinction of this gesture from others. That is, a gesture model takes one to ten trajectory models for as many different finger tips as its constituents. In the initial version trajectory models are purely spatial, leaving out conditions of time.

A trajectory model consists of two elements, both adhering to absolute locations in physical 3D space:

- a 3D spatial tunnel-of-validity, i.e. a volume in 3D physical space which the trajectory of the tip is modelled to progress within and always confine itself to,
- a set of points, termed crooks, defined along a so-called core trajectory modelled for the tip. Ideally the core trajectory is the single best-modelling model trajectory, i.e. the curve of closest fit for some defined metric, in average for all trajectories the tip ever traverses in performances of the gesture in question. The definition of crooks base on these points to share the property that they are salient for the shape of the trajectory in some and the same respect. E.g. as points do, that are located at the centres of sharp bends on the trajectory. In line with this, a suggestion is to use points at local maxima of curvature as the set of crooks. In the gesture recognition the crooks act as checkpoints: To be recognised being an instantiation of a model, an observed tip trajectory must pass the crooks of the model in correct order, and further, exhibit crooks of its own exactly and only at these model crook locations, too. Some tolerance may be allowed both for model crook passing and observed crook location.

The set of crooks and their use as checkpoints request the observed trajectory to have shape essentials similar to that of the core trajectory both in number and in spatial

---

[1] Strictly, observed tip trajectories are calculated, and as such reside in a "calculation space", possibly slightly different from true physical space due to imperfections in calculation basis (measurement values, hand model), whereby the gesture models also strictly truly reside in this "calculation space".

distribution, with respect to the definition of crooks. And even stronger, each shape essential of the observation must occur in close vicinity of the absolute position in 3D space of the corresponding shape essential of the core. The tunnel-of-validity states complementary requests upon the trajectory segments in between the crooks by setting out a requested bounding volume for the entire trajectory shape and its absolute location in 3D space. Altogether this yields a firm stating designed to separate different gestures, of a path an observed trajectory must follow to be accepted as originating from an ongoing, modelled gesture.

Then to accommodate the innate variation noted earlier in the exact outline of the same gesture, a set of guidelines is introduced to allow a some spatial transformation of an observed trajectory, in principle prior to evaluation against a model. Ideally any observed trajectory that truly originates from a performance of a certain gesture, can and will through proper transformation in accordance with the guidelines be brought in compliance with the corresponding trajectory model included in the model for this gesture, that is, the trajectory model for the same finger tip as the observation concerns. The guideline contents are further explained by the following pictorial expression of the criteria for an observed trajectory to comply with a model: It must be possible to lift up the observation trajectory by its crooks as if bent in wire, from its original location in absolute 3D space, and hang it on the model core trajectory, placing the crooks of the observation on the crooks of the core in proper order of traversal. Possibly the observation must be rotated differently than originally. In this handling the observation is treated as a rigid object except that each crook-to-crook segment may be uniformly adjusted, compressed or expanded, separately as to scale up or down the size of the segment as necessary but not deform its shape. And in the end, at no place the observation must exceed the tunnel-of-validity volume.

Accept of an observed trajectory to comply with a trajectory model is said to produce a piece of evidence for the hypothesis that a performance of the gesture, the model for which includes this trajectory model, is ongoing or just recently finished.

3.5   Gesture Recognition

The overall scheme for gesture recognition proceeds as a multiple concurrent hypotheses checking. Each hypothesis is an assumption that a different gesture from the set of gestures of interest is being or has just been performed. A hypothesis renders true when sufficient evidence for it is provided. As stated above, a piece of evidence is contributed by each currently observed trajectory found to be an acceptable realisation of a trajectory model which is one of the constituents for the model of the gesture in question. Thus to validate to the strongest level the hypothesis that performance of a certain gesture is ongoing or just recently finished, the observed trajectories for all finger tips for which the model for this gesture includes a trajectory model, must comply with their correspondence respectively amongst these.

In the comparison of an observed trajectory to a trajectory model it is crucial that the two are properly synchronised. This is not trivial, recalling the model is finite whereas the observation may in principle be infinite as continuous operation is intended. For correct comparison the two must be aligned, so that the start of the model is compared to the point on the observation, that the finger tip traverse at the state of the gesture performance which corresponds to the model start. That the location of this point on the observation is unknown is overcome by evaluating multiple comparisons of the model to the observation using multiple starting points for the alignment: Whenever a new crook is exposed on the observation, it is hypothesised that this is exactly the crook to match the first crook encountered on the model; and evaluation of a comparison with the model aligned at this position is initiated. This is done concurrently for all trajectory models defined for the finger tip in question. And it is done in parallel with the concurrent further evaluations of the comparisons initiated at preceding exposures of crooks on the observation. For a given comparison, eventually the observed crooks exhaust the crooks of the model. Then either the comparison finally fails, or is accepted to validate that an instantiation of a model has been found. Measures may be put up to allow to reject or to accept a comparison before counterparts to all crooks of the compared model have been exposed on the observation.

4   Conclusion

The paper presents a gesture recogniser to realise the one input channel in a bi-modal human natural dialogue interface where the other is realised by a speech recogniser.

For the gesture recognition a model-based approach is suggested. The approach is based on a novel scheme for gesture modelling. The scheme adheres distinctly to the spatiality of human hand gestures as this manifests itself in the trajectories in specific finger tips traverse in the 3D physical space of the gesturer when performing the gestures. The paper gives a detailed conceptual description of the gesture model grounds and use. Also thorough considerations are put forth for the ideal choice of a limited vocabulary of gestures, more precisely gesture forms, to be the set of interest for the recogniser. The rationale behind construction of a human-computer interface based on communicative capabilities on level of the human for natural speech and in specific natural gesture is discussed as well. The discussion includes a pointing to the ideal conjuncture of tasks of the gesture recogniser and of the discourse manager or similar module in a multi-modal system.

At current state, implementation of the approach into a prototype is still in progress. Theoretic work on methodological planning and algorithmic derivations has been elaborated. Gesture capture devices of a data glove and a magnetic tracker have been installed. But the realisation of the recogniser procedures into program code has not been finished yet. Hence, so far it has not been possible to carry out any empirical tests, neither regarding the chosen set of gestures of interest, nor the feasibility of the suggested approach at large.

Once fully implemented, the prototype plug-in gesture recogniser will provide an interesting test-bed in the narrow scope for the capability of the suggested approach to do the intended recognition. Further it will support investigation of different gestures of interest for suitability for recognition by an automatic (in specific this) gesture recogniser, and for how artificial or natural humans subjectively experience it is to use these gestures. Proper evaluation of the latter requires that the full intended context of operation is established, that is, setting up the total bi-modal interface system into which the gesture recogniser is intended to be plugged. Feedback in parallel on the supported expressive channels possibly employed in parallel by the gesturer, that is, gesture and speech, must be enabled since impact of this feedback on the gesturer's experience must be taken into account.

In the broader scope the total bi-modal set-up will support usability tests for clarifying the value of a restricted vocabulary gesture interface when in combination with speech, in particular natural gestures combined with natural speech. These tests may point to gestures it is crucial to include in the set of interest. At the top level such studies address the fundamental position underlying the work presented. Previous investigations by others [2-5,15] have indicated that addition of a gesture recogniser to a human-computer natural dialogue speech-based interface does mean a benefit. With the suggested approach, we believe it is also feasible, tractable, and worth the cost.

Acknowledgements

The work presented is a contribution to the Danish national research project "The Staging of Inhabited, Virtual, 3D-Spaces", funded by the Danish Research Councils.

References

1. Andersen, C. S., A survey of Gloves for Interaction with Virtual Worlds. Tech. Rep. LIA 98-8, Lab. for Image Analysis, Aalborg University, Denmark. Aug. 1998.
2. Bavelas, J. B., Chovil N.,Coates, L., Roe, L. Gestures Specialized for Dialogue. Personality and Social Psychology bulletin, V. 21, No. 4, pp. 394-405. April 1995.
3. Cassell, J., Bickmore, T., Billinghurst, M., et al. Embodiment in Conversational Interfaces: Rea. In: CHI'99 Proc., Ass. for Computing Machinery, Inc., pp. 520-527. 1999.
4. Cassell, J., Bickmore, T., Campbell, L., et al. Requirements for an Architecture for Embodied Conversational Characters. In: Computer Animation and Simulation '99 (Eurographics Series), Springer Verlag, Vienna, Austria. 1999.
5. Cassell, J., Stone, M. Living Hand to Mouth: Psychological Theories about Speech and Gesture in Interactive Dialogue Systems. In: AAAI 1999 Fall Sym. on Narrative Intel., American Ass. for Art. Intel. 1999.
6. Cédras, C., Shah, M. Motion-based recognition: a survey. Image and Vision Computing, V. 13, No. 2, pp. 129-155. March 1995.
7. Gavrila, D. M., The Visual Analysis of Human Movement: A Survey. Computer Vision and Image Understanding, V. 73, No.1, pp. 82-98. Jan. 1999.
8. Klesen, M., Szatkowski, J., Lehmann, N. The Black sheep – Interactive Improvisation in a 3D Virtual World. (To app.) In: Proc. i3 Ann. Conf. 2000, Jönköping, Sweden. Sept. 2000.
9. Lee, J., Kunii, T.L. Model-Based Analysis of Hand Posture. IEEE Computer Graphics and Applications, V. 15, No. 5, pp. 77-86. Sept. 1995.

10. Madsen, C. B., Granum, E. Aspects of Interactive Autonomy and Perception. In: Virtual Interaction: Interaction in Virtual Inhabited 3D Worlds, Qvortrup, L. (ed), Springer Verlag, London. 2001.
11. McNeill, D. Hand and Mind: What Gestures Reveals about Thought. University of Chicago Press, Chicago. 1992.
12. Moeslund, T. B. Interacting with a Virtual World Through Motion Capture. In: Virtual Interaction: Interaction in Virtual Inhabited 3D Worlds, Qvortrup, L. (ed), Springer Verlag, London. 2001.
13. Moeslund, T. B., Granum, E. A Survey of Computer Vision-Based Human Motion Capture. (Subm. to:) Int. Journal on Computer Vision and Image Understanding. Feb. 2000.
14. Nölker, C., Ritter, H. Detection of Fingertips in Human Hand Movement Sequences. In: Gesture and Sign Language in Human-Computer Interaction, Proc. Int. Gesture Ws.'97, pp. 209-218, Lecture Ns. in Art. Intel. Ser., No. 1371, Springer Verlag, Berlin Heidelberg. 1998.
15. Paggio, P., Jongejan, B. Representing Multimodal Input in a Unification-based System: the Staging Project. In: Proc. Ws. on Integrating Information from Different Channels in Multi-Media-Contexts, ESSLLI MM, Holmqvist, K., Kühnlein, P., Rieser, H. (eds), paper 1. 1999.
16. Paggio, P., Jongejan, B. Unification-Based Multimodal Analysis in a 3D Virtual World: the Staging Project. In: Learning to Behave: Interacting Agents. Proc. Twente Ws. on Language Technology 17 (TWLT   17), Enschede, The Netherlands. Oct. 2000.
17. Paggio, P., Music, B. Linguistic Interaction in Staging – a Language Engineering View. In: Virtual Interaction: Interaction in Virtual Inhabited 3D Worlds, Qvortrup, L. (ed), Springer Verlag, London. 2001.
18. Pavlovic, V. I., Sharma, R., Huang, T.S. Visual Interpretation of Hand Gestures for Human-Computer Interaction: A Review. IEEE Trans. PAMI, V. 19, No. 7, pp. 677-695. July 1997.
19. Rehg, J., Kanade, T. DigitEyes: Vision-Based Human Hand Tracking. Tech. Rep. CMU-CS-93-220, School of Comp. Science, Carnegie Mellon University, Pittsburgh. Dec. 1993.
20. Sturman, D. J., Zeltzer, D. A Survey of Glove-based Input. IEEE Computer Graphics and Applications, V. 14, No. 1, pp. 30-39. Jan. 1994.
21. 5DT Data Glove 16, 5DT Data Glove 16-W, Data gloves for the fifth dimension, User's manual V. 1.00. 5DT Fifth Dimension Technologies, Pretoria, South Africa. Feb. 2000.
22. 3SPACE ISOTRAK II User's Manual. Polhemus Inc., Colchester, Vermont, USA. 1993.

A Natural Interface
to a Virtual Environment through Computer Vision-Estimated Pointing Gestures

Thomas B. Moeslund, Moritz Störring, and Erik Granum

Laboratory of Computer Vision and Media Technology, Aalborg University
Niels Jernes Vej 14, DK-9220 Aalborg East, Denmark
{tbm,mst,eg}@cvmt.auc.dk

Abstract. This paper describes the development of a natural interface to a virtual environment. The interface is through a natural pointing gesture and replaces pointing devices which are normally used to interact with virtual environments. The pointing gesture is estimated in 3D using kinematic knowledge of the arm during pointing and monocular computer vision. The latter is used to extract the 2D position of the user's hand and map it into 3D. Off-line tests show promising results with an average errors of $8cm$ when pointing at a screen $2m$ away.

1 Introduction

A virtual environment is a computer generated world wherein everything imaginable can appear. It has therefore become known as a virtual world or rather a virtual reality (VR). The 'visual entrance' to VR is a screen which acts as a window into the VR. Ideally one may feel immersed in the virtual world. For this to be believable a user is either to wear a head-mounted display or be located in front of a large screen, or even better, be completely surrounded by large screens.

In many VR applications [4] the user needs to interact with the environment, e.g. to pinpoint an object, indicate a direction, or select a menu point. A number of pointing devices and advanced 3D mouses (space mouses) have been developed to support these interactions. These interfaces are based on the computer's terms which many times are not natural or intuitive to use.

In this paper we propose to replace such pointing devices with a computer vision system capable of recognising natural pointing gestures of the hand. We choose to explore how well this may be achieved using just one camera and we will focus on interaction with one of the sides in a VR-CUBE, see figure 1 a), which is sufficient for initial feasibility and usability studies.

2 The Approach

The pointing gesture belongs to the class of gestures known as *deictic gestures* which MacNeill [3] describes as "gestures pointing to something or somebody

I. Wachsmuth and T. Sowa (Eds.): GW 2001, LNAI 2298, pp. 59–63, 2002.

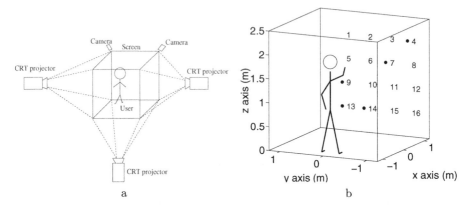

Fig. 1. VR-CUBE: a) Schematic view of the VR-CUBE. The size is 2.5 x 2.5 x 2.5m. b) Experimental setup. 16 points in a .5cm raster are displayed

either concrete or abstract". The use of the gesture depends on the context and the person using it [2]. However, it has mainly two usages: to indicate a direction or to pinpoint a certain object.

When the object pointing to is more than approximately one meter away, which is usually the case when pointing in a virtual environment, the pointing direction is indicated by the line spanned by the hand (index finger) and the visual focus (defined as the centre-point between the eyes). Experiments have shown that the direction is consistently (for individual users) placed just lateral to the hand-eye line [5].

The user in the VR-CUBE is wearing stereo-glasses and a magnetic tracker is mounted on these glasses. It measures the 3D position and orientation of the user's head which is used to update the images on the screen from the user's point of view. The 3D position of the tracker can be used to estimate the visual focus and therefore only the 3D position of the hand needs to be estimated in order to calculate the pointing direction.

Since we focus on the interaction with only one side we assume that the user's torso is fronto-parallel with respect to the screen. That allows for an estimation of the position of the shoulder based on the position of the head (glasses). The vector between the glasses and the shoulder is called the displacement vector in the following. This is discussed further in section 3.

The 3D position of the hand can usually be found using multiple cameras and triangulation. However, experiments have shown that sometimes the hand is only visible in one camera. Therefore we address the single-camera problem. We exploit the fact that the distance between the shoulder and the hand (denoted R), when pointing, is rather independent of the pointing direction. This implies that the hand, when pointing, will be located on the surface of a sphere with radius R and centre in the user's shoulder.

The camera used in our system is calibrated [6] which enables us to map an image point (pixel) to a 3D line in the VR-CUBE coordinate system. By estimating the position of the hand in the image we obtain an equation of a straight line in 3D and the 3D position of the hand is found as the point where the line intersects the sphere.

2.1   Estimating the 2D Position of the Hand in the Image

The following is done to segment the user's hand and estimate its 2D position in the image. Firstly the image areas where the user's hand could appear when pointing are estimated using the 3D position and orientation of the user's head (from the magnetic tracker), a model of the human motor system and the kinematic constraints related to it, and the camera parameters (calculating the field of view). Furthermore, a first order predictor [1] is used to estimate the position of the hand from the position in the previous image frame.

The histogram of the intensity image can be approximated by bimodal distribution, the brighter pixels originate from the background whereas the darker originate from the user. This is used to segment the user from the background. The optimal threshold between the two distributions can be found by minimising the weighted sum of group variances [4].

The colour variations in the camera image are poor. All colours are close the the gray vector. Therefore the saturation of the image colours is increased by an empirical factor. The red channel of the segmented pixels has maxima in the skin areas as long as the user is not wearing clothes with a high reflectance in the long (red) wavelengths. The histogram of the red channel can be approximated as a bimodal distribution, hence it is also thresholded by minimising the weighted sum of group variances. After thresholding the group of pixels belonging to the hand can be found [4].

3   Pointing Experiments without Visual Feedback

Five users were each asked to point to 16 different points displayed on the screen, see figure 1 b). No visual feedback was given during these experiments, hence the users should be unbiased and show a natural pointing gesture. An image of each pointing gesture was taken together with the data of the magnetic head tracker. The displacement vector between the head tracker and the shoulder was measured for each user.

Figure 2 a) shows the results of a representative pointing experiment. The circles (o) are the real positions displayed on the screen and the asterisks (*) connected by the dashed lines are the respective estimated positions where the user is pointing to. The error in figure 2 a) is up to $0.7m$. There are no estimates for the column to the left because there is no intersection between the sphere described by the hand and the line spanned by the camera and the hand of the user.

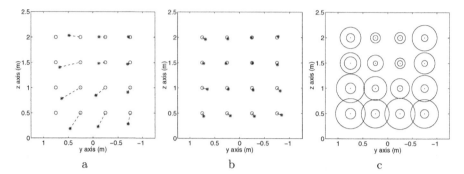

Fig. 2. Results from pointing experiments. See text

The error is increasing the more the user points to the left. This is mainly due to the incorrect assumption that the displacement vector is constant. The direction and magnitude of the displacement vector between the tracker and shoulder is varying.

For each user a lookup table (LUT) of displacement vectors as a function of the head rotation was build using the shoulder position in the image data and the tracker data. Figure 2 b) shows the result of a representative pointing experiment (same as used in figure 2 a) using a LUT of displacement vectors to estimate the 3D position of the shoulder. Notice that after the position of the shoulder has been correction estimates for the left column is available. In figure 2 c) the average result of all experiments are shown. Each inner circle illustrates the average error while each outer circle illustrates the maximum error. The total average is $76mm$ and the maximum error to $30mm$.

4   Discussion

In this paper we have demonstrated that technical interface devices can be replaced by a natural gesture, namely finger pointing. During off-line tests we showed the average error to be $76mm$ and the maximum error to $308mm$. This we find to be a rather accurate result given the user is standing two meters away. However, whether this error is too large depends on the application.

In the final system the estimated pointing direction will be indicated by a bright 3D line seen through the stereo glasses starting at the finger of the user and ending at the object pointed to. Thus, the error is less critical since the user is part of the system loop and can correct on the fly.

Currently we are deriving explicit expressions for the error sources presented above. Further experiments will be done in the VR-CUBE to characterise the accuracy and usability as soon as the real time implementation is finished. The experiments will show whether the method allows us to replace the traditional pointing devices as is suggested by our off-line tests.

References

1. Y. Bar-Shalom and T. E. Fortmann. *Tracking and Data Association.* Academic Press, INC., 1988. 61
2. E. Littmann, A. Drees, and H. Ritter. Neural Recognition of Human Pointing Gestures in Real Images. *Neural Processing Letters*, 3:61–71, 1996. 60
3. D. MacNeill. *Hand and mind: what gestures reveal about thought.* University of Chicago Press, 1992. 59
4. M. Störring, E. Granum, and T. Moeslund. A Natural Interface to a Virtual Environment through Computer Vision-estimated Pointing Gestures. In *Workshop on Gesture and Sign Language based Human-Computer Interaction*, London, UK., April 2001. 59, 61
5. J. Taylor and D. McCloskey. Pointing. *Behavioural Brain Research*, 29:1–5, 1988. 60
6. R. Y. Tsai. A versatile camera calibration technique for high-accuracy 3d machine vision metrology using off-the-shelf tv cameras and lenses. *IEEE Journal of Robotics and Automation*, 3(4):323–344, August 1987. 61

Towards an Automatic Sign Language Recognition System Using Subunits

Britta Bauer and Karl-Friedrich Kraiss

Department of Technical Computer Science
Aachen University of Technology (RWTH), Germany
Phone: +49-241-8026105, Fax: +49-241-8888308
Ahornstrasse 55, D-52074 Aachen
bauer@techinfo.rwth-aachen.de
http://www.techinfo.rwth-aachen.de

Abstract. This paper is concerned with the automatic recognition of German continuous sign language. For the most user-friendliness only one single color video camera is used for image recording. The statistical approach is based on the Bayes decision rule for minimum error rate. Following speech recognition system design, which are in general based on subunits, here the idea of an automatic sign language recognition system using subunits rather than models for whole signs will be outlined. The advantage of such a system will be a future reduction of necessary training material. Furthermore, a simplified enlargement of the existing vocabulary is expected. Since it is difficult to define subunits for sign language, this approach employs totally self-organized subunits called fenone. K-means algorithm is used for the definition of such fenones. The software prototype of the system is currently evaluated in experiments.

1 Introduction

Sign language is the natural language of deaf people. Although different in form, they serve the same function as spoken languages. Spread all over the world, it is not a universal language. Regionally different languages have been evolved as e.g. GSL (German Sign Language) in Germany or ASL (American Sign Language) in the United States. Sign languages are visual languages and they can be characterised by manual (handshape, handorientation, location, motion) and non-manual (trunk, head, gaze, facial expression, mouth) parameters. Mostly one-handed and two-handed signs are used, where for the one-handed signs only the so called dominant hand performs the sign. Two-handed signs can be performed symmetrically or non-symmetrically. The 3D space in which both hands act for signing represents the signing space. If two signs only differ in one parameter they are called minimal pair [12].

1.1 Motivation

There are mainly two different motivations for studying automatic sign language recognition. A first aspect is the development of assistive systems for deaf people.

I. Wachsmuth and T. Sowa (Eds.): GW 2001, LNAI 2298, pp. 64–75, 2002.

For example, the development of a natural input device for creating sign language documents would make such documents more readable for deaf people. Moreover, hearing people have difficulties to learn sign language and likewise the majority of those people who were born deaf or who became deaf early in live, have only a limited vocabulary of the accordant spoken language of the community in which they live. Hence, a system for translating sign language to spoken language would be of great help for deaf as well as for hearing people. These are only two examples of possible assistive systems for deaf people for which a sign language recognition system would be of great value. A second aspect is that sign language recognition serves as a good basis for the development of gestural human-machine interfaces. Many assistive systems not only for deaf but also for e.g. elderly people are thinkable. In addition commercial application could base on the results of sign language recognition.

1.2   Related Difficulties and Recognition Approach

Currently, the developed system uses a single colour video camera. Hence, the following problems in terms of sign language recognition must be taken into account:

- While signing some fingers or even a whole hand can be occluded.
- The position of the signer of front of the camera may vary. Movements, like shifting in one direction or rotating around the body axis must be considered.
- The projection of the 3D scene on a 2D plane results in loss of depth information. The reconstruction of the 3D-trajectory of the hand is not always possible.

In particular, further problems occur concerning the recognition part:

- Each sign varies in time and space. Even if the same person is performing the same sign twice, small changes in speed and position of the hand will occur.
- Generally, a sign is affected by the preceding and the subsequent sign (co-articulation)
- The System is able to detect sign boundaries automatically, thus the user is *not* required to segment the sign sentence into single signs.

Hidden Markov Models (HMMs) based on statistical methods are capable to solve these problems [7]. The ability of HMMs to compensate time and amplitude variances of signals has been proven for speech and character recognition. Due to these characteristics, HMMs are an ideal approach to sign language recognition. A block-scheme is depicted in figure 1. Like speech, sign language can be considered as a non-deterministic time signal. Instead of words or phonemes we here have a sequence of signs. Most of the speech recognition systems available are based on phonemes rather than whole words. Problematic for the recognition with whole signs is the training corpus. Every time a new word is added to the vocabulary, training material for that word has to be added in order to train

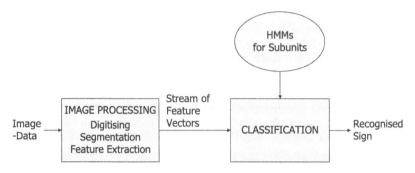

Fig. 1. System Architecture

the accordant model. Hence, the definition of a limited set of subunits of signs is required in order to estimate Hidden Markov Models for these, rather than for a whole sign. The models for the individual signs belonging to the basic vocabulary will be made up as a concatenation of these subunits, which are calculated by the data itself. The advantage of such a method will be as follows:

– The amount of necessary training material will be reduce, as every new sign of the database consists of a concatenation of limited subunits.
– A simplified enlargement of the vocabulary is expected.
– A person-independent recognition of sign language will be feasible.

This paper will outline a concept of an automatic video-based recognition system, which is based on subunits. The presented recognition system concentrates on manual sign parameters. Non-manual sign parameters are at this stage not used for this task. Properties, such as incorporation and indication of sign language will be ignored at the moment. The training- and test corpus consists of GSL–sentences which are meaningful and grammatically well-formed. All sentences are built according to the grammar of GSL. Furthermore, there are no constraints regarding a specific sentence structure. Continuous sign language recognition in this context means that signs within a sentence are not separated by a pause.

The structure of the paper is as follows: Section two gives an overview of previous work in the field of automatic continuous sign language recognition. The following section describes the system architecture, in particular HMMs and the data-driven construction of subunits. Our approach for continuous sign language recognition is given in section four. A description of the experiments undertaken and their results can be found in section five. In the last section the conclusion and the description of further work in this field are drawn.

2 Related Work in the Field of Automatic Continuous Sign Language Recognition

The aim of this section is to briefly discuss systems which are concerned with the recognition of continuous sign language.

Braffort [3] presented a recognition system for French sign sentences. Data acquisition is based on a dataglove. Signs are divided into conventional signs, non-conventional signs and variable signs. The recognition system is based on HMMs. Two modules are used for recognition. One for classifying conventional signs and another one for recognizing the remaining two classes. Models are trained for whole signs. The system achieves a recognition rate of 96% and 92% for the two modules on a lexicon of 44 sign sentences (vocabulary of 7 signs).

Liang et al. [8] used linear ten-states HMMs for the recognition of continuous Taiwanese sign language. Data acquisition is carried out by using a dataglove. The system is designed to classify 250 signs. Here, signs are divided into subunits by a hold-and-movement classification. A recognition rate of 80.4% is reached.

Starner and Pentland [13] presented a video-based system for the recognition of short sentences of ASL, with a vocabulary of 40 signs. Signs are modeled with four-states HMM. A single camera is used for image recording. The recognition rate is 75% and 99%, allowing syntactical structures.

In 1997, Vogler and Metaxas [16] described a HMM-based system for continuous ASL recognition with a vocabulary of 53-signs. Three video cameras are used interchangeably with an electromagnetic tracking system for obtaining 3D movement parameters of the signer's arm and hand. The sentence structure is unconstrained and the number of signs within a sentence is variable. Vogler et al. performed two experiments, both with 97 test sentences: One, without grammar and another with incorporated bigram probabilities. Recognition accuracy ranges from 92.1% up to 95.8% depending on the grammar used.

Another interesting work was presented by Vogler at the Gesture Workshop in 1999 [15]. Here, signs are divided into subunits according to a description of Liddell and Johnson [9], i.e. movements and holds. Movements are characterized by segments during which some aspects of the signer's configuration changes. Holds are defined as those segments during which all aspects of the signer's configuration remain stationary. Experiments were performed with a vocabulary size of 22 signs. The achieved recognition accuracy is 91,8%, similar to a similar experiment with models for whole signs.

3 Theory of Hidden Markov Models

This section briefly discusses the theory of HMMs. A more detailed description of this topic can be found in [10,11].

Given a set of N states s_i we can describe the transitions from one state to another at each time step t as a stochastic process. Assuming that the state-transition probability a_{ij} from state s_i to state s_j only depends on preceding states, we call this process a Markov chain. The further assumption, that the actual transition only depends on the very preceding state leads to a first order Markov chain. We can now define a second stochastic process that produces at each time step t symbol vectors x. The output probability of a vector x only depends on the actual state, but not on the way the state was reached. The output probability density $b_i(x)$ for vector x at state s_i can either be discrete or continuous.

This double stochastic process is called a Hidden Markov Model (HMM) λ which is defined by its parameters $\lambda = (\Pi, A, B)$. Π stands for the vector of the initial state-transition probabilities π_i, the $N \times N$ matrix A represents the state-transition probabilities a_{ij} from state s_i to state s_j, and B denotes the matrix of the output densities $b_i(x)$ of each state s_i.

Given the definition of HMMs, there are three basic problems to be solved [10]:

- The evaluation problem: Given the observation sequence $O = O_1, O_2, \dots, O_T$, and the model $\lambda = (\Pi, A, B)$, the problem is how to compute $P(O \mid \lambda)$, the probability that this observed sequence was produced by the model. This problem can be solved with the Forward-Backward algorithm.
- The estimation problem: This problem covers the estimation of the model parameters $\lambda = (\Pi, A, B)$, given one or more observation sequences O. No analytical calculation method is known to date, the Viterbi training represents a solution, that iteratively adjusts the parameters Π, A and B. In every iteration, the most likely path through an HMM is calculated. This path gives the new assignment of observation vectors O_t to the states s_j.
- The decoding problem: Given the observation sequence O, what is the most likely sequence of states $S = s_1, s_2, \dots, s_T$ according to some optimality criterion. A formal technique for finding this best state sequence is called the Viterbi algorithm.

4   HMM-Based Approach for Continuous Sign Language Recognition

In the previous section, the basic theory of HMMs has been introduced. This section details an approach of an HMM-based automatic recognition system for continuous sign language based on whole words models. After recording, the sequence of input images is digitized and the images are segmented. In the next processing step features regarding size, shape and position of the fingers, hands and body of the signer are calculated. Using this information a feature vector is built that reflects the manual sign parameter. Classification is performed by using HMMs. For both, training and recognition, feature vectors must be extracted from each video frame and input into the HMM.

4.1   Feature Extraction

Since HMMs require feature vectors, an important task covers the determination and extraction of features. In our approach the signer wears simple colored cotton gloves, in order to enable real-time data acquisition and to retrieve easily information about the performed handshape. Taking into account the different amount of information represented by the handshape of the dominant and non-dominant hand and the fact that many signs can be discriminated only by looking at the dominant hand, different gloves have been chosen: one with seven colors - marking each finger, the palm, and the back of the dominant hand - and a second glove in a eighth color for the non-dominant hand. A threshold algorithm generates Input/Output-code for the colors of the gloves, skin, body and background. In the next processing step the size and the center of gravity (COG) of the colored areas are calculated and a rule-based classifier estimates the position of the shoulders and the central vertical axis of the body silhouette. Using this information we build a feature vector that reflects the manual parameters of sign language, without explicitly modeling them [1].

Table 1. Feature vector for coding manual sign parameters

Manual parameter	Feature
Location	x-coordinate, relative to central body axis
	y-coordinate, relative to height of the right shoulder
	Distance of the COGs of the dominant and non-dominant hand
Handshape	Distances of all COGs of the coloured markers of the hand to each other (dominant hand only)
Handshape/ Orientation	Size of coloured areas
Orientation	Angles of the fingers relative to the $x - axis$ (dominant hand only)

4.2   Problems of Recognition Based on Models for Each Sign

Modelling sign language with one HMM for each sign as described in [1], the following problems occur. For a limited set of vocabulary recognition based on models for each sign is an ideal approach. However, since the possibility to enlarge of the vocabulary size is limited, this approach is problematic. Hence, the definition of a limited set of subunits or base forms of signs is required in order to estimate Hidden Markov Models for these, rather than for whole signs. The models for the individual signs belonging to the basic vocabulary will be made up as a concatenation of these subunits, which are calculated by the data itself. The advantage of such a method will be as follows:

- The amount of necessary training material will be reduced, as every new sign of the database consists of a concatenation of limited subunits.
- A simplified enlargement of the vocabulary is expected.
- A person-independent recognition of sign language will be feasible.

In general, in speech recognition exists two ways of constructing an acoustic model that the system may use to compute the probabilities $P(A|W)$ for any acoustic data string $A = a_1, a_2, \cdots, a_m$ and hypothesized word string $W = w_1, w_2, \cdots, w_n$. The two types of acoustic models differ fundamentally. One method is the phonetic acoustic model, which is based on an intuitive linguistic concept. The second possible breakdown is the so called *fenonoic* model [7], which is completely self-organized from data. The following section gives a brief introduction of possible subunits in sign language in an intuitive linguistic view. The possibilities of these serving as a phonetic acoustic model in respect of HMM-based sign language recognition and the resulting problems are detailed.

4.3 Phoneme Modeling in Sign Language

Currently, phoneme modeling in sign language is one main focus in sign language research. A good overview can be found in [4]. In this section we expose two approaches, which will be briefly discussed in the following as these two are fundamental different:

For Stokoe [13] each sign can be broken down into three different parameters, e.g. the location of a sign, the handshape and the movement. These parameters are performed simultaneously. Notation systems such as HamNoSys are based on this system. Stokoe's approach of phonemes in sign language is very difficult to adopt for sign language recognition since these parameters are performed simultaneously while the specific modeling with Hidden Markov Models requires a sequential order of models.

A different assumption leads Liddel and Johnson [9] to a movement and hold model of sign language. Here each sign is partitioned in a sequential way into movements and holds, where for a movement aspects of the signer's configuration changes in contrast to holds, where the configuration remains as it was before. This approach of Liddel and Johnson seems to be more appropriate for the recognition of sign language, as here sign language is partitioned in a sequential way. However, problematic is the transcription of sign language phonemes. No unified lexicon of transcription based on this approach exists for sign language, hence the transcription has to be undertaken manually, which is for a reasonable vocabulary size not feasible. In the following section the introduction of a totally self organized breaking down of sign language into subunits will be shown.

5 Self Organizing Subunits for Sign Language

Generally, in speech recognition two different methods exists to perform a recognition by means of subunits. The commonly used method is based on a phonetic acoustic model which uses an intuitive linguistic concept. The second methods is relatively unknown and not often used in speech recognition systems. The so called fenomic model is completely self-organized from the data itself [7]. They do not presuppose any phonetic concepts, which makes them an ideal approach for sign language. They are also a useful tool for modeling new words that do not belong to the prepared vocabulary.

5.1   K-Means Clustering Algorithm

The aim is to divide all existing feature vectors in a set of k clusters. K-means clustering is performed by moving a feature vector to the cluster with the centroid closest to that example and then updating the accordant centroid vector. A centroid vector is a set of different feature vectors, where each feature of the centroid vector is the average of values for that feature of all feature vectors. For the initial partition one possibility is to take the first k feature vectors and define these as the first k centroid vectors. After this step, take all remaining feature vectors and assign these to the closest centroid. After each assignment, the centroid vectors has to be recalculated. Afterwards, again take every existing feature vector and recalculate its distance to all possible centroid vectors. Assign these to the centroid with the closest distance and afterwards recalculated the new centroid vector and the one losing the accordant feature vetor. This step has to be repeated until no more new assignments take place [7].

5.2   Fenonic Baseform

The structure of a fenonic baseform is depicted in figure 2. A fenonic base-form consists of two states s_1 and s_2 and three different transitions between these states, namely a_{00} the loop-transition, and a_{01} the transition between both states. The third transition is the so called null transition, which changes state but result in no output. Our approach to get a fenonic baseform for whole sign language sentences is as follows: Take for each sign of the vocabulary all available feature vectors. With these perform the above described k-means algorithm. After the convergence of the k-means algorithm an assignment of each feature vector to a specific cluster is clear. Hence, all feature vectors belonging to the same cluster are in a way similar. Thus, it makes sense to make use of these similarities and take each cluster as a fenonic baseform. The models for the individual words belonging to the basic vocabulary will themselves be made up as a concatenation of these fenonic baseforms. After the performance of the k-means algorithm, a codebook is generated, which trace back the assignments of the feature vectors of all different signs. An example of how to establish fenonic base form is depicted in figure 3. Here, the extracted feature vectors are shown for the sentence 'DAS WIEVIEL KOSTEN' ('How much does that cost?').

After recording, the stream of input images is digitized and segmented. In the next processing step features are extracted form each frame in order to build a feature vector. The first row of the figure shows two images of the sign DAS, four images of the sign WIEVIEL and three images of the sign KOSTEN as recorded by the video camera. For each image a feature vector X is calculated. To find out, which feature vectors are similar to others, the k-means clustering algorithm is used and the extracted feature vectors are assigned to the accordant cluster. Here three different clusters are determined ($k = 3$). Here, the two feature vectors of the sign DAS were assigned to cluster $A1$ and $A2$. Hence, the sign DAS will be handled as a concatenation of these two fenonics.

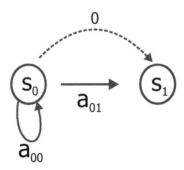

Fig. 2. Possible structure of a fenone HMM

5.3   Training HMMs on Continuous Sign Language

Training HMMs on continuous sign language based on fenonics base forms is very similar to training based on word models. One of the advantages for hidden Markov modeling is that it can absorb a range of boundary information of models automatically for continuous sign language recognition. Given a specific number of observation (training) sequences O, the result of the training are the model parameters $\lambda = (\Pi, A, B)$. These parameters are later used for the recognition procedure. Since the entire sentence HMM is trained on the entire observation sequence for the corresponding sentence, sign boundaries are considered automatically. Variations caused by preceding and subsequent signs are incorporated into the model parameters. The model parameters of the single sign must be reconstructed from this data afterwards. The overall training is partitioned into the following components: the estimation of the model parameters for the complete sentence, the detection of the sign boundaries and the estimation of the model parameters for each fenonic baseform. For both, the training of the model parameters for the entire sentence as well as for fenones, the Viterbi training is employed. By means of the generated Codebook after the execution of the k-means algorithm an assignment of feature vectors to fenonic baseforms is after performing the training step on sentences clearly possible.

The number of states for a sentence is equal to the sum of states of the fenonic baseforms within the sentence, i.e. the number of fenone times two as an HMM for a fenone consist of two states. Next, the system assigns the vectors of each sequence evenly to the states and initializes the matrix A, i.e., all transitions are set equally probable. Using the initial assignment the mean and deviation values of all components of the emission distributions of each state can be calculated. After viterbi alignment the transition probabilities a_{ij} are re-calculated and the split criterion is examined. With a sufficient convergence the parameters of an HMM $\lambda = (\Pi, A, B)$ are available, otherwise the next iteration is requested.

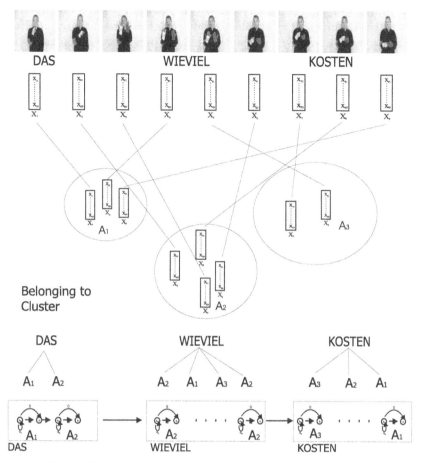

Fig. 3. General construction of fenonic baseforms for signs

5.4 Recognition of Continuous Sign Language

In continuous sign language recognition, a sign and in particular a fenone may begin or end anywhere in a given observation sequence. As the sign boundaries cannot be detected accurately, all possible beginning and end points have to be accounted for. Furthermore, the number of signs and with this the number of subunits of signs within a sentence are unknown at this time. This converts the linear search, as necessary for isolated sign recognition, to a tree search. Obviously, an optimal full search is not feasible because of its computational complexity for continuous sign recognition. Instead of searching all paths, a threshold is used to consider only a group of likely candidates. These candidates are selected in relation to the state with the highest probability. Depending on that value and a variable B_0 a threshold for each timestep is defined [7]. Every state with a calculated probability below this threshold will be discarded from further considerations. The variable B_0 influences the recognition time. Having many likely candidates, i. e., a low threshold is used, recognition needs more time than considering less candidates. B_0 must be determined by experiments. After this, the order of recognized subunits is clear, but not yet the order of signs within a sentence. Hence, by means of the generated codebook the order of signs will be decoded.

6   First Results and Future Work

On the input side a color video camera is used. A Pentium-II 300 PC with an integrated image processing system allows the calculation of the feature vectors at a processing rate of 13 frames per second. In order to ensure correct segmentation, there are few restrictions regarding the clothing of the signer. The background must have a uniform color. First experiments are undertaken with a very limited set of signs. The training- and testcorpus comprises of 12 different signs and these consists of 10 determined subunits (k=10). The achieved recognition rate is in the moment 80.8%. Analyzing these results it can be stated that the vocabulary size is currently far too limited to make general statements about the performance of this recognition system. With a vocabulary of 12 signs only a limited set of subunits is available. Due to these subunits the partitioning of the data is presumably not adequate enough.

It is expected that with an increasing vocabulary size and thereby an increasing set of subunits the calculated HMM parameters of the fenonic baseform will be significantly better as it is evaluated by common speech recognition systems. These are usually based on a set of 200 subunits [7]. Future work will have a main focus on this issue.

7   Summary

In this paper we introduced an HMM-based continuous sign language recognition system using subunits, which are self-organized derived by the data itself. The system is equipped with a single color video camera for image recording. Real-time image segmentation and feature extraction are achieved by using simple colored cotton gloves. The extracted sequence of feature vector reflects the manual sign parameters. The k-means algorithm was introduced to produce a general set of subunits. Signs of the vocabulary are made up by the concatenation of these subunits. Compared to future systems 2 the enlargement of the vocabulary size is simplified by this approach. Moreover the training material for a large vocabulary will be reduced.

References

1. Bauer B. and H. Hienz: *Relevant Features for Video-Based Continuous Sign Language Recognition* In: Proceedings of the 4th International Conference on Automatic Face and Gesture Recognition FG 2000 , pp. 440-445, March 28-30, Grenoble, ISBN 0-7695-0580-5 69

2. Boyes Braem, P.: *Einführung in die Gebärdensprache und ihre Erforschung.* Signum Press, Hamburg, 1995.

3. Braffort, A.: *ARGo: An Architecture for Sign Language Recognition and Interpretation.* In P.Harling and A. Edwards (Editors): Progress in Gestural Interaction, pp. 17–30, Springer, 1996. 67

4. Coulter, G. R.: *Phonetics and Phonology, VOLUME3 Current Issues in ASL Phonology.* Academic Press, Inc. SanDiego, California, ISBN 0-12-193270. 70

5. Hienz, H. and K. Grobel: *Automatic Estimation of Body Regions from Video Images.* In Wachsmuth, I. and M. Fröhlich (Editors): Gesture and Sign Language in Human Computer Interaction, International Gesture Workshop Bielefeld 1997, pp. 135–145, Bielefeld (Germany), Springer, 1998.

6. Jelinek, F.: *Self-organized Language Modeling for Speech Recognition.* In A. Waibel and K.-F. Lee (Editors): Readings in Speech Recognition, pp.450–506, Morgan Kaufmann Publishers, Inc., 1990.

7. Jelinek, F.: *Statistical Methods For Speech Recognition.* MIT Press 1998, ISBN 0262-10066-5 65, 70, 71, 73, 74

8. Liang, R. H. and M. Ouhyoung:*A Real-Time Continuous Gesture Recognition System for Sign Language.* In Proceedings of the Third Int. Conference on Automatic Face and Gesture Recognition, Nara (Japan), pp. 558-565 1998. 67

9. Liddel, S. K. and R. E. Johnson.*American Sign Language The phonological base.*In: Sign Language Studies, 64: 195-277, 1989 67, 70

10. Rabiner, L. R. and B. H. Juang: *An Introduction to Hidden Markov Models.* In IEEE ASSP Magazin, pp. 4–16, 1989. 67, 68

11. Schukat-Talamazzini, E.G: *Automatische Spracherkennung.* Vieweg Verlag, 1995. 67

12. Starner, T., J. Weaver and A. Pentland: *Real-Time American Sign Language Recognition using Desk- and Wearable Computer-Based Video.*In IEEE Transactions on Pattern Analysis and Machine Intelligence, 20(12):1371-1375, 1998 64

13. Stokoe, W. C.:*Sign Language Structure: An Outline of the Visual Communication System of the American Deaf.*Studies in Linguistics: Occasional Papers Linstok Press, Silver Spring, MD, 1960, Revised 1978 67, 70

14. Stokoe, W., D. Armstrong and S. Wilcox: *Gesture and the Nature of Language.* Cambridge University Press, Cambridge (UK), 1995.

15. Vogler, C. and D. Metaxas: *Toward Scalability in ASL Recognition: Breaking Down Signs into Phonemes.*In: Int. Gesture Workshop Gif-sur-Yvette, France 1999 67

16. Vogler, C. and D. Metaxas: *Adapting Hidden Markov Models for ASL Recognition by using Three-Dimensional Computer Vision Mehtods.*In Proceedings of IEEE Int. Conference of Systems, Man and Cybernetics, pp. 156-161, Orlando (USA), 1997. 67

Signer-Independent Continuous Sign Language Recognition Based on SRN/HMM

Gaolin Fang[1], Wen Gao[1,2], Xilin Chen[1], Chunli Wang[3], and Jiyong Ma[2]

[1] Department of Computer Science and Engineering
Harbin Institute of Technology, Harbin, 150001, China
[2] Institute of Computing Technology, Chinese Academy of Sciences
Beijing, China
[3]Department of Computer Science, Dalian University of Technology
Dalian, China
{fgl,wgao,xlchen,chlwang}@ict.ac.cn

Abstract. A divide-and-conquer approach is presented for signer-independent continuous Chinese Sign Language(CSL) recognition in this paper. The problem of continuous CSL recognition is divided into the subproblems of isolated CSL recognition. The simple recurrent network(SRN) and the hidden Markov models(HMM) are combined in this approach. The improved SRN is introduced for segmentation of continuous CSL. Outputs of SRN are regarded as the states of HMM, and the Lattice Viterbi algorithm is employed to search the best word sequence in the HMM framework. Experimental results show SRN/HMM approach has better performance than the standard HMM one.

Keywords: Simple recurrent network, hidden Markov models, continuous sign language recognition, Chinese sign language

1 Introduction

Sign language as a kind of gestures is one of the most natural means of exchanging information for most deaf people. The aim of sign language recognition is to provide an efficient and accurate mechanism to transcribe sign language into text or speech so that communication between deaf and hearing society can be more convenient. Sign language recognition has emerged as one of the most important research areas in the field of human-computer interaction. In addition, it has many other applications, such as controlling the motion of a human avatar in a virtual environment (VE) via hand gesture recognition, multimodal user interface in virtual reality (VR) system.

Attempts to automatically recognize sign language began to appear in the literature in the 90's. Previous work on sign language recognition focuses primarily on finger spelling recognition and isolated sign recognition. There has been very little work on continuous sign language recognition. Starner [1] used a view-based approach with a single camera to extract two-dimensional features as input to HMMs. The correct rate

I. Wachsmuth and T. Sowa (Eds.): GW 2001, LNAI 2298, pp. 76–85, 2002.

was 91% in recognizing the sentences comprised 40 signs. By imposing a strict grammar on this system, an accuracy of 97% was possible with real-time performance. Liang and Ouhyoung [2] used HMMs for continuous recognition of Taiwan Sign language with a vocabulary between 71 and 250 signs with Dataglove as input devices. However, their system required that gestures performed by the signer be slow to detect the word boundary. Vogler and Metaxas [3] used computer vision methods to extract the three-dimensional parameters of a signer's arm motions, coupled the computer vision methods and HMMs to recognize continuous American sign language sentences with a vocabulary of 53 signs. They modeled context-dependent HMMs to alleviate the effects of *movement epenthesis*. An accuracy of 89.9% was observed. In addition, they used phonemes instead of whole signs as the basic units [4], experimented with a 22 sign and achieved recognition rates similar to sign-based approaches. A system has been described in our previous work [5], which used Dataglove as input device and HMMs as recognition method. It can recognize 5177 isolated signs with 94.8% accuracy in real time and recognize 200 sentences with 91.4% word accuracy.

As the previous work showed, most researches on continuous sign language recognition were done within the signer-dependent domain. No previous work on signer-independent continuous sign language recognition was reported in the literature. For continuous sign language recognition, a key problem is the effect of *movement epenthesis*, that is, transient movements between signs, which vary with the context of signs. For signer-independent recognition, different signers vary their hand shape size, body size, operation habit, rhythm and so on, which bring about more difficulties in recognition. Therefore recognition in the signer-independent domain is more challenging than in the signer-dependent one. Signer-independent continuous sign language recognition is first investigated in this paper. Our experiments show that continuous sign language has the property of segments, according to which a divide-and-conquer approach is proposed. It divides the problem of continuous CSL recognition into the subproblems of isolated CSL recognition. The SRN modified to assimilate context information is regarded as the segment detector of continuous CSL. The Lattice Viterbi algorithm is employed to search the best word sequence path in the output segments of SRN.

The organization of this paper is as follows. In Section 2 we propose the improved SRN and present SRN-based segmentation for continuous sign language. In Section 3 we introduce the Lattice Viterbi algorithm in the HMM framework and discuss the computation of segment emission probability. In Section 4 we show experimental results and comparisons. The conclusion is given in the last section.

2   SRN-Based Segmentation

2.1 Simple Recurrent Network

Elman proposed a simple recurrent network[6] in 1990 on the basis of the recurrent network described by Jordan. Networks have the dynamic memory performance because of the introduction of recurrent links. They have been successfully applied to speech recognition [7], [8], handwriting recognition [9], isolated sign language recognition [10]. The SRN in Figure 1 has four units. The input units receive the input vector I_t at time t, and the corresponding outputs are hidden units H_t, feedback units C_t, output units O_t, target output units R_t. Defining W_C^H, W_I^H, W_H^O as the weight matrices of feedback units to hidden units, input units to hidden units and hidden units to output units respectively.

In the network the feedback units is connected one-to-one corresponding hidden units through a unit time-delay.

$$C_t = \begin{cases} H_{t-1} & t \geq 2 \\ \overrightarrow{0.5} & t = 1 \end{cases} \tag{1}$$

The input units and the feedback units make up of the combined input vector. Let $D_t = \begin{bmatrix} C_t & I_t \end{bmatrix}$, $W_D^H = \begin{bmatrix} W_C^H \\ W_I^H \end{bmatrix}$, Φ, Ψ is respectively the bias of hidden units, output units.

$$H_t = f\left(C_t \cdot W_C^H + I_t \cdot W_I^H \ - \ \Phi\right) = f\left(D_t \cdot W_D^H \ - \ \Phi\right) \tag{2}$$

$$O_t = f\left(H_t \cdot W_H^O \ - \ \Psi\right) \tag{3}$$

The activation function $f(\cdot)$ is the standard sigmoid one, $f(x) = \left(1 + e^{-x}\right)^{-1}$. The errors between the network's outputs and the targets are propagated back using the generalized delta rule [11].

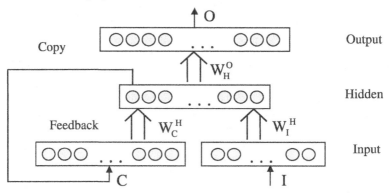

Fig. 1. Simple recurrent network

Because of the introduction of feedback units, the outputs of network depends not only on the external input but also on the previous internal state that relies on the result of the preceding all external inputs. So the SRN can memorize and utilize a relatively larger preceding context [6].

The SRN can memorize the preceding context, but it cannot utilize following context information. Thus the SRN is modified to efficiently utilize the following context information in two ways. One is that the following context vector is regarded as one of the input. Thus $I_t = \begin{bmatrix} I_t & I_{t+1} \end{bmatrix}$, the rest of calculations are same as the standard SRN. The other is that training samples are respectively in turn and in reverse turn fed into the SRN with the same architecture, and then one forward SRN and one backward SRN are trained. So the context information can be assimilated through two SRNs. The experiment of segmentation of continuous sign language is performed in two improved ways. But in the latter outputs of the forward SRN conflict with those of the backward SRN so that experimental results are inferior to the former. Thus, we employ the former method as the improved SRN.

2.2 Segmentation of Continuous Sign Language

Input preprocess: We use two 18-sensor Datagloves and three position trackers as input devices. Two trackers are positioned on the wrist of each hand and another is fixed at back (the reference tracker). The Datagloves collect the variation information of hand shapes with the 18-dimensional data at each hand, and the position trackers collect the variation information of orientation, position, movement trajectory. Data from position trackers can be converted as follows: the reference Cartesian coordinate system of the trackers at back is chosen, and then the position and orientation at each hand with respect to the reference Cartesian coordinate system are calculated as invariant features. Through this transformation, the data are composed of the three-dimensional position vector and the three-dimensional orientation vector for each hand. Furthermore, we calibrate the data of different signer by some fixed postures because everyone varies his hand shape size, body size, and operation habit. The data form a 48-dimensional vector in total for two hands. However, the dynamic range of each component is different. Each component value is normalized to ensure its dynamic range is 0-1.

Original experiment regards the 48-dimensional data as the input of SRN, chooses 30 hidden units and 3 output units. Training this network costs 28 hours in the PIII450(192M Memory) PC. But the result is not satisfying and the segment recall is only 87%. This approach is unsuccessful in the scalability and accuracy. This only considers the case of detecting all actual segments and doesn't care the case of detecting several segments for an actual segment. The latter will be solved by the Lattice Viterbi algorithm in the HMM framework.

$$\text{Segment recall} = \frac{\text{Number of correct segments}}{\text{Number of all actual segments}} \times 100\% \qquad (4)$$

Thus the above approach is modified. We use self-organizing feature maps(SOFM) as the feature extraction network. The outputs of SOFM have the strong segment properties (Fig. 2 shows an example). The output of SOFM is regarded as the input of SRN by the encoding. We select 48 input units and 256 output units for the SOFM, and 16 input units and 15 hidden units and 3 output units for the SRN through trial and error. Training the SRN costs 45 minutes in the PIII450(192M Memory) PC. The segment recall of SRN is 98.8%. The training of SOFM is implemented by the toolbox in Matlab. The input, output, training and recognition of SRN will be discussed in the following section.

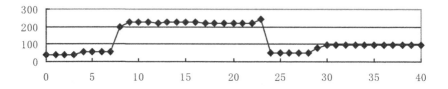

Fig. 2. The segments property of sign language"我们什么时候走"(when will we leave)

Input: The encoding of 256 output units of SOFM requires 8 units, and the introduction of the following context also demands 8 units. In total there are 16 input units. The value of the input $I_t^i \in \{0,1\}$, $i = 1, 2, \cdots, 16$.

Output: We define 3 output units: the left boundary of segments 1, the right boundary of segments 2, the interior of segments 3, and the corresponding units o_t^1, o_t^2, o_t^3.

$$O_t = \begin{cases} [1 \quad 0 \quad 0] & \text{Output is 1} \\ [0 \quad 1 \quad 0] & \text{Output is 2} \\ [0 \quad 0 \quad 1] & \text{Output is 3} \end{cases} \tag{5}$$

Training: We can't straightforward find the SRN output of training data because sign language is continuous. Thus automatic segmentation is employed to find the target segments. Let the sample sentence in the training $W = w_1 w_2 \cdots w_k$, the corresponding frame sequence $T = t_1 t_2 \cdots t_i$, we decide frame t_i belongs to word w_m or w_{m+1}, and if $t_i \in w_m$, $t_{i+1} \in w_{m+1}$, then frame t_i is the right boundary of segments and frame t_{i+1} is the left boundary of segments. Each state probability of frame t_i belonging to word w_m ($m = 1 \cdots k$) is calculated by the approach of isolated sign language recognition with the isolated sign language model parameters (see Sect. 4). Then the constrained Viterbi algorithm is used to search the best segment sequence. The constrained refers to the search sequence followed by $w_1 w_2 \cdots w_k$. This segment sequence is regard as the target output. Training data are composed of two group data from two signers.

Back-propagation through time is introduced as the learning algorithm of SRN. The samples in the training set are transformed by the SOFM, and the SOFM outputs by the encoding together with the following ones are fed into the SRN, then the errors between the SRN outputs and the targets are propagated back using back-propagation and changes to the network weights are calculated. At the beginning of learning, the weight matrices, the bias of hidden units and output units are initialized the random value (-1,+1), the feedback units are initialized to activations of 0.5.

Recognition: Continuous sign language in the test set is firstly fed into the SOFM. The quantized outputs are formed by the feature extraction of SOFM. The SOFM outputs by the encoding together with the following ones are fed into the SRN. The segmentation result of SRN is $i^* = \arg\max_i(o_t^i)$ at time t. The adjacency property of the left and right boundary of segments is used as constraint in the segmentation.

3 HMM Framework

The segmentation result of SRN is fed into the HMM framework. The state of HMM is composed of one or several segments (2—4). Because different signs vary their lengths and may be composed of several segments, we should search the best path in those segments through the Lattice Viterbi algorithm. The best refers to two respects: the recombined segments sequence is the best, and the word sequence selected from the best segment sequence is the best. The standard Viterbi algorithm is modified in order to adapt itself to the search on lattices, and the modified algorithm is referred to as the Lattice Viterbi algorithm.

To illustrate more conveniently, we define two Viterbi algorithms: one is the isolated sign language Viterbi algorithm. For the data frame sequence of the input which is known in advance as only one word component, the algorithm will search all states of the word for each frame in recognition, and get the best state sequence.

The other is the continuous sign language Viterbi algorithm. For the data frame sequence of the input whose component cannot be known beforehand, the algorithm will search not only all states of this word but also the states of the rest words for each frame in recognition, and get the best state sequence. It requires more time and has less accuracy than the isolated sign language Viterbi algorithm in recognizing the same data frame sequence.

Both the isolated sign language Viterbi algorithm and the continuous sign language Viterbi algorithm in essence belong to the standard one, because they search the frame one by one. But the Lattice Viterbi algorithm that is different from the standard one can span one or more segments to search. It will be discussed in detail in the following section.

3.1 Lattice Viterbi Algorithm

We define an edge as a triple $< t, t', q >$, starting at segment t, ending at segment t' and representing word q, where $0 \le t < T$, $t < t' \le T$. All triple $< t, t', q >$ form the set L. We introduce accumulator $\delta(t, t', q)$ that collects the maximum probability of covering the edge $< t, t', q >$. To keep track of the best path, we define the auxiliary argument $\psi(t, t', q)$ as the previous triple argument pointer of the local maximum $\delta(t, t', q)$. We denote $b(t, t', q)$ as the emission probability of word q covering segments from position t to t'. $P(q \mid q')$ is defined as the transition probability from word q' to q, which is estimated in some 3000 million Chinese words corpus from the Chinese newspapers in the 1994-1995 year. The Lattice Viterbi algorithm is as follows.

1) Intialization:

$$\delta(0, t, q) = b(0, t, q) \tag{6}$$

$$\psi(0, t, q) = \text{NULL} \tag{7}$$

2) Recursion:

$$\delta(t, t', q) = \max_{<t', t, q'> \in L} \delta(t'', t, q')P(q \mid q')b(t, t', q) \tag{8}$$

$$\psi(t, t', q) = \arg\max_{<t', t, q'> \in L} \delta(t'', t, q')P(q \mid q') \tag{9}$$

3) Termination:

$$P^* = \max_{<t, T, q> \in L} \delta(t, T, q) \tag{10}$$

$$< t_1^*, T, q_1^* > = \arg\max_{<t, T, q> \in L} \delta(t, T, q) \tag{11}$$

4) Path backtracking:

Let $T = t_0^*$, we iterate the function $< t_{i+1}^*, t_i^*, q_{i+1}^* > = \psi(t_i^*, t_{i-1}^*, q_i^*)$ until $< t_{k+1}^*, t_k^*, q_{k+1}^* > = \text{NULL}$, and the word sequence $q_k^* \cdots q_1^*$ is the best path.

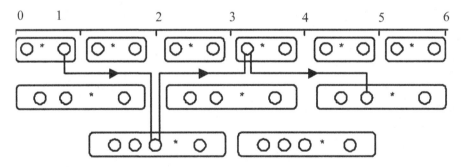

Fig. 3. The sketch map of the Lattice Viterbi algorithm

Figure 3 illustrates the search procedure of the Lattice Viterbi algorithm with a continuous sign language sentence of 6 segments. For example: when $t = 4$, $t' = 6$, $t'' = 3$ or $t'' = 2$ in Fig. 3. Through the computation of formula (8), $t'' = 3$ is the local maximum of $\delta(4, 6, q)$, thus $\psi(4, 6, q) = < 3, 4, q' >$. The search result is 4 recombined segments: <0 1> <1 3> <3 4> <4 6>, and the corresponding word in each segment forms the best word sequence.

3.2   Computation of Segment Emission Probability

The computation of segment emission probability is similar to the one of isolated sign language recognition. But it stores the probabilities of all possible candidate words that are looked on as the emission probability $b(t, t', q)$ in the HMM framework. We get the best path by the Lattice Viterbi algorithm at last. The SOFM/HMM method is employed for isolated sign language recognition [12]. This method uses the output of SOFM to construct the state probability of HMM, combines the powerful self-organizing capability of SOFM with excellent temporal processing properties of HMM to improve the performance of HMM-based sign language recognition, and enhances the SOFM/HMM discrimination with the posteriori probability modifying the current state probability density functions in recognition. So it increases the recognition accuracy.

We define embedded training and un-embedded training recognition according to the model in the computation of segment emission probability. The sentences in the training set are segmented into words and corresponding frame sequences by the automatic segmentation introduced in Section 1.2. Using the segmented words as the training set, the model parameters trained is referred to as the embedded model parameters. The combination of the embedded model parameters with the isolated sign language model parameters is used as the candidate model in recognition, and this recognition procedure is regarded as embedded training recognition. If we use only the isolated sign language parameters as the candidate model in recognition, this recognition procedure is referred to as un-embedded training recognition.

4   Experiments and Comparisons

The data are collected from 7 signers with each performing 208 isolated signs 3 times. We at random select 5 from 7 signers, and then select 2 group data (i.e. all signs performed once) from each signer. In total 10 group data are regarded as the isolated sign language training set. The isolated sign language model parameters are trained by the SOFM/HMM method with the isolated sign language training set. Meanwhile, the standard HMM parameters are trained using the same training set. We select 2 from 5 signers in the isolated sign language training set, respectively represented with A, B, and select 1 from the rest 2 signers represented with C. There are 3 signers in total. The 100 continuous sentences composed of a vocabulary of 208 signs are respectively performed twice by 3 signers with the natural speed. There are 6 group data marked with A_1, A_2, B_1, B_2, C_1, C_2 .We choose A_1, B_1 as the training set which is used in the SOFM, SRN and embedded training. One of A_2, B_2 is referred to as the registered test set(Reg.), and one of C_1, C_2 is referred to as the unregistered test set(Unreg.). We compare the performances of SRN/HMM with those of HMM in signer-independent continuous sign language recognition and the results are as follows.

Table 1. Un-embedded training recognition, test set is Unreg

Method	Recognition accuracy(%)	Recognition time(s/word)
HMM	68.1 (S=58, I=40, D=19)	0.503
SRN/HMM	73.6 (S=57, I=8, D=32)	0.241

Table 2. Embedded training recognition, test set is Unreg

Method	Recognition accuracy(%)	Recognition time(s/word)
HMM	81.2 (S=35, I=25, D=9)	1.016
SRN/HMM	85.0 (S=33, I=6, D=16)	0.479

Table 3. Embedded training recognition, test set is Reg

Method	Recognition accuracy(%)	Recognition time(s/word)
HMM	90.7 (S=13, I=18, D=3)	1.025
SRN/HMM	92.1 (S=12, I=5, D=12)	0.485

All experiments are performed with the Bigram language model in the PIII450(192M Memory) PC. S, I, D denote respectively the error number of substitution, insertion and deletion. The total number of signs in the test set is 367. Compared with the standard HMM, the SRN/HMM has higher recognition accuracy and shorter recognition time. The possible reasons for the results are as follows. Firstly, the HMM uses the continuous sign language Viterbi algorithm which is liable to be influenced by the *movement epenthesis*, but the SRN/HMM alleviates the effects of *movement epenthesis* by discarding the transient frames near the sentences segmentation judged according to the output of SOFM. Secondly, unlike the HMM which searches the best state sequence, the SRN/HMM gets the best word sequence. Thirdly, the SRN/HMM employs the isolated Viterbi algorithm that can get the higher accuracy than the continuous Viterbi algorithm. However, the divide-and-conquer approach in the SRN/HMM may lead to the accumulation of errors. Thus we introduce the soft-segmentation instead of the fixed-segmentation in the SRN segmentation and decide the boundary of words in the Lattice Viterbi algorithm so as to improve the performance of SRN/HMM.

5   Conclusion

While the HMM employs the implicit word segmentation procedure, we present a novel divide-and-conquer approach which uses the explicit segmentation procedure for signer-independent continuous CSL recognition, and which, unlike the HMM, isn't liable to the effects of *movement epenthesis*. The divide-and-conquer approach is as follows. Firstly, the improved SRN is introduced to segment the continuous sentences, and the results of segmentation are regarded as the state inputs of HMM. Secondly, the best word sequence is gotten through the search of Lattice Viterbi algorithm in the sentence segments. This approach alleviates the effects of *movement epenthesis*, and experimental results show that it increases the recognition accuracy and reduces the recognition time. In addition, the improved SRN, the Lattice Viterbi algorithm and the segment property of continuous sign language found in the experiment are not only used in this approach but also provide the foundation for further researches.

Acknowledgment

This work has been supported by National Science Foundation of China (contract number 69789301), National Hi-Tech Development Program of China (contract number 863-306-ZD03-01-2), and 100 Outstanding Scientist foundation of Chinese Academy of Sciences.

References

1. Starner, T., Pentland A.: Visual Recognition of American Sign Language Using Hidden Markov Models. International Workshop on Automatic Face and Gesture Recognition, Zurich, Switzerland, 1995, pp. 189-194
2. Liang, R.H., Ouhyoung, M.: A Real-time Continuous Gesture Recognition System for Sign Language. In Proceeding of the Third International Conference on Automatic Face and Gesture Recognition, Nara, Japan, 1998, pp. 558-565
3. Vogler, C., Metaxas, D.: ASL Recognition Based on a Coupling between HMMs and 3D Motion Analysis. In Proceedings of the IEEE International Conference on Computer Vision, 1998, pp. 363-369
4. Vogler, C., Metaxas, D.: Toward Scalability in ASL Recognition: Breaking Down Signs into Phonemes. In Proceedings of Gesture Workshop, Gif-sur-Yvette, France, 1999, pp. 400-404
5. Gao, W., Ma, J.Y., et al.: HandTalker: A Multimodal Dialog System Using Sign Language and 3-D Virtual Human. Advances in Multimodal Interfaces-ICMI 2000, pp. 564-571
6. Elman, J.L.: Finding Structure in Time. Cognitive Science, 1990, Vol. 14, No. 2, pp. 179-211
7. Robinson, T.: An Application of Recurrent Nets to Phone Probability Estimation. IEEE Transactions on Neural Networks, 1994, Vol. 5, pp. 298-305
8. Kershaw, D.J., Hochberg, M., Robinson, A.J.: Context-Dependent Classes in a Hybrid Recurrent Network-HMM Speech Recognition System. Advances in Neural Information Processing Systems, 1996, Vol. 8, pp. 750-756
9. Senior, A., Robinson, A.J.: Forward-Backward Retraining of Recurrent Neural Networks. Advances in Neural Information Processing Systems, 1996, Vol. 8, pp. 743-749
10. Murakami, K., Taguchi, H.: Gesture Recognition using Recurrent Neural Networks. In CHI'91 Human Factors in Computing Systems, 1991, pp. 237-242
11. Rumelhart, D.E., Hinton, G. E., Williams, R. J.: Learning Internal Representations by Error Propagation. In Parallel Distributed Processing: Explorations in the Microstructure of Cognition, Vol. 1, 1986, pp. 318-362
12. Fang, G.L., Gao, W.: A SOFM/HMM System for Person-Independent Isolated Sign Language Recognition, INTERACT2001 Eight IFIP TC.13 Conference on Human-Computer Interaction, Tokyo, Japan, 2001, pp. 731-732

A Real-Time Large Vocabulary Recognition System for Chinese Sign Language

Wang Chunli[1], Gao Wen[2], and Ma Jiyong[2]

[1] Department of Computer, Dalian University of Technology
Dalian 116023
[2] Institute of Computing Technology, Chinese Academy of Science
Beijing 100080
{chlwang,wgao,jyma}@ict.ac.cn
Clwang@mail.dlptt.ln.cn

Abstract. The major challenge that faces Sign Language recognition now is to develop methods that will scale well with increasing vocabulary size. In this paper, a real-time system designed for recognizing Chinese Sign Language (CSL) signs with a 5100 sign vocabulary is presented. The raw data are collected from two CyberGlove and a 3-D tracker. An algorithm based on geometrical analysis for purpose of extracting invariant feature to signer position is proposed. Then the worked data are presented as input to Hidden Markov Models (HMMs) for recognition. To improve recognition performance, some useful new ideas are proposed in design and implementation, including modifying the transferring probability, clustering the Gaussians and fast matching algorithm. Experiments show that techniques proposed in this paper are efficient on either recognition speed or recognition performance.

1   Introduction

The hand gesture recognition that can contribute to a natural man-machine interface is still a challenging problem. Closely related to the field of gesture recognition is that of sign language recognition. Sign language, as a kind of structured gesture, is one of the most natural means of exchanging information for most deaf people. The aim of sign language recognition is to provide an efficient and accurate mechanism to transcribe sign language into text or speech.

Attempts to automatically recognize sign language began to appear in the 90's. Charaphayan and Marble [1] investigated a way using image processing to understand American Sign Language (ASL). This system can recognize correctly 27 of the 31 ASL symbols. Starner[2] reported a correct rate for 40 signs achieved 91.3% based on the image. The signers wear color gloves. By imposing a strict grammar on this system, the accuracy rates in excess of 99% were possible with real-time performance. Fels and Hinton[3,4] developed a system using a VPL DataGlove Mark

I. Wachsmuth and T. Sowa (Eds.): GW 2001, LNAI 2298, pp. 86–95, 2002.
© Springer-Verlag Berlin Heidelberg 2002

II with a Polhemus tracker attached for position and orientation tracking as input devices. The neural network was employed for classifying hand gestures. Takahashi and Kishino[5] investigated a system for understanding the Japanese kana manual alphabets corresponding to 46 signs using a VPL dataGlove. Their system could correctly recognize 30 of the 46 signs. Y. Nam and K. Y. Wohn[6] used three–dimensional data as input to HMMs for continuous recognition of a very small set of gestures. They introduced the concept of movement primes, which make up sequences of more complex movements. R. H. Liang and M. Ouhyoung[7] used HMM for continuous recognition of Taiwan Sign language with a vocabulary between 71 and 250 signs based Dataglove as input devices. Kisti Grobel and Marcell Assan [8] used HMMs to recognize isolated signs with 91.3% accuracy out of a 262-sign vocabulary. They extracted the features from video recordings of signers wearing colored gloves. C. Vogler and D. Metaxas[9] used HMMs for continuous ASL recognition with a vocabulary of 53 signs and a completely unconstrained sentence structure. C. Vogler and D. Metaxas[10][11] described an approach to continuous, whole-sentence ASL recognition that uses phonemes instead of whole signs as the basic units. They experimented with 22 words and achieved similar recognition rates with phoneme-based and word-based approaches. Wen Gao[12] proposed a Chinese Sign language recognition system with a vocabulary of 1064 signs. The recognition accuracy is about 93.2%.

So far, the challenge that sign language recognition faces is to develop methods that will resolve well large vocabulary recognition. There are about 5100 signs included in Chinese Sign Language (CSL) dictionary. Therefore the task of CSL recognition becomes very challenging. We proposed a Chinese Sign language recognition system with a vocabulary of 1064 signs in [12]. An ANN-DP combined approach was employed for segmenting subwords automatically from the data stream of sign signals. To tackle the epenthesis movement problem, a DP-based method has been used to obtain the context-dependent models. The recognition accuracy is about 93.2%. This paper is an extension of the work done in [12]. Our goal is to realize the 5100-sign vocabulary CSL recognition.

Two CyberGlove and a Pohelmus 3-D tracker with three receivers positioned on the wrist of CyberGlove and the waist are used as input device in this system. The raw gesture data include hand postures, positions and orientations. An algorithm based on geometrical analysis for purpose of extracting invariant feature to signer position is proposed in this paper.

In large vocabulary continuous speech recognition, phonemes are used as the basic units; the number of them is much smaller compared with those in CSL. There are 50~60 phonemes in speech and about 2500~3000 basic units in sign language. Because there is not the description about all the basic units in CSL at hand and the possible number of them is quite large, we do not build HMMs for the basic units. A HMM is built for each sign in this system, and there are 5100 HMMs. The numbers of basic gestures included in different signs are variable, so these HMMs include different number of states. It is difficult to realize real-time recognition because of too much computational cost if CHMMs (Continuous HMMs) were employed in large vocabulary recognition. Clustering the Gaussians is proposed to reduce the computational cost, so the parameters of Gaussians on the states need to be classified. But the number of possible combinations of the six data streams in sign language could be approximately an order of 10^8. If they were dealt with together, the code words will be too many. In this system, six codebooks are set for these six parts.

Another key factor for speeding up the recognition procedure and reducing the memory resources is how to pruning unlikely hypothesis as soon as possible. Selecting the candidate signs are used during the decoding process in this system. The experiments show that the signs can be recognized in real-time.

The organization of this paper is as follows: Section 2 describes the outline of the designed system. Section 3 presents a feature extraction method of hand gesture. Section 4 discusses recognition approaches of sign language recognition. Section 5 demonstrates the performance evaluation of the proposed approaches. The summary and discussion are given in the last section.

2   System Architecture

The architecture of this designed system is shown in Fig.1. The sign data collected by the gesture-input devices is fed into the feature extraction module, the output of feature vectors from the module is then input into the training module, in which a model is built for each sign. The number of the states of a model is set according to the transformation of the sign's signal. In order to decrease the computational cost, the Gaussians on the states are classified, then the index of the code word that is the nearest to the Gaussian on each state is recorded. The details are discussed in Sec.3. The search algorithm we used will be described in Sec.4. When the sign is output from the decoder, the sign drives the speech synthesis module to produce the voice of speech.

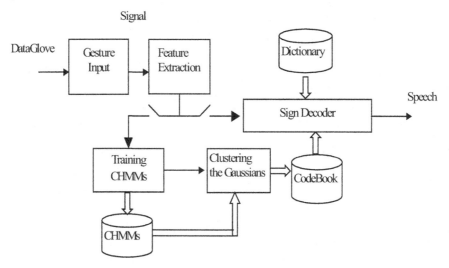

Fig.1. Sign Language Recognition

3   Feature Extraction

To calculate all gesture data from the left hand, right hand and body parts in a well-defined space, we need to consider the modeling of the relative 3D motion of three receivers working with a transmitter. The timely motion of the transmitter will be also considered. 3D motion of receivers can be viewed as the rigid motion. It is well known that 3D displacement of a rigid object in the Cartesian coordinates can be modeled by an affine transformation as the following,

$$X' = R(X - S) \tag{1}$$

where R is a 3×3 rotation matrix,

$$
R = \begin{pmatrix} 1 & 0 & 0 \\ 0 & \cos\alpha & -\sin\alpha \\ 0 & \sin\alpha & \cos\alpha \end{pmatrix} \begin{pmatrix} \cos\beta & 0 & \sin\beta \\ 0 & 1 & 0 \\ -\sin\beta & 0 & \cos\beta \end{pmatrix} \begin{pmatrix} \cos\gamma & -\sin\gamma & 0 \\ \sin\gamma & \cos\gamma & 0 \\ 0 & 0 & 1 \end{pmatrix} \tag{2}
$$

$$
= \begin{pmatrix} \cos\beta\cos\gamma & \cos\beta\sin\gamma & -\sin\beta \\ \sin\alpha\sin\beta\cos\gamma - \cos\alpha\sin\gamma & \sin\alpha\sin\beta\sin\gamma + \cos\alpha\cos\gamma & \sin\alpha\cos\beta \\ \cos\alpha\sin\beta\cos\gamma + \sin\alpha\sin\gamma & \cos\alpha\sin\beta\sin\gamma - \sin\alpha\cos\gamma & \cos\alpha\cos\beta \end{pmatrix}
$$

$X = (x_1, x_2, x_3)^t$ and $X' = (x_1', x_2', x_3')^t$ denote the coordinates of the transmitter and receiver respectively, S is the position vector of the receiver with respect to Cartesian coordinate systems of the transmitter.

The receiver outputs the data of Eulerian angles, namely, α, β, γ, the angles of rotation about X_1, X_2 and X_3 axes. Normally these data cannot be used directly as the features because inconsistent reference might exist since the position of the transmitter might be changed between the processing of training and that of testing. Therefore, it is necessary to define a reference point so that the features are invariant wherever the positions of transmitter and receivers are changed. The idea we propose to fix this problem is as follows. There is a receiver on each hand, and the third receiver is mounted at a fixed position on the body, such as the waist or the back. Suppose that S_r, S_l, and S are the position vectors of the receivers at right hand, left hand and on the body. It is clear that the product RR_r^t, RR_l^t, $R(S_r - S)$ and $R(S_l - S)$ are invariant to the positions of the transmitter and the signer, where R_l, R_r, and R are the rotation matrix of the receivers at right hand, left hand and on the body. R_r^t is the transpose matrix of R_r that is the rotation matrix of the receiver at the right hand respect to Cartesian coordinate systems of the transmitter.

The raw gesture data, which in our system are obtained from 36 sensors on two datagloves, and three receivers mounted on the datagloves and the waist, are formed as 48-dimensional vector. A dynamic range concept is employed in our system for satisfying the requirement of using a tiny scale of data. The dynamic range of each element is different, and each element value is normalized to ensure its dynamic range 0-1.

4   Sign Language Recognition

Hidden Markov Models (HMMs)[13] have been used successfully in speech recognition, handwriting recognition, etc. A HMM is a doubly stochastic state machine that has a Markov distribution associated with the transitions across various states, and a probability density function that models the output for every state. A key assumption in stochastic gesture processing is that the signal is stationary over a short time interval.

CHMMs is used in the system proposed in [12], which can recognize 1064 signs. But the system is not applicable to 5100-sign vocabulary because the computational cost is too much. In order to improve the performance of the system, some useful techniques are used.

4.1 CHMMs with Different Number of States

In our system, a HMM is built for each sign. But the numbers of basic gestures included in different words are variable. Some signs are simple, such as "chair"(Fig.2(a)) and "Room"(Fig.2(b)). There is only one basic gesture in each one of these signs. It is enough to set 3 states in their HMMs. But some signs are complicated, such as Chinese idioms "zuo-jing-guan-tian"(Fig.2(c)) and "hu-tou-she-wei"(Fig.2(d)). There are at least four basic gestures in these signs.

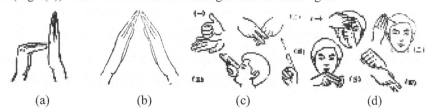

(a)   (b)   (c)   (d)

Fig.2. (a) "Chair" ;(b) "Room"; (c) "zuo-jing-guan-tian" (d) "hu-tou-she-wei"

If the number of the states in HMMs is set to 3, each vector on the states in the complicated signs does not correspond to a gesture. The accuracy will be affected. On the other hand, if the number of the states in HMMs is set to the maximum number of the basic gestures included in one sign, the computational cost will be too much. The number of states in a HMM should be consistent with the number of basic gestures included in the sign. In our system, an approach based on dynamic programming is used to estimate the number of the states.

There is a problem of matching of long signs and short ones if the number of states is different. The minimum number of states is 3 and the maximum number is 5, and a_{ij} is the probability of transferring from the state i to j. For the long signs, $a_{ij} < 1$ when $i, j \geq 3$. For the short signs, $a_{33} = 1$.

Given the observation sequence $O = (o_1 o_2 ... o_T)$, the best state sequence is $q = (q_1 q_2 ... q_T)$, where q_1 is the initial state. The probability of $O = (o_1 o_2 ... o_T)$ is obtained over the best state sequence q can be written as

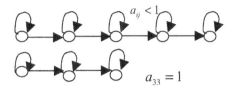

Fig. 3. The HMMs of long signs and short ones

$$p(O \mid \lambda) = \pi_{q_1} b_{q_1}(o_1) a_{q_1 q_2} b_{q_2}(o_2)...a_{q_{T-1} q_T} b_{q_T}(o_T)\qquad(3)$$

The interpretation of the computation in the above equation is the following. Initially (at time $t = 1$) we are in state q_1 with probability π_{q_1}, and generate the symbol o_1 with probability $b_{q_1}(o_1)$. The clock changes from time t to $t+1$ and we make a transition to state q_2 from state q_1 with probability $a_{q_1 q_2}$, and generate the symbol o_2 with probability $b_{q_2}(o_2)$. This process continues in this manner until the last transition (at time T) from state q_{T-1} to state q_T with probability $a_{q_{T-1} q_T}$. Because $a_{ij} < 1$ (i,j>3) for long sign, the decrease of the probability of long signs is larger than that of short signs. A long sign is easily recognized as the short sign that is similar to the later half of the long one. To match the long signs and short ones, the transferring probabilities a_{ij} ($i, j \geq 3$) in long signs are set to 1, namely the transfers between the later states are set to null.

4.2 Clustering the Gaussians

In systems that use a time-synchronous beam search, only a few percent of the states may be active at a given time, but even so, this implies a lot of Gaussian likelihoods need to be evaluated, which is still substantial. A common technique for a "fast match" for efficient Gaussian calculation is to pre-cluster all the Gaussians on the states into a relatively small number of clusters. During recognition, the likelihoods of the clusters are first evaluated, and these clusters whose scores are higher than the threshold are set active. Only those Gaussians corresponding to the active clusters are evaluated. The amount of computation that can be saved is good-sized. There are many variations for computing the clusters, for example, VQ-based methods, and hyperplane-based methods. In our system, the vector is consisted of the left hand shape, position, orientation, right hand shape, position and orientation. Because the number of possible combinations of these six parts can be approximately 10^8, in this system, hand shape, position and orientation of the mean vectors in Gaussians are classified respectively to reduce the number of codewords. There are six streams totally. As the probability in log domain can be computed by the summation of all the streams probabilities in log domain as follows:

$$\log p(x) = \sum_{l=1}^{6} \log p_l(x_l) \tag{4}$$

Where $p(x)$ is the Gaussian probability, and $p_l(x_l)$ is the observation probability of the l'th data stream. The clustering algorithm used here is k-means[13].
For each data stream, the stream state probabilities are clustered. The probability of a class needs to be computed only once. If $p_l(x_l)$ is belong to the i'th class, the probability of $p_l(x_l)$ is equal to that of the i'th class. As the number of distinguishable patterns in each data stream is relatively small, for a given observation vector, after these six observation probabilities have been computed, the log likelihood of each signs can be easily gotten by 5 times addition operations. For the case of continuous sign recognition, because the computation time for the state observation probabilities is relatively small, the probably active model candidates can be quickly determined. The likelihood computation can be reduced by a factor of 10 or more with minimal loss in accuracy.

4.3 Fast Matching Algorithms

Given the observation sequence $O = (o_1 o_2 ... o_T)$ and all models $\lambda = \{\lambda_1, \lambda_2, \cdots, \lambda_V\}$ (V is the number of the models), compute the probability that each model λ_v outputs the observation sequence.

$$P(O \mid \lambda_v) = P(O, Q^* \mid \lambda_v) \tag{5}$$

where Q^* is the best state sequence. Then find the sign according to the following expression:

$$v^* = \arg \max_{1 \le v \le V} P(O \mid \lambda_v) \tag{6}$$

In order to find v^*, the probabilities of all the models need to be calculated. Viterbi can be used. Viterbi[13] search and its variant forms belong to a class of breadth-first search techniques.

There are more than 5000 models, so the computational cost is still large even if "clustering the Gaussian" is used. In order to conserve the computing and memory resources, the fast matching method is proposed in time-synchronous search. The idea is as followed. For the first several frames, the probabilities of all the codeword of each stream data are calculated. For each stream, a threshold according to the maximum one is set to select the active codeword. If the probability of one codeword is larger than the threshold, the codeword is set active, otherwise inactive. For each HMM, if there is a state on which the six codewords are active and the probability that the sign begins with this state is higher than a threshold, this model is set active. Only the probabilities $P(O \mid \lambda_v)$, in which λ_v is an active model, need to be calculated. From these probabilities, the model that has the maximum probability is selected, and the sign corresponding to this model is the recognition result.

5   Experiment

The hardware environment is Pentium III 700MHz, with two CyberGloves and three receivers of 3D tracker; each Cyberglove is with 18 sensors. The baud rate for both CyberGlove and 3D tracker is set to 38400.

5100 signs included in Chinese sign language dictionary are used as evaluation vocabularies. Each sign was performed five times by one signer, four times are used for training and one for testing. The minimum number of states in HMM of each sign is 3. The maximum number of states in HMM of each sign is 3~7. The recognition rates with different maximum number of states are shown in Table.1. According to Table.1, when the maximum number of states is 5l the best recognition accuracy is 95.0% and the average number of states is only 3.27. This result is very encouraging.

Table.1. The recognition rates with different maximum number of states

The maximum number of states	Average number of states	Rate
3	3	93.9%
4	3.162658	94.3%
5	3.268776	95.0%
6	3.349578	94.8%
7	3.418354	94.78%

The above result is with modifying the transferring probability. The comparison is given in Table 2. The recognition rate with modifying the transferring probability is 95% and the recognition rate without modifying the transferring probability is 93.8%. The results show that modifying the transferring probability has effect on the recognition rate.

Table.2. The recognition rates of large vocabulary signs (5100 signs)

Without modifying the transferring probability	93.8%
With modifying the transferring probability	95.0%

The recognition rates of isolated signs with different numbers of the codewords are shown in Table3. The test data are collected when the signs were recognized online.

Table.3. The recognition rates of signs for different numbers of codewords

The number of codewords						Recognitio n rates(%)
Right positio n	Right orientation	Left positio n	Left orientatio n	Left hand shape	Right hand shape	
256	256	256	256	256	256	89.33
128	128	128	128	256	256	92.73
128	128	128	128	300	300	93.19
128	128	128	128	350	350	93.74
128	128	128	128	400	400	93.61
128	128	128	128	512	512	93.91

According to Table 3, considering both the accuracy and the speed, the numbers of codewords of right position, right orientation, left position, left orientation, left hand shape and right hand shape are set to 128, 128, 128, 128, 350, 350. It spends no more than 1 second to recognition a sign online and the recognition rate is more than 90%. Real-time recognition has been realized.

6   Conclusion

In this paper, a real-time CSL recognition system based on large vocabulary is presented using HMM based technology. Our contributions within this system are three aspects: modified the transition probability, clustering the Gaussian and search algorithm. Experimental results have shown that the proposed techniques are capable of improving both the recognition performance and speed. This system is for signer dependent isolated signs recognition task. The system will be extended for the task of continuous and signer independent recognition in the future.

Acknowledgment

This research is sponsored partly by Natural Science Foundation of China (No.69789301), National Hi-Tech Program of China (No.863-306-ZD03-01-2), and 100 Talents Foundation of Chinese Academy of Sciences.

References

1.   C. Charayaphan, A. Marble: Image processing system for interpreting motion in American Sign Language. Journal of Biomedical Engineering, 14(1992) 419-425.
2.   T. Starner: Visual recognition of American Sign Language using hidden Markov models. Master's thesis, MIT Media Laboratory, July. 1995.

3. S. S. Fels, G. Hinton: GloveTalk: A neural network interface between a DataDlove and a speech synthesizer. IEEE Transactions on Neural Networks 4(1993) 2-8.
4. S. Sidney Fels: Glove –TalkII: Mapping hand gestures to speech using neural networks-An approach to building adaptive interfaces. PhD thesis, Computer Science Department, University of Torono, 1994.
5. Tomoichi Takahashi, Fumio Kishino: Gesture coding based in experiments with a hand gesture interface device. SIGCHI Bulletin (1991) 23(2) 67-73.
6. Yanghee Nam, K. Y. Wohn: Recognition of space-time hand-gestures using hidden Markov model. To appear in ACM Symposium on Virtual Reality Software and Technology (1996).
7. R. H. Liang, M. Ouhyoung: A real-time continuous gesture recognition system for sign language. In Proceeding of the Third International Conference on Automatic Face and Gesture Recognition, Nara, Japan (1998) 558-565.
8. Kirsti Grobel, Marcell Assan: Isolated sign language recognition using hidden Markov models. In Proceedings of the International Conference of System,Man and Cybernetics (1996) 162-167.
9. Christian Vogler, Dimitris Metaxas: Adapting hidden Markov models for ASL recognition by using three-dimensional computer vision methods. In Proceedings of the IEEE International Confference on Systems, Man and Cybernetics, Orlando, FL (1997) 156-161.
10. Christian Vogler, Dimitris Metaxas: ASL recognition based on a coupling between HMMs and 3D motion analysis. In Proceedings of the IEEE International Conference on Computer Vision, Mumbai, India (1998) 363-369.
11. Christian Vogler, Dimitris Metaxas: Toward scalability in ASL Recognition: Breaking Down Signs into Phonemes. In Proceedings of Gesture Workshop, Gif-sur-Yvette, France (1999) 400-404.
12. Wen Gao, Jiyong Ma, Jiangqin Wu, Chunli Wang: Large Vocabulary Sign Language Recognition Based on HMM/ANN/DP. International Journal of Pattern Recognition and Artificial Intelligence, Vol. 14, No. 5 (2000) 587-602.
13. L. Rabiner, B. Juang: Fundamentals of Speech Recognition. Publishing Company of TsingHua University.

The Recognition of Finger-Spelling
for Chinese Sign Language

Wu Jiangqin[1] and Gao Wen[2]

[1] Department of CS, Zhejiang University, Hangzhou, 310027
wu_jiangqin@yahoo.com
[2] Chinese Science Academy, ICT, Beijing, 100080
wgao@ict.ac.cn

Abstract. In this paper 3-layer feedforward network is introduced to recognize Chinese manual alphabet, and Single Parameter Dynamic Search Algorithm(SPDS) is used to learn net parameters. In addition, a recognition algorithm for recognizing manual alphabets based on multi-features and multi-classifiers is proposed to promote the recognition performance of finger-spelling. From experiment result, it is shown that Chinese finger-spelling recognition based on multi-features and multi-classifiers outperforms its recognition based on single-classifier.

1   Introduction

Chinese Sign Language(CSL) is primarily composed of gestures, complemented by finger-spelling. There is 30 manual alphabets [1] in finger-spelling. One finger posture stands for one manual alphabet. Any character, word and sentence could be spelled by manual alphabets. Especially, some proper nouns, name of person and place can only be communicated by finger-spelling.

Finger-spelling recognition system is an assistant for a deaf to communicate with their surroundings through computer. So Chinese finger-spelling recognition is discussed here.

Artificial Neural Network(ANN) proves to be one of the main recognition technique for finger-spelling postures. Beale and Edwards [2]use a neural network to recognize five different American Sign Language finger-spelling postures. H. Hienz and K. Grobel [3] recognized video-based handshape using Artificial Neural Networks because of the network's learning abilities.

Here 3-layer feedforward network is also introduced into our paper to recognize Chinese manual alphabets. As it is plagued by significant time required for training, a kind of fast learning algorithm – Single Parameter Dynamic Search Algorithm (SPDS) [4] is used to learn net parameters. In addition, multi-classifiers based on multi-features are constructed and a recognition algorithm for recognizing manual alphabets based on multi-features and multi-classifiers is proposed to promote the recognition performance of finger-spelling.

I. Wachsmuth and T. Sowa (Eds.): GW 2001, LNAI 2298, pp. 96-100, 2002.

2   Data Input and Pre-processing

As the manual alphabet is one-handed gesture and the finger posture standing for each alphabet is still, the 18-D raw sensor data from one CyberGlove and the 3-D data from one Polhemus 3-D tracker with one receiver positioned on the wrist of glove jointly constitute the feature vector.

Because there exist unnecessary frames generated by transition from one alphabet to another when gesturing one word or one sentence, it is necessary to extract the *effective frames* characterizing the alphabets included in the word. By analysis of the raw sensor data of one word or the sentence, it is found that the effective frame corresponds to a steady span of raw sensor data and the transition frame corresponds to the changed span between two adjacent effective frames. Effective frame is determined by detecting steady and varied span of raw sensor data. The mean of the steady span is that effective frame.

3   Chinese Manual Alphabet Recognition Based on SPDS

We apply 3-layer feedforward network with strong discrimination and anti-jamming for Chinese manual alphabets recognition. In order to increase the speed of training net parameters, SPDS is used in the process of training. In each iteration only one parameter is allowed change and one-dimension precise search is done then. According to the features of both multi-layer neural network and error function, for each adjustment only the computation related to changed part of error function is needed, so the computation complexity is reduced greatly and the training speed promotes a lot.

In our experiment. 300 samples of 30 manual alphabets(A,B,C,D,E,F,G,H,I,J,K, L,M,N,O,P,Q,R,S,T,U,V,W,X,Y,Z,CH,SH,ZH,NG) are collected as training samples. The net parameters corresponding to each letter are learned by SPDS. For letters from A to NG, 100 times on-line test are done, statistical experiment results are shown in table 1.

From the results, it is shown that most of the letters can be recognized both accurately and effectively, but the recognition rate of some letters are relatively low. What makes the letters to be wrongly recognized is that there exist a few letters close to them in the 21-D feature space. For example, the reason that the recognition rate of X is relatively small is that there exists a letter H close to it in the feature space. If a sub-classifier is built based on primary features that discriminate X and H such as ring-middle abduction, middle-index abduction and thumb abduction, which is used to recognize X, the recognition rate of X(100%) is much higher than the previous recognition rate (90%, as shown in table 1). According to the above idea, a recognition algorithm based on multi-features and multi-classifiers is proposed in the next section to recognize manual alphabet.

4   Manual Alphabet Recognition Based on Multi-features and Multi-classifiers

4.1   A Recognition Algorithm Based on Multi-Features and Multi-classifiers

The idea of the algorithm is that some feature subsets that make letters discriminated more easily are selected firstly, subclassifiers based on these feature subsets are then built, and finally the recognition results of all the subclassifiers are fused to improve the recognition performance.

The Construction of Subclassifier. Assuming that the number of elements in feature set M is $|M|$, and the number of recognized letters is P. According to the posture and orientation of each letter, feature set M is divided into K feature subsets M_l $l=1,...,K$. To retain the recognition performance of each letter, feature sub-sets are allowed intersected. Apparently $|M_l|\le|M|$, $l=1,...,K$ and $\bigcup_{l=1}^{K}M_l = M$. The net parameters $\{\omega_{ij}^{(l)}\}$ $\left(i=1,...,|M_l|;j=1,...,H^{(l)}\right)$, $\{\upsilon_{jk}\}$ $\left(j=1,...,H^{(l)};k=1,...,P\right)$, $\{\theta_j^{(l)}\}$ $\left(j=1,2,...,H^{(l)}\right)$, $\{\gamma_k^{(l)}\}(k=1,...,P)$ for subclassifier C_l based on $M_l, l=1,2,...,K$ is learned by SPDS algorithm, where $\{\omega_{ij}\}$ is the connection weight from input node to hidden node, $\{\upsilon_{jk}\}$ is the connection weight from hidden node to output node, $\{\theta_j^{(l)}\}$ is the output threshold of hidden nodes, $\{\gamma_k^{(l)}\}$ is the output threshold of output nodes, $|M_l|, l=1,...,K$ is the number of input nodes for each subclassifier, $H^{(l)}$ is the number of hidden nodes for l-th classifier, P is the number of output nodes.

Recognizing Algorithm. Firstly, any raw sensor data flow G from input device is segmented into effective frame series $x(1),x(2),...,x(T)$. The data for each frame $x(t)=\left(x_1(t),x_2(t),...,x_{|M|}(t)\right)$, $t=1,...,T$ based on $M_l, l=1,...,K$ is notated as $x^{(l)}(t)=\left(x_1^{(l)}(t),...,x_{|M_l|}^{(l)}(t)\right)$. The error $E_q^{(l)}(t)$ between its output corresponding to classifier C_l and target output for q-th pattern ($q=1,...,P$) is calculated, according to equation

$$E_q^{(l)}(t)=\sum_{j=1}^{P}\left(\left(f\left(\sum_{s=1}^{H^{(l)}}v_{sj}^{(l)}\left(f\left(\sum_{i=1}^{|M_l|}\omega_{is}^{(l)}x_i^{(l)}(t)\right)+\theta_s^{(l)}\right)\right)+\gamma_j^{(l)}\right)-t_{qj}\right)^2 \tag{1}$$

where $t_q=\left(t_{q1},t_{q2},...,t_{qP}\right)$ is the target output for q-th pattern, f is Sigmoid function.

If ε is selected to satisfy that when $E_q^{(1)}(t) \leq \varepsilon$ and $x(t)$ belongs to q-th pattern, then $\{1 : E_q^{(1)}(t) \leq \varepsilon, 1 = 1, \ldots, K\}$ represents the set of subclassifiers that $x(t) = \left(x_1(t), x_2(t), \ldots, x_{|M|}(t)\right)$, $t = 1, \ldots, T$ belongs to pattern q, and $\left|\{1 : E_q^{(1)}(t) \leq \varepsilon, 1 = 1, \ldots, K\}\right|$ represents the number of subclassifiers that $x(t) = \left(x_1(t), x_2(t), \ldots, x_{|M|}(t)\right)$, $t = 1, \ldots, T$ belongs to pattern q.

$$A_q(x(t)) = \frac{\left|\{1 : E_q^{(1)}(t) \leq \varepsilon\}\right|}{K} \tag{2}$$

is the degree that $x(t)$ belongs to pattern q. The pattern that maximizes $A_q(x(t))$

$$q_0(t) = \arg\max_q A_q(x(t)) \tag{3}$$

is the pattern that $x(t)$ belongs to, $t = 1, \ldots, T$.

4.2   Experiment Result

Under the same experiment environment as in section 3, the recognition algorithm based on multi-features and multi-classifiers is used to test letters from A to NG on-line (here 4 classifiers are used, one is orientation classifier), results are also shown in table 1.

From the experiment results, it is shown that the recognition performance based on multi-classifiers is much better than that based on single classifier. For example, the recognition rate of X promotes from 90% to 98%.

5   Conclusion

In this paper Artificial Neural Network (ANN) is introduced into finger-spelling recognition, and a kind of fast learning algorithm-- Single Parameter Dynamic Search Algorithm (SPDS) is used to learn net parameters. A recognition algorithm for manual alphabets based on multi-features and multi-classifiers is proposed to promote the performance of recognizing letters. From experiment results, it is shown that Chinese manual alphabets recognition based on multi-features and multi-classifiers outperforms its recognition based on single-classifier.

Table 1. The Experiment Result for Single Classifier and Multi-classifiers

L	A	B	C	D	E	F	G	H	I	J
MC	100	100	98	100	100	98	100	97	100	100
SC	100	100	97	100	100	90	100	96	100	100

L	K	L	M	N	O	P	Q	R	S	T
MC	100	100	97	100	96	100	100	100	98	100
SC	100	100	95	98	90	100	100	100	90	100

L	U	V	W	X	Y	Z	CH	SH	ZH	NG
MC	100	97	100	98	100	100	98	100	100	100
SC	100	80	92	90	100	100	90	100	100	100

References

1. Chinese deaf association, Chinese Sign Language. Huaxia publishing company, 1994
2. R. Beale and A Edwards, Recognizing postures and gestures using neural networks. Neural Networks and Pattern Recognition in Human Computer Interaction.E.Horwood,1992
3. H. Hienz, K. Grobel and G. Beckers, Video-based handshape recognition using Artificial Neural Networks, 4th European Congress on Intelligent Techniques and Soft Computing, Aachen, September 1996, PP. 1659-63
4. Wang Xuefeng, Feng Yingjun, A New Learning Algorithm of Multi-Layer Neural Network, Jouranl of Harbin Institute of Technology, Vol 29, No. 2 (1997)

Overview of Capture Techniques
for Studying Sign Language Phonetics

Martha E. Tyrone

Department of Language and Communication Science, City University
Northampton Square, London EC1V0HB
m.e.tyrone@city.ac.uk

Abstract. The increased availability of technology to measure human movement has presented exciting new possibilities for analysing natural sign language production. Up until now, most descriptions of sign movement have been produced in the context of theoretical phonology. While such descriptions are useful, they have the potential to mask subtle distinctions in articulation across signers or across sign languages. This paper seeks to describe the advantages and disadvantages of various technologies used in sign articulation research.

1   Background

As the field of sign linguistics has become more established and grown to include more atypical signers, the need to describe and categorise signing deficits caused by neurological or other sensorimotor impairments has increased. In particular, deficits that are articulatory rather than linguistic have been increasingly documented, though there is no clear means of assessing them. While linguistics has well-developed methods for measuring the articulation and acoustics of speech, sign linguistics has no comparable methodology. Traditionally, sign language research has focused on theoretical aspects of sign language structure (i.e. syntax and phonology) rather than the physical output mechanism. As a result, no widely used technique has developed to examine the phonetics of sign language. Also, while some motor control research has focused on speech articulation, researchers' only interest in limb movements has been in non-linguistic tasks, such as pointing or grasping, which are motorically very different from signing. So standardised techniques for collecting human movement data cannot be directly applied to sign language. For these reasons, it is difficult to analyse sign articulation in a way that allows comparisons across studies, much less across sign languages. We believe that new research on sign articulation deficits is currently leading to the development of a useful methodology for examining sign language phonetics.

Although there has been a great deal of research on sign language phonology, it is only recently that researchers have examined *fluid* signing from the standpoint of phonetics or motor control [1-4]. Early work on the relationship between sign phonology and anatomy or physiology was based on inventories of canonical

I. Wachsmuth and T. Sowa (Eds.): GW 2001, LNAI 2298, pp. 101-104, 2002.
© Springer-Verlag Berlin Heidelberg 2002

handshapes in signed languages [5, 6, 7]. While establishing the existence of specific handshapes is important for determining phonetic constraints on sign forms, one cannot rely solely on dictionary forms to describe the phonetics of signed language. It is necessary to look at fluid signing to uncover how signs are actually produced, by different signers and in different linguistic contexts. Doing so is difficult, though, because it requires the researcher to determine the positions, orientations, velocities and trajectories of many independent articulators. While it is clear that these are the fundamental physical features that need to be measured, it is not yet clear whether those features have consistent properties analogous to fundamental frequency or voice onset time in speech. Nor can this point be addressed in the absence of large amounts of data drawn from a consistent methodology.

2   Videotape Recording

Most descriptions of articulation in fluid signing have been motivated by the discovery of linguistic and motor control deficits caused by stroke [8] or Parkinson's disease [3, 9, 10]. The most basic technique used in studies of sign articulation is to videotape subjects signing, look at the videotape, and notate errors in some type of coding scheme. Videotaping is relatively unintrusive, can be used anywhere (an important consideration in studies of impaired subjects), and the collected data are easily stored and retrieved. In terms of analysis, however, videotaped data leave much to be desired. Both the temporal and spatial resolution are greatly limited, given the speed and precision with which signs are executed. Additionally, the actual process of coding and analysing the data is slow and labour-intensive.

3   Optical Motion Tracking

One alternative to videotape is an optical motion-tracking system, which uses a set of digital cameras to record the positions and movements of infrared light-emitting diodes placed on the subject's body [11]. Optical tracking systems are useful because the data from them are stored onto computer as they are collected, making them easier to analyse. In addition, optical tracking systems have high spatial and temporal resolution (~3mm and 100Hz, respectively). However, because the system is optical, if anything opaque comes between the cameras and a diode, the system cannot identify the diode's position. One implication of this for signing is that when the subject's hand or arm is oriented so that the diodes are not facing the camera, the data from those articulators are occluded. This is especially problematic for finger and wrist movements, e.g. in fingerspelling, because the fingers frequently cross over each other and the fingertips can move at approximately a 160 degree angle relative to the palm.

4   Datagloves

The problem of occlusions in optical tracking systems can be partially addressed by using a set of datagloves combined with diodes placed on parts of the body that are not likely to be occluded (e.g., the elbow). Datagloves can be used to collect information on the relative, but not absolute, positions of articulators; and because they record the fingers' pressure against the glove and not their position viewed by the cameras, data can be collected even when the fingers are not visible [12]. When datagloves are used with an optical tracking system and the sampling rates of the two datastreams are co-ordinated, the two sets of data can be combined to determine the absolute positions of the fingers, as well as their positions relative to each other. As with the optical system alone, the resolution of the system is high and the data can be analysed quickly and easily. However, occlusion of wrist position is still a problem when datagloves are used in combination with optical systems. Apart from the question of measurement accuracy which is necessary for *phonetic* analysis, wrist position underlies the *phonological* feature of orientation in sign languages.

5   Magnetic Tracking and Datagloves

Finally, a method that is being developed to record sign articulation data uses a magnetic tracking system combined with datagloves [13]. This operates on the same principle as the optical system: 16 independent markers are placed on the subject's body and their positions are tracked through space; and datagloves, each with 18 pressure sensors, are placed on the subject's hands. The magnetic system uses magnets rather than diodes as markers, and a magnetic field ($\sim 10^{-6}$ T) to track the positions of the magnets. Like the optical system, the magnetic system has high spatial and temporal resolution ($\sim$3mm and 60Hz); and the collected data are easy to analyse. However, the most significant advantage of a magnetic system is that data are not occluded when articulators are turned away from view. This is of great importance in allowing accurate measurements of sign articulation as it occurs naturally. The main disadvantage of this system is that the separate streams of data can be difficult to integrate.

6   Conclusions

Technology has become available to allow researchers to study real-time articulation of sign language with great accuracy. This should pave the way for new research on motor control deficits and their effect on signing, gesture, and other motor tasks. Plans are underway to use magnetic tracking to study signing deficits in British Sign Language resulting from Parkinson's disease. In principle, magnetic tracking can be used to examine other motor control deficits affecting sign output. Additionally, having a viable method for measuring sign articulation will allow collection of large amounts of normative data, not only from British Sign Language, but also from other signed languages. It is only by the extensive collection of sign articulation data that a well-defined, empirically-driven field of sign phonetics can develop.

References

1. Brentari, D., Poizner, H., Kegl, J.: Aphasic and Parkinsonian signing: differences in phonological disruption. Brain and Language 48 (1995) 69-105
2. Poizner, H., Kegl, J.: Neural basis of language and behaviour: perspectives from American Sign Language. Aphasiology 6 (1992) 219-256
3. Tyrone, M. E., Kegl, J., Poizner, H.: Interarticulator co-ordination in Deaf signers with Parkinson's disease. Neuropsychologia 37 (1999) 1271-1283
4. Cheek, A.: Handshape variation in ASL: Phonetics or phonology? Poster presented at the 7th International Conference on Theoretical Issues in Sign Language Research (2000)
5. Ann, J.: On the relation between ease of articulation and frequency occurrence of handshapes in two sign languages. Lingua 98 (1996) 19-42
6. Mandel, M. A.: Natural constraints in sign language phonology: Data from anatomy. Sign Language Studies 8 (1979) 215-229
7. Loncke, F.: Sign phonemics and kinesiology. In: Stokoe, W.C., Volterra, V. (eds.): SLR' 83. Proceedings of the 3rd International Symposium on Sign Language Research. CNR/Linstok Press, Rome SilverSpring (1985) 152-158
8. Poizner, H., Klima, E. S., Bellugi, U.: What the Hands Reveal About the Brain. MIT Press, Cambridge, MA (1987)
9. Loew, R. C., Kegl, J. A., Poizner, H.: Flattening of distinctions in a Parkinsonian signer. Aphasiology 9 (1995) 381-396
10. Poizner, H., Kegl, J.: Neural disorders of the linguistic use of space and movement. In: Tallal, P., Galaburda, A., Llinas, R., vonEuler, C. (eds.): Temporal Information Processing in the Nervous System. New York Academy of Science Press, New York (1993) 192-213
11. Kothari, A., Poizner, H., Figel, T.: Three-dimensional graphic analysis for studies of neural disorders of movement. SPIE 1668 (1992) 82-92
12. Fels, S. S., Hinton, G. E.: Glove-TalkII-- A neural network interface which maps gestures to parallel formant speech synthesizer controls. IEEE Transactions on Neural Networks 9 (1998) 205-212
13. Bangham, J. A., Cox, S. J., Lincoln, M., Marshall, I., Tutt, M., Wells, M.: Signing for the deaf using virtual humans. Paper presented at the IEEE Colloquium on Speech and Language Processing for Disabled and Elderly People (2000)

Models with Biological Relevance
to Control Anthropomorphic Limbs: A Survey

Sylvie Gibet, Pierre-François Marteau, and Frédéric Julliard

Valoria, Université de Bretagne Sud, Campus de Tohannic
Rue Yves Mainguy, F-56000 Vannes, France
{sylvie.gibet,pierre-francois.marteau,frederic.julliard}@univ-ubs.fr

Abstract. This paper is a review of different approaches and models underlying the voluntary control of human hand-arm movement. These models, dedicated to artificial movement simulation with application to motor control, robotics and computer animation, are categorized along at least three axis: Direct vs. Inverse models, Dynamics vs. Kinematics models, Global vs. Local models. We focus on sensory-motor models which have a biologically relevant control scheme for hand-arm reaching movements. Different methods are proposed with various points of view, related to kinematics, dynamics, theory of control, optimization or learning theory.

1 Introduction

Movement has always a particular meaning. It is integrated into an action and oriented towards a physical or intentional goal. For instance, the surgery aims at inserting instruments inside the human body; a violin player aims at reaching some particular musical phrase. To achieve these tasks, the central nervous system (CNS) uses all its sensory information: vision, tactile, proprioceptive and kinesthesia senses and also auditory sense. Beyond these cooperation capabilities of the motor control system, is a prediction necessity: the brain has only a few milliseconds to react to a given situation and to select the appropriate sensors to achieve a movement. Considering skilled actions such as piano playing or juggling, it becomes difficult to ignore the role of anticipation in movement production and perception. Prediction capabilities are possible if internal representations or internal models are coded in the brain.

Over recent years, the concept of internal model, defined as internally-coded model that mimic the behavior of motor systems, has emerged as an important theoretical concept in motor control. These models can be approached by the simulation of human motion. The objectives of such simulation models are twofold: first, they provide new insights into motor system neuro-physiology and bio-mechanics. In particular, they can improve the understanding via simulation of hypothetical strategies that the CNS uses to control movements of the limbs. Second, they allow the designing of artificial arms according to biological principles. One of the advantages of building such biologically-inspired devices is that the resulting movements might exhibit prop-

I. Wachsmuth and T. Sowa (Eds.): GW 2001, LNAI 2298, pp. 105-119, 2002.
© Springer-Verlag Berlin Heidelberg 2002

erties inherent to natural movements in response to simple patterns of activation. Among these properties, some of them can be expressed in the 3D-Cartesian space, as for instance the smoothness of the endpoint motion of the hand-arm system, or the endpoint velocity profile which is uni-modal and bell-shaped. Many computer applications involving motion simulation have been developed in the past decades. They above all concern the animation of virtual characters, intelligent robotics and gesture-driven applications, requiring control gestures which have a certain degree of expressivity.

Within the scope of motion simulation, a basic question can be raised: "what muscle variables does the CNS control?"; several candidate variables can be proposed: force or torque, length, velocity, stiffness, viscosity, etc. But if the movement of a limb can be defined in terms of the contraction of individual muscles, it is clear that complex motor behaviors cannot be understood as a simple extrapolation of the properties of its elementary components. One of the most convincing argument that supports the idea of distributed control and non specific muscle commands is that of motor equivalence. That is, a similar movement pattern can be attained through the use of several different muscle combinations. For example, subjects can retain the same style of writing whether writing with a pen and paper or with chalk and board [1]. Similar observations have been made regarding speech, where it has been shown that intelligible speech can be produced when the proper articulators are obstructed, requiring the use of different vocal-tract configurations to achieve the desired phonation [2]. Motor equivalence suggests that the mapping between CNS command variables and muscles command variables consists of several levels of control, with control variables distributed among a number of control structures, thus allowing plasticity and flexibility in the organization of muscle synergies. Flexibility refers also to the fact that motor programs remain the same, whereas cinematic or dynamic skeleton might differ. For example, the cinematic structure would change when adding a new segment at the extremity of a cinematic chain, such as when using a stick for a reaching movement, but the gesture would not necessarily need to be learnt again.

All these properties, i.e. motor equivalence, flexibility, anticipation and prediction that characterize biological motion influence the nature of information internally coded in the CNS. The problem is not so much to know what muscle variables are involved when a movement is performed, but to find out the structure of the internal model, to identify the internal control variables and the variables at the command level. Moreover, it is necessary, when performing skilled movements, to have a representation of motion which is rather invariant and close to the task representation. Physiologists suggest that spatial representation expressed in the 3D space is more invariant than the representations such as force-time patterns or joint rotations. Kawato [3] and Bullock [4] for example highlight the need for spatial representation in the control of motor behaviors.

After drawing a general framework for the representation of the motor apparatus, we will present in this paper a review of different models and algorithmic methods that address the control motion problem with various points of view related to kinematics, dynamics, theory of control, optimization or learning theory.

2   Forward and Inverse Internal Models

Forward models simulate the behavior of the motor system in response to outgoing motor commands. These models give a causal representation of the motor system. By contrast, inverse models invert the causal flow of the motor system, e.g. they generate from inputs about the state and the state transitions an output representing the causal events that produced that state. A representation of internal forward models of sensory-motor systems is given in figure 1, where an estimated sensory feedback is generated, given the state of the system and the motor command.

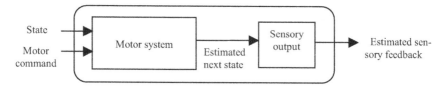

Fig. 1. Forward internal model

Most of the time, the state refers to the angular position and angular velocity of the motor system. More generally, this can be a set of parameters that condition the future behavior of the system. This state is not necessarily known by the internal model; this is why the state and the sensory information must be distinguished. The sensory output can be defined as information ending in muscles, joints and skin. They are related to proprioceptive, kinesthetic, tactile and visual information. Miall and al. [5] propose a forward sensory output model that predicts the sensory signals which would be caused by a particular state transition.

This internal representation can be approximated by several cascaded transformation systems, involving motor command, state space coordinates (joint coordinates) and Cartesian space coordinates. Classically, two main transformations are considered, as shown in figure 2: a forward dynamics transformation that links the motor command to the state space and a forward kinematics transformation that links the state space to the Cartesian space.

Forward models are generally well-defined systems, with a redundancy in the articulated systems characterized by an excess of degrees of freedom. Both of the dynamics and kinematics transformations are many-to-one mappings, since multiple motor means can produce the same effect. That is to say, given the current state of the system and the knowledge of the motor command, it is possible to calculate the next state of the system. The solution is unique, and depends on the initial data. On the contrary, a particular state of the system can be reached with many different motor means. For example, a seven degree of freedom arm moving a finger along a desired path in 3-D space can realize this task with different sequences of arm configurations.

With inverse transformations, it becomes possible to infer the appropriate motor command responsible for a desirable movement specified in the Cartesian space or the joint space. Identifying inverse problems intrinsically depends on the properties of the muscular-skeleton system.

When modeling an anthropomorphic limb, we do not try to reproduce a faithful representation of a human limb, which is a highly non linear system with an excess of degrees of freedom. Most of the time it is represented by functional blocks expressing the different transformation stages involved in motor control, from the neural signals to the motor commands used by the muscular-skeleton apparatus. In figure 2 a model of the transformation between motor command signals that activate the muscles and sensory output signals is represented.

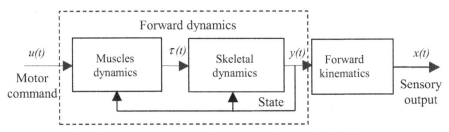

Fig.2. Forward dynamics model (represented by the dashed line box)

The muscles, activated by the muscular activation signal, $u(t)$, produce efforts which are distributed into equivalent torque applied to the joints, $\tau(t)$, according to the geometrical model of the limb. The forward-dynamics model predicts the next state values $y(t+1)$ (for instance q, dq/dt), according to the current values of $u(t)$ and $y(t)$. The forward kinematics block diagram transforms the state signal $y(t)$ into the sensory output, given for example by the Cartesian signal $x(t)$.

According to this representation, at least three inverse problems can be identified. The first one characterizes the trajectory formation and can be stated as follows: if we consider a multiple-joint limb, the endpoint of which has to go through a starting point, a via-point and an endpoint, there are an infinite number of trajectories satisfying these conditions. The second inverse problem, generally called inverse kinematics is to determine the joint angles of a multiple-joint arm with excess degrees of freedom when the hand position in the Cartesian space is given. Because of the redundancy, even when the time course of the hand position is strictly determined, the time course of the n-joint angles cannot uniquely be determined. The third inverse problem is the inverse dynamics problem when the arm is modeled with pairs of agonist-antagonist muscles. This consists in determining the time courses of agonist and antagonist muscle tensions when the joint angle time course is determined. Even when the time course of the joint angle is specified, there are an infinite number of tension waveforms of the muscles that realize the same joint angle time course. A feature of biological motor system is that joint torques are generated by a redundant set of muscles. Therefore, we might either choose to find out muscle activation using muscle-tendon models [6-7], or find out equivalent torques applied to joints [8].

Two inverse problems are generally considered in robotics, animation and motor control communities: the inverse kinematics problem and the inverse dynamics problem. Classical kinematics-based or dynamics-based control methods are presented in the next two sections.

3   Kinematics Control

Forward kinematics is represented by a non linear function f, that carries out the transformation between the n-dimensional vector of joint angle q and the position and orientation of the endpoint as the m-dimensional vector x:

$$x = f(q) \tag{1}$$

Direct kinematics techniques are widely used in computer animation industry, where the control of motion is based on the specification of the evolution with time of the trajectory of each degree of freedom of the articulated structure. The animator has in charge the specification of key-frames. Interpolation algorithms provide trajectories for the different angular coordinates. The realism of the produced motion can be improved if the number of key-frames increases. Determining, from spatial representation the internal state of the system is called inverse kinematics. If we consider the formulation given by the equation (1), the inverse kinematics is to compute the state q from the spatial representation x:

$$q = f^{-1}(x) \tag{2}$$

For redundant arms, i.e. $n > m$, f is a many-to-one mapping (there are many qs which lead to the same x). Then, the inverse transformation of f does not exist in a mathematical sense: solutions are generally not unique or do not exist at all (degenerate case), and even for $n = m$ multiple solutions can exist. Most techniques developed to solve the inverse kinematics problem aim at controlling in real-time the endpoint of an articulated system. The algorithms need to address how to determine a particular solution to (2) in face of multiple solutions.

Using the taxonomy of Tolani and Badler [9], inverse kinematics techniques are categorized into analytical or numerical solutions. In contrast to the analytical methods, which find all possible solutions, but generally use kinematic chains with constrained structures, numerical methods iteratively converge to a single solution from an initial condition. As we are interested here in redundant systems, adapted to model arms, legs or any other articulated structure, we only address numerical techniques. Two main approaches are considered: the approaches which require the inversion of the Jacobian, and other optimization approaches.

3.1   Inverse Jacobian

Two generic approaches, derived from robotics, can be used to calculate an approximation of the inverse of the Jacobian [10]. Global methods find an optimal path of q with respect to the entire trajectory. They are usually applied in computationally expensive off-line calculations. In contrast, local methods, feasible in real time, only compute an optimal change in q, δq for a small change in x, δx, and then integrate δq to generate the entire joint space path. Local methods can be solved for redundant manipulators, using pseudo-inverse computing [11-12].

To accelerate the resolution of the equations, improved algorithms have been developed. They are both used in robotics [13] and in computer animation [14]. However, these algorithms fail in either generality or performance, when operating on highly articulated figures. Moreover, they are generally based on the definition of a criterion which does not ensure that the produced movements look natural. Therefore if these methods cannot be ignored since they are largely used in most graphical software, they are not of prime importance according to biological relevance, nor are they efficient to cope with changes in environment or with the anatomical variability among individuals.

3.2   Optimization-Based Approaches

The inverse kinematics problem can also be regarded as a nonlinear optimization problem, thus avoiding the calculation of the pseudo inverse Jacobian. Examples of optimization-based approaches can be found in computer graphics, by minimizing a potential function expressing the goal to reach [15]. Since these methods do not attempt to generate realistic joint trajectories, they partially respond to our problem.

Other approaches better fit biomechanical analysis of movement [16-17]. Among these approaches, the Sensory-Motor Model (SMM), described in section 5, integrates biological plausible control functions.

The main interest of the inverse kinematics models is that the motor command can be expressed in terms of spatially defined goals, or targets.

4   Dynamics Control

The main interest of using dynamics to simulate internal models is the possibility to produce movements that satisfy the laws of physics, and thus have a realistic behavior. The inputs of the forward dynamical model of the limb are both the current state (e.g., joint angles and angular velocities) of the system, and the motor commands expressed in terms of torques applied to each joint.

If the joint position and velocities co-ordinates are simultaneously specified, it is known that the state of the system is completely determined and its subsequent motion can, in principle, be calculated. The equations of motion, expressed in terms of relation between the accelerations, velocities and position, can thus be derived. They are second-order differential equations for the vector q and its derivatives, and their integration makes possible the determination of the functions $q(t)$ and so of the path of the system.

If we represent the arm dynamics by a multi-articulated mechanical system characterized by its kinematics structure, the matrix of inertia and the nature of the links, we can use the Lagrangian formalism which expresses the motion equations from the energy of the system:

$$\frac{d}{dt}\left(\frac{\partial L}{\partial \dot{q}_i}\right) - \frac{\partial L}{\partial q_i} = \Gamma_i + Q_i \tag{3}$$

The function L, called the Lagrangian of the system, represents the sum of the cinetic and potential energies of the system. The set $\{q_i\}$ represents a set of n independent variables joint coordinates of the articulated system ($1 \cdot i \cdot n$), n being the number of degrees of freedom of the system. Q_i is the torque exerted by gravity and Γ_i is the torque exerted by the outside forces on the i-th joint variable.. A set of n non linear differential equations can then be derived from the Lagrangian formalism and numerically resolved by different integration algorithms schemes, for instance the Newton-Raphson formalism or the Runge-Kutta algorithm.

The objective of control methods is to provide a means to automatically calculate at each time step the forces and torque exerted by muscles to achieve a desired action. After presenting formally the control problem, we describe three control approaches: control by constraints, optimization principle and nonlinear control.

4.1   Formalization of the Control Problem

The system can be described by a mathematical model characterized by a function f establishing a relation between the state variables $y(t)$, and the command function $u(t)$:

$$\dot{y}(t) = f(t, y(t), u(t)) \tag{4}$$

The objectives of the system to control can be expressed in terms of:

- constraints: as a constraint, we can impose for instance that the system tracks a specified trajectory in the state space or in the Cartesian space.
- criteria to minimize: we might want to minimize the energy of the system during movement execution.

Control laws depending exclusively on time, refer to open-loop control. On the contrary, control laws depending on time and on the state variable y, refer to closed-loop control (Fig. 3).

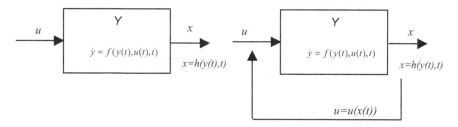

Fig. 3. Open and closed loop control model

4.2   Control by Constraints

This consists in adding to the set of motion equations describing the evolution of the system, other equations, which are additional kinematics constraints to be satisfied. Non constrained parts of the system can move freely.

The interest of this approach is that it combines a kinematics specification (constraints) and the simulation of the motion equations, thus avoiding the search of a command law. However, the introduction of efforts that have not necessary a mechanical reality can yield non natural behaviors. Moreover, the simulation cost is rather heavy, since the principle consists in increasing the dimension of the system by the adjunction of constraints.

Various methods are proposed to solve the set of equations. Among them, the use of Lagrange multipliers or penalization methods [18-19] can be retained.

4.3   Control by Optimization

The problem consists in finding a control trajectory $u^*(t)$ which minimizes the scalar function:

$$J(u) = \int_i^{t_f} \eta(t, y(t), u(t))dt + C(y(t_f)), \qquad (5)$$

while satisfying equality and non equality constraints.

This kind of methods solves the control problem in a global fashion, by proposing a control trajectory linking the initial state to the final state of the system, while minimizing a specific criterion. The interest is to globally calculate the motion, thus taking into account anticipation effects. However, it remains difficult to determine an accurate criterion which expresses some kind of naturalness of the generated movements. Moreover, these methods have heavy computing costs that limit there use in real time control.

There are numerous works exploiting optimal control and nonlinear optimization tools to simulate motion in computer animation [20-23].

4.4   Applying Control Theory to Motion Control

Most control concepts used in neurophysiology are based on regulation concepts. Feedback-controlled mechanisms applied to motor behavior are indeed concerned with the design and implementation of control algorithms capable of regulating a variable to a desired value.

Note that most control theory is based on linear mathematics. This reflects more a facility to solve problems rather than a tool in accordance with physics. In particular, it is difficult to understand the vast repertory of motor behavior with the rigid and restricted capabilities of linear control and servomechanisms. And if linear command theory is well-known today, the control of non linear systems remains a theoretical difficult problem. Non linearity of the system can be modeled by adding a non linear transfer function in the loop. Command laws must be elaborated so that a correct evolution of the system output is ensured.

Control theory has been largely explored in robotics, with the use of Proportional Derivative controllers, and more rarely in computer graphics [24-25]. However, the objectives of robotics are to move articulated structures which respect hard constraints such as stability or mechanical feasibility.

Adaptive control is one of the more recently emerging aspects of control technology. It applies to control systems whose behavior or control strategy changes as the task changes, and it may be an alternative to the description of motor behavior. Two forms of adaptive control have emerged. The first form is model-referenced adaptive control, in which a model of the controlled system behavior is used to generate future inputs to the system based on a comparison of observed behavior with predictions from the model. The second form is parameter-adaptive control, in which one or more of the controlled system parameters (such as stiffness or viscosity) may be changed to suit task needs. Adaptive control theory assumes that there exists some objective function to optimize when performing a task. The difficulty can be to find a physiologically relevant function. But it provides a powerful modeling technique that adds a pertinent criterion (such as energy or smoothness) to the resolution of a set of differential equations.

5   Our Approach: The Sensory-Motor Model (SMM)

In our approach, the sensory-motor systems can be seen as non linear mappings between sensory signals and motor commands, which can be considered as basic components involved in the control of complex multi-articulated chains. The state of the articulatory system q is computed on the basis of the scalar potential function $E(q)$ defined by:

$$E(q) = (f(q) - x_t)^T . (f(q) - x_t) \qquad (6)$$

and the evolution of q is given by:

$$\frac{\partial q}{\partial t} = -g(E(x, x_t, t)).\nabla(E(x, x_t, t)), \qquad (7)$$

where x is the current location and x_t the target location expressed in the Cartesian space ; f is the transformation which links the state observation vector x to the state vector q: $x = f(q)$; g is a gain function and $\bullet$ the gradient operator.

With SMM, the inverse kinematics problem is solved by using an optimization approach, according to a gradient descent strategy. In order to ensure the stability of the system and to generate damped behaviors, a nonlinear function and a second order filter have been introduced. The nonlinear function has a sigmoid shape: the gain of this function increases significantly when the error between the observable position and the reference target position goes towards zero. Stability and asymptotic properties of such a model have already been studied in [26]. This model is applied successfully to the control of a hand-arm system [27]. A gesture is associated with an internal command which consists in assigning a succession of goal positions to the arm endpoint.

To execute motion, SMM deals with symbolic data as well as continuous data in an adaptation process. Symbolic data includes the representation of tasks and goals, expressed in the Cartesian space. Continuous data are sensory signals used jointly with the motor command in a feedback process to update the state variables. This model yields natural behaviors, and can be seen as a biologically plausible functional model.

However, in the scope of complex artificial system design, analytical equations that drive the dynamics and the kinematics of the system could be difficult to extract, and the solution to the corresponding set of differential equations fastidious to estimate. Setting up control strategies for complex system control is consequently not a simple task. In this context, learning part of the control strategy from the observation of the system behavior is an appealing and efficient approach.

6   Learning Sensory-Motor Maps Involved in Internal Models

As developed in previous sections, sensory-motor systems exhibit non linear mappings between sensory signals and motor commands, that are in many respects basic components involved in the control of complex multi-articulated chains. However, to cope with the need for plasticity (adaptation to changes), generic control performance (similar control principles for various kinds of mappings, various kinds of articulated chains or neuro-anatomical variability among individuals) and anticipation (predictive capability and optimization of movements) it is more or less accepted in the neuro-physiology community that these mappings are learned by biological organisms rather than pre-programmed.

Two distinct and competing approaches are available when facing the problem of learning non linear transforms (NLT) and in particular non linear mappings involved in multi-joint control system: parametric learning (PL) and non parametric learning (NPL) (Cf. [28] for a pioneer and detailed synthesis on PL and NPL, and [29] for a more recent review of PL and NPL models with biological relevance arguments regarding internal sensory-motor maps). The fundamental difference between PL and NPL is that PL addresses the learning essentially globally while NPL addresses it much more locally. In other words, PL methods try to learn the non linear transform on the whole domain where the transform is defined. That means that if a change in the environment occurs locally, it will potentially affect the learning process everywhere in the definition domain of the transform. Conversely, NPL learns the properties of the transform in the neighborhood of each point of interest within the definition domain of the transform. Neuromimetic networks are an instance of the PL class with synaptic weights as parameters, while near neighbors derived methods are instances of the NPL class.

Biological relevance can be found for the two kind of approaches. Nevertheless, local characteristic of NPL is undoubtedly a great advantage when addressing incremental learning or continuous learning in variable environments, since the local modification resulting from any change does not affect the overall structure of the NLT already learned.

Two related approaches have been proposed in the context of learning algorithms, that allow flexibility and adaptability to the context of hand-arm movement tasks. The first one uses Parameterized Self-Organizing Maps which can be rapidly built on the basis of a small number of data examples[30]. The authors show their ability to identify the position of finger tips in 3D-hand shapes images. The second PL approach uses a recurrent network (MMC) which relaxes to adopt a stable state corresponding to a geometrically correct solution [31], with relatively few learning samples and parameters.

To exemplify the flexibility and the generic characteristics of NPL methods we present a simple experiment [32-33] carried on an simulated arm system submitted to a reaching task. The arm is composed with three joints and integrates 7 degrees of freedom. The arm is controlled by a SMM where the gradient descent strategy has been estimated using the learning of a non linear mapping f characterized as follows:

$$\widehat{\delta q} = f(q, \delta x), \tag{8}$$

where δx is the 3D direction vector towards the target position specified in the 3D Cartesian space, q the vector of the seven joint variables and $\widehat{\delta q}$ the estimated elementary modification within the joint space that will minimize the distance between the arm end point and the position of the target.

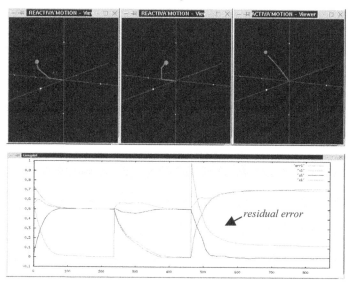

Fig. 4. Three targets are successively activated within the 3D Cartesian space. The two first targets are effectively reached. The last one cannot be reached since the location is out of hand of the simulated arm

In the first experiment (Fig. 4) three spatial targets t1[0.5, 0.5, 0], t2[0, 0.5, 0.5] and t3[0.8, 0, 0.8] are successively activated. The two first targets are asymptotically reached, while the third one, out of hand, cannot be reached by the three-joint arm. In the second experiment (Fig. 5), the same targets are activated, but the last link length of the arm is doubled to simulate the adjunction of a stick. Without any update of the learning, all three targets are now reached, showing the flexibility of the learned map f.

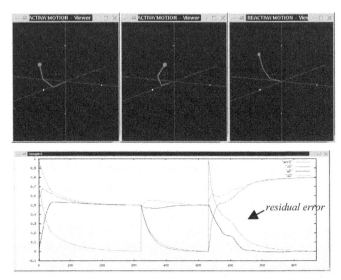

Fig. 5. Same experiment with a stick that double the length of the third (the last) joint. The last target is now reached and the corresponding residual error goes towards zero. No further learning had been required to cope with the stick adjunction

7   Conclusion

In this paper we reviewed various models dedicated to artificial movement generation with application to motor control, robotics and computer animation. These models have been categorized along at least three axis : Direct/Inverse, Kinematics/Dynamics, Global/Local. If it is easy to express the differential equations that describe the direct kinematics or dynamics models, and to derive the equations of motion, the inverse control problem is more difficult to formalize. The inverse kinematics problem is solved by mathematical computation. Control theory and optimization principles are an alternative to the control problem of mechanical or dynamical systems. A Sensory-motor model, combining linear and non linear biologically relevant functions is also proposed to solve in an iterative fashion the inverse kinematics problem.

Some of these models find some kind of relevance within the neuro-physiological community. They are labeled "internal model" since they are intended to mimic the behavior of the real articulated system to estimate further evolutions, anticipate changes or optimize the overall movement according to the context or to a dynamical environment. Internal model - either dynamical or kinematic, direct or inverse, local or global - need to be learned by internal neurological structures.

The learning issue has been briefly addressed within the paper, through the presentation of two competing approaches : Parametric Learning and Non Parametric Learning. Parametric approaches prove to be efficient when the issue is to learn a global trajectory formation or a fixed sensory-motor mapping. Conversely, Non parametric approaches seem to be more attractive when local strategy of movement formation or dynamical environment are considered. To exemplify these last arguments, we have shown how a Non Parametric Learning approach could efficiently learn the gradient descent strategy (minimal joint displacements) exploited in a Sensory-Motor Model.

The research material referenced or presented throughout the paper shows the large multidisciplinary field and the high productivity of movement control disciplines. Important results are provided in neurophysiology, biomecanics, or psychomotricity, robotics and computer animation communities. Promising expectations are still ahead in all of these disciplines either at a research level or at an application level.

References

1. Stelmach, G.E. and Digles V.A.: Control theories in motor behavior. Acta psychologica (1982)
2. Abbs, J.H. and Gracco, V.L.: Control of Complex motor gestures: Orofacial muscle responses to load perturbations on lip during speech. Journal of Neurophysiology, 51(4) (1984) 705-723
3. Kawato, M., Maeda Y., Uno, Y., Suzuki R.: Trajectory Formation of Arm Movement by Cascade Neural Network Model Based on Minimum Torque Criterion. Biological Cybernetics, vol. 62 (1990) 275-288
4. Bullock, D., Grossberg, S., Guenther, F.H.: A Self-Organizing Neural Model of Motor Equivalent Reaching and Tool Use by a Multijoint Arm. Journal of Cognitive Neuroscience, vol. 54 (1993) 408-435
5. Miall, R.C., Wolpert, D.M.: Forward Models for Physiological Motor Control. Neural Networks, vol. 9, n°8 (1996) 1265-1279
6. Hill A.V.: The heat of shortening and the dynamic constants of muscle. In proc. Royal Society of London, 1938, vol. B126, pp. 136-195
7. Winters J.M., and Stark L.: Muscle models: What is gained and what is lost by varying model complexity. Biological Cybernetics, vol. 55 (1987) 403-420
8. Zajac, F.E. and Winters, J.M.: Modeling Musculoskeletal Movement Systems: Joint and Body Segmental Dynamics, Musculoskeletal actuation, and Neuromuscular control; in Multiple Muscle Systems, J.M. Winters and S.L-Y Woo Eds., New York, Springer Verlag (1990) 121-148
9. Tolani, D., Goswami, A., Badler, N.: Real-Time Inverse Kinematics Techniques for Anthropomorphic Limbs. Graphical Models, vol. 62 (2000) 353-388
10. Baillieul, J. and Martin, D.P.: Resolution of kinematic redundancy, in Proceedings of Symposia in Applied Mathematics, American Mathematical Society[2] (1990) 49-89
11. Whitney, D.E.: Resolved motion rate control of manipulators and human prostheses. IEEE Transactions on Man-Machine systems, vol. 7(12), (1969) 47-53
12. Klein, C. and Huang, C.: Review of pseudoinverse control for use with kinematically redundant manipulators. IEEE Trans. Systems Man Cybernetics, vol. 7 (1997) 868-871
13. Nakamura, Y.: Advanced Robotics: Redundancy and Optimization, Addison-Wesley, Reading, MA (1991)

14. Badler, N., Phillips, C., Webber, B.: Simulating Hulmans – Oxford University Press, Inc, 200 Madison Avenue, New York, 10016 (1993)
15. Zhao, J. and Badler, N.: Inverse Kinematics Positioning Using Nonlinear Programming for Highly Articulated Figures. Transactions of Computer Graphics, 13(4), (1994) 313-336
16. Koga, Y., Kondo, K., Kuffner, J., Latombe, J.C.: Planning motions with intentions. In ACM Computer Graphics, Annual conference series (1994) 395-408
17. Gibet S. & Marteau P.F.: NonLinear feedback Model of sensori-motor system. International Conference on Automation, Robotics and Computer Vision, Singapore (1992).
18. Kearney, J.K., Hansen, S., Cremer, J.F. Programming mechanical simulations. The Journal of Visualization and Computer Animation, vol. 4 (1993) 113-129
19. Arnaldi, B., Dumont, G., Hégron, G.: Animation of physical systems from geometric, kinematic and dynamic models. Modeling in Computer Graphics, Springer Verlag. IFIP Working Conference 91 (1991) 37-53
20. Witkin, A., Kass, M.: Spacetime constraints. Computer Graphics, SIGGRAPH'88 Proceedings, vol. 22, n°4, (1988) 159-168
21. Cohen, M.: Interactive spacetime control for animation. Computer Graphics, SIGGRAPH'92 Proceedings, vol. 24, n°2, (1992) 293-302
22. Liu, Z., Gortler, S., Cohen, M.: Hierarchical spacetime control. Computer Graphics, Proceedings of ACM SIGGRAPG'94, (1994) 35-42
23. Brotman, L., Netravali, A.N.: Motion interpolation by optimal control. Computer Graphics, 22 (4) (1988) 309-315
24. Van de Panne, M., Fiume, E., Vranesic, Z.: Reusable motion synthesis using state-space controllers, ACM Computer Graphics (1990) 225-234
25. Lamouret, A., Gascuel, M.P., Gascuel, J.D.. Combining physically-based simulation of colliding objects with trajectory control. The journal of Visualization and Computer Animation, vol. 5 (1995) 1-20
26. Gibet, S., Marteau, P.F.: A Self-Organized Model for the Control, Planning and Learning of Nonlinear Multi-Dimensional Systems Using a Sensory Feedback, Journal of Applied Intelligence, vol.4 (1994) 337-349
27. Lebourque, T., Gibet, S.: A complete system for the specification and the generation of sign language gestures, in Gesture-Based Communication in Human-Computer Interaction. Proc. GW'99, Gif sur Yvette (France), A. Braffort, R. Gherbi, S. Gibet, J. Richardson, D. Teil Eds., Springer-Verlag Pub. (1999) 227-251
28. Duda, R. O., Hart, P. E.: Pattern classification and scene analysis. New York, NY: Wiley (1973)
29. Schaal, S.: Nonparametric regression for learning nonlinear transformations. In: Ritter, H., Holland, O., Möhl, B. (eds.). Prerational Intelligence in Strategies, High-Level Processes and Collective Behavior. Kluwer Academic (2000)
30. Walter, J., Ritter, H.: Rapid learning with Parametrized Self-Organizing Maps. Neurocomputing, vol. 12 (1996) 131-153
31. Steinkühler, U., Cruse, H.: A holistic model for an internal representation to control the movement of a manipulator witrh redundant degrees of freedom. Biological Cybernetics, 79, 457-466 (1998)

32. Marteau, P.F., Gibet, S., Julliard, F.: Non parametric learning of Sensory Motor Maps. Application to the control of multi joint systems. WSES/IEEE International Conference on Neural, Fuzzy and Evolutionary Computation, CSCC 2001, Rethymnon, Crete (2001)
33. Marteau, P.F., Gibet, S.: Non Parametric Learning of Sensory-Motor Mappings Using General Regression Neural Networks. 3rd WSES International Conference on Neural Networks and Application, Feb. 11-15, Interlaken, Switzerland (2002)

Lifelike Gesture Synthesis and Timing for Conversational Agents

Ipke Wachsmuth and Stefan Kopp

Artificial Intelligence Group
Faculty of Technology, University of Bielefeld
D-33594 Bielefeld, Germany,
{ipke,skopp}@techfak.uni-bielefeld.de

Abstract. Synchronization of synthetic gestures with speech output is one of the goals for embodied conversational agents which have become a new paradigm for the study of gesture and for human-computer interface. In this context, this contribution presents an operational model that enables lifelike gesture animations of an articulated figure to be rendered in real-time from representations of spatiotemporal gesture knowledge. Based on various findings on the production of human gesture, the model provides means for motion representation, planning, and control to drive the kinematic skeleton of a figure which comprises 43 degrees of freedom in 29 joints for the main body and 20 DOF for each hand. The model is conceived to enable cross-modal synchrony with respect to the coordination of gestures with the signal generated by a text-to-speech system.

1 Introduction, Previous Work, and Context

Synthesis of lifelike gesture is finding growing attention in human-computer interaction. In particular, synchronization of synthetic gestures with speech output is one of the goals for embodied conversational agents which have become a new paradigm for the study of gesture and for human-computer interfaces. Embodied conversational agents are computer-generated characters that demonstrate similar properties as humans in face-to-face conversation, including the ability to produce and respond to verbal and nonverbal communication. They may represent the computer in an interaction with a human or represent their human users as "avatars" in a computational environment [3].

The overall mission of the Bielefeld AI lab is interacting with virtual reality environments in a natural way. Three things have been important in our previous work toward incorporating gestures as a useful input modality in virtual reality: (1) measuring gestures as articulated hand and body movements in the context of speech; (2) interpreting them by way of classifying features and transducing them to an application command via a symbolic notation inherited from sign language; (3) timing gestures in the context of speech in order to establish correspondence between accented behaviors in both speech and gesture channels.

I. Wachsmuth and T. Sowa (Eds.): GW 2001, LNAI 2298, pp. 120–133, 2002.

An important aspect in the measuring of gestures is to identify cues for the gesture stroke, i.e. the most meaningful and effortful part of the gesture. As indicators we screen the signal (from trackers) for pre/post-stroke holds, strong acceleration of hands, stops, rapid changes in movement direction, strong hand tension, and symmetries in two-hand gestures. To give an idea, Figure 1 shows the signal generated from hand movement in consecutive pointing gestures. We have developed a variety of methods, among them HamNoSys [13] descriptions and timed ATNs (augmented transition networks), to record significant discrete features in a body reference system and filter motion data to object transformations which are put into effect in the 3D scene. The things we have learned from investigating these issues help us to advance natural interaction with 3D stereographic scenes in a scenario of virtual construction.

In previous years we have dealt with pointing and turning gestures accompanying speech, commonly classified as deictics and mimetics, e.g. [8]. In the DEIKON project ("Deixis in Construction Dialogues"), we have now started to research into more sophisticated forms of deictics that include features indicating shape or orientation, which lead into iconic gesture [16]. The DEIKON project was begun in the year of 2000 and is concerned with the systematic study of referential acts by coverbal gesture. In a scenario setting where two partners cooperate in constructing a model aeroplane, we investigate how complex signals originate from speech and gesture and how they are used in reference. One goal is to elucidate the contribution of gestural deixis for making salient or selecting objects and regions. Another goal is to make an artificial communicator able to understand and produce coverbal gestures in construction dialogues.

In this context, this contribution focusses on an approach for synthesizing lifelike gestures for an articulated virtual agent. Particular emphasis lies on how to achieve temporal coordination with external information such as the signal

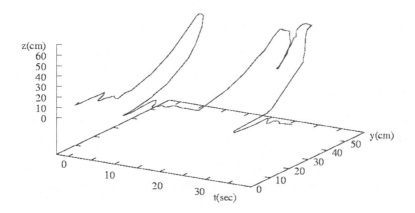

Fig. 1. Tracked hand movement in consecutive pointing gestures

Fig. 2. Articulated communicator (target interaction scenario)

generated by a text-to-speech system. The mid-range goal of this research is the full conception of an "articulated communicator" (cf. Figure 2), i.e. a virtual agent that conducts multimodal dialogue with a human partner in cooperating on a construction task.

The remainder of the paper is organized as follows. Having sketched some of our previous work and the context in which it is carried out, we next turn to the issue of lifelike gestures synthesis which is a core issue of our most recent work. In Section 3 we describe some details of the articulated communicator. The focus of Section 4 is timing, in particular, with respect to the gesture stroke. In Section 5 we give an outlook on how we proceed to synchronize synthetic speech with gesture.

2   Lifelike Gesture Synthesis

The rationales for our research on lifelike gesture synthesis are twofold. On the one hand, we seek for a better understanding of biologic and of cognitive factors of communication abilities through a generative approach ("learning to generate is learning to understand"). That is, models of explanation are to be provided in the form of "biomimetic" simulations which imitate nature to some extent. On the other hand, the synthesis of lifelike gesture is finding growing attention in human-computer interaction. In the realm of a new type of advanced application interfaces, the generation of synthetic gesture in connection with text-to-speech systems is one of the goals for embodied conversational agents [3].

If we want to equip a virtual agent with means to generate believable communicative behaviors automatically, then an important part in this is the production of natural multimodal utterings. Much progress has been made with respect to combining speech synthesis with facial animation to bring about lip-synchronous speech, as with so-called talking heads [10]. Another core issue is the skeletal animation of articulated synthetic figures for lifelike gesture synthesis that resembles significant features of the kinesic structure of gestural movements along with synthetic speech. Especially the achievement of precise timing for accented behaviors in the gesture stroke as a basis to synchronize them with, e.g., stressed syllables in speech remains a research challenge.

Although promising approaches exist with respect to the production of synthetic gestures, most current systems produce movements which are only parametrizeable to a certain extent or even rely on predefined motion sequences. Such approaches are usually based on behavioral animation systems in which stereotyped movements are identified with a limited number of primitive behaviors [14] that can be pieced together to form more complex movements, e.g. by means of behavior composition [1] or script languages [2,11]. The REA system by Cassell and coworkers (described in [3]) implements an embodied agent which is to produce natural verbal and nonverbal outputs regarding various relations between the used modalities (cf. Figure 3, middle). In the REA gesture animation model, a behavior is scheduled that, once started, causes the execution of several predefined motor primitives which in turn employ standard animation techniques, e.g. keyframe animation and inverse kinematics. Although the issue of exact timing of spoken and gestural utterances is targetted in their work, the authors state that it has not yet been satisfactorily solved.

A more biologically motivated approach is applied in the GeSSyCa system (cf. Figure 3, left) by Gibet et al. [6]. This system is able to produce (French SL) sign language gestures from explicit representations in which a set of parametrizeable motion primitives (pointing, straight line, curved, circle, wave form movements) can be combined to more complex gestures. The primitives define targets for an iterative motion control scheme which has been shown to reproduce quite natural movement characteristics. However, since the movements depend entirely on the convergence of the control scheme, stability problems may arise for certain targets. Furthermore, temporal or kinematic features of a movement can only be reproduced within certain limits.

Fig. 3. Gesture production in the GeSSyCa (left), REA (middle), and MAX systems (right)

The goal of our own approach, demonstrable by the MAX system (cf. Figure 3, right; kinematic skeleton exposed) is to render real-time, lifelike gesture animations from representations of spatio-temporal gesture knowledge. It incorporates means of motion representation, planning, and control to produce multiple kinds of gestures. Gestures are fully parametrized with respect to kinematics, i.e. velocity profile and overall duration of all phases, as well as shape properties. In addition, emphasis is given to the issue of "peak timing", that is, to produce accented parts of the gesture stroke at precise points in time that can be synchronized with external events such as stressed syllables in synthetic speech. In more detail this is described in the following sections.

3   Articulated Communicator

In earlier work we have developed a hierarchical model for planning and generating lifelike gestures which is based on findings in various fields relevant to the production of human gesture, e.g. [7]. Our approach grounds on knowledge-based computer animation and encapsulates low-level motion generation and control, enabling more abstract control structures on higher levels. These techniques are used to drive the kinematic skeleton of a highly articulated figure – "articulated communicator" – which comprises 43 degrees of freedom (DOF) in 29 joints for the main body and 20 DOF for each hand (cf. Figure 4, left). While it turned out to be sufficient to have the hands animated by key-framing, the arms and the wrists are driven by model-based animation, with motion generators running concurrently and synchronized.

To achieve a high degree of lifelikeness in movement, approaches based on control algorithms in dynamic simulations or optimization criteria are often considered a first method, since they lead to physically realistic movements and provide a high level of control. But due to high computational cost they are usually not applicable in real-time. Starting from the observation that human arm movement is commonly conceived as being represented kinematically, we employ a single representation for both path and kinematics of the movement. The representation is based on B-splines which in particular make it possible to have smooth arm gestures that may comprise several subsequent, relative guiding strokes.

Our model (see Figure 6 for an overview) incorporates methods for representing significant spatiotemporal gesture features and planning individual gestural animations, as well as biologically motivated techniques for the formation of arm trajectories. The fundamental idea is that, in planning a gestural movement, an image of the movement is created internally. It is formed by arranging constraints representing the mandatory spatial and temporal features of the gesture stroke. These are given either as spatiotemporal descriptions from previous stages of gesture production, e.g., location of a referent for deictic gestures, or they are retrieved from stored representations, e.g., for the typical hand shape during pointing. Therefore, our model comprises a gesture lexicon or *gestuary* as postulated by deRuiter [4]. It contains abstract frame-based descriptions of gestural movements in the stroke phase, along with information about their usage for transferring communicative intent. The entries can hence be considered as defining a mapping from a communicative function to explicit movement descriptions of the gesture stroke.

In the gestuary, gestures are described in terms of either postural features (static constraints) or significant movement phases (dynamic constraints) that occur in the gesture stroke. Features that can be defined independently are described using a symbolic gesture notation system which is built on HamNoSys [13], while others must be determined for each individual gesture. To this end, the description further accommodates entries which uniquely refer to specific values of the content the gesture is to convey, e.g., quantitative parameters for deictic or iconic gestures. The gesture's course in time is defined by arranging the constraint definitions in a tree using PARALLEL and SEQUENCE nodes which can optionally be nested. While means for defining parallel and sequential actions were provided with original HamNoSys, Version 2.0 [13], further HamNoSys-style conventions were introduced for our animation model which allow it to describe repetitions and symmetries in gestural movement.

An example gesture template from the gestuary is shown in Figure 4; right (the concrete syntax is actually denoted in XML). The communicative function of the pointing gesture specified is to refer to a location plus indicating a pointing direction. By the use of these two parameters it is, for instance, possible to have the finger point to a location from above or from the side. The pointing gesture stroke is defined to have movement constraints that describe a target posture of the hand to be reached at the apex, namely, that – in parallel – the handshape is basic shape index finger stretched, the hand location is directed to the referenced location, the extended finger orientation is to the referenced direction, and the palm orientation is palm down. The symbols used in Figure 4 are ASCII equivalents of selected HamNoSys symbols that we use for gesture description.

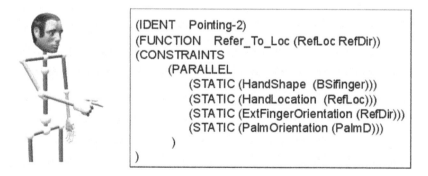

Fig. 4. Articulated Communicator (left); template of a pointing gesture (right)

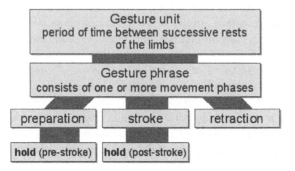

Fig. 5. Hierarchical kinesic structure of gestural movements (after [9])

In the following section we explain how gesture templates are instantiated to include timing constraints. These lead to individual movement plans that are executed by the articulated communicator's motor system.

4   Timing

As was said earlier, our gesture generation model is based on a variety of findings of gesture production and performance in humans which is a complex and multi-stage process. It is commonly assumed that representational gestural movements derive from spatiotemporal representations of "shape" in the working memory on cognitively higher levels. These representations are then transformed into patterns of control signals which are executed by low-level motor systems. The resulting gesture exhibits characteristical shape and kinematic properties enabling humans to distinguish them from subsidiary movements and to recognize them as meaningful [4]. In particular, gestural movements can be considered as composed of distinct movement phases which form a hierarchical kinesic structure (cf. Figure 5). In coverbal gestures, the stroke (the most meaningful and effortful part of the gesture) is tightly coupled to accompanying speech, yielding semantic, pragmatic, and even temporal synchrony between the two modalities [9]. For instance, it was found that indexical gestures are likely to co-occur with the rheme, i.e. the focused part of a spoken sentence, and that the stroke onset precedes or co-occurs with the most contrastively stressed syllable in speech and covaries with it in time.

4.1   Prerequisites

In Figure 6, an outline of the main stages of the gesture animation process (left) and the overall architecture of the movement planning and execution (right) are shown. In the first gesture planning stage – planning – an image of the movement is created in the way indicated in the previous section, by arranging constraints

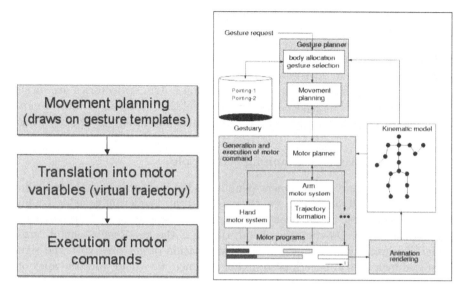

Fig. 6. Main stages and overall architecture of the gesture animation system

representing the mandatory spatial and temporal features of the gesture stroke. In the second planning stage, these ordered lists of constraints are separated and transferred to specialized hand, wrist, and arm motor control modules. These modules produce submovements for preparation, stroke, and retraction phases of the corresponding features that occur in a gesture phrase (cf. Figure 5).

We briefly describe how motor commands are put into effect in the third (execution) stage of the animation system. The overall movement is controlled by a motor program which is able to execute an arbitrary number of local motor programs (LMPs) simultaneously (for illustration see bottom of Figure 5; right). Such LMPs employ a suited motion generation method for controlling a submovement (affecting a certain set of DOFs) over a designated period of time. LMPs are arranged in a taxonomy and share some functionality necessary for basic operations like creating, combining, and coordinating them. In detail, each LMP provides means for (1) self-activation and self-completion and (2) concatenation.

Since different movement phases need to be created by different motion generators, LMPs are arranged in sequences during planning (concatenation). The specializied motor control modules create proper LMPs from the movement constraints at disposal and define concatenations by assigning predecessor, resp. successor relationships between the LMPs. At run-time, the LMPs are able to activate and complete themselves as well as to pass control over to one another. In order to guarantee continuity in the affected variables, each LMP connects itself fluently to given boundary conditions, i.e., start position and velocity. The devised method accounts for co-articulation effects, e.g., fluent gesture transitions emerge from activation of the subsequent gesture (resp. its LMPs) before the preceding one has been fully retracted.

According to the distribution of motor control, independent LMPs exist for controlling wrist, hand, and arm motions. Any of the defining features may thus be left unspecified and does not affect the submovements within the complementary features (e.g. hand shape or wrist location). Arm trajectories are formed explicitly in the working space based on a model that reproduces natural movement properties and, in addition, allows to adjust the timing of velocity peaks.

4.2 Timed Movement Planning – Example

The most crucial part in movement planning is the issue of timing the individual phases of a gestural movement. On the one hand this is relevant to achieve a high degree of lifelikeness in the overall performance of a gesture and, in particular, the gesture stroke. On the other hand, timing is the key issue to enable cross-modal synchrony with respect to the coordination of gestures with the signal generated by a text-to-speech system. For instance, we would want the apex of a gesture stroke to be coordinated with peak prosodic emphasis in spoken output.

The gesture planner forms a movement plan, i.e. a tree representation of a temporally ordered set of movements constraints, by (1) retrieving a feature-based gesture specification from the gestuary, (2) adapting it to the individual gesture context, and (3) qualifying temporal movement constraints in accordance with external timing constraints. The movement planning modules for both hand and arm motor systems are able to interpret a variety of HamNoSys symbols (selected with respect to defining initial, intermediate and target postures), convert them into position and orientation constraints with respect to an egocentric frame of reference, and generate a movement which lets the arm/hand follow an appropriate trajectory. It is possible to specify movement constraints with respect to static or dynamic features of the gesture stroke (see Section 3) and, further, to set timing constraints with respect to start, end, and peak times of each feature incorporated in the gesture stroke. Hence, roughly, a movement plan is generated by a specification of the following details:

HamNoSys + movement constraints + timing constraints
(selected)   {STATIC, DYNAMIC}   {Start, End, Manner}

To give an example of a movement plan, Figure 7 shows the instantiated gesture template of the stroke phase of a left-hand pull gesture which starts with an open flat hand, stretched out with the palm up, and ends with a fist-shaped hand near the left shoulder, palm still up (relative to the forearm). Dynamic movement constraints are specified for arm and hand motion, with initial and target postures given in symbolic HamNoSys descriptions. From the way the timing constraints are instantiated (Start, End, and Manner), the stroke would be performed within the period of 310 ms, with a velocity peak close to the end of that period. The peak can be placed at any time within the stroke phase. Having achieved this flexibility, all timing constraints can be instantiated automatically

```
Pull-2
  □―(PARALLEL (Start 1.1)(End 1.41))
      ―(DYNAMIC (Start 1.1)(End 1.41)(HandLocation ((LocShoulder LocLeftBeside LocStretched)
                 (LocShoulder LocLeftBeside LocNear))) (Manner ((Peak 1.39))))
      ―(DYNAMIC (Start 1.1)(End 1.41)(HandShape ((BSflato)(BSfist)))(Manner ((Peak 1.39))))
      ―(STATIC (Start 1.1)(End 1.41)(PalmOrientation (PalmU)))
```

Fig. 7. Example of pull gesture description – stroke phase – with instantiated start, end, and peak times (relative to utterance onset)

in the planning of a multimodal utterance (cf. Section 5). In the example, the apex of the arm motion and the apex of the hand motion are synchronous because they are carried out in parallel and their peaks are specified at identical times. Due to a STATIC movement constraint, the palm orientation remains the same over the full gesture stroke.

Preparation and retraction of the pull gesture are supplied from the motor planner automatically, with smooth (C^1-continuous) transitions and tentative, but not full-stop, pre- and post-stroke holds automatically generated. The resulting velocity profile for the pull gesture as specified in Figure 7 is shown in Figure 8, and snapshots from the preparation, stroke, and retraction phases are shown in Figure 9. Similarly, we have sucessfully specified a variety of further gestures, among them pointing gestures with and without peak beats, iconic two-handed gestures, and gestures that include several guiding strokes such as the outlining of a rectangular shape. Lab experience has shown that any specific gesture in such a realm can be specified in the effort of not much more than a minute and put into effect in real-time.

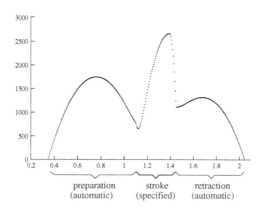

Fig. 8. Velocity profile for a pull gesture with a timed peak

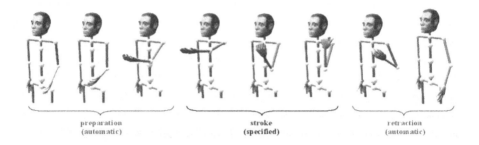

preparation   stroke   retraction
(automatic)   (specified)   (automatic)

Fig. 9. Preparation, stroke and retraction phase for pull gesture

5 Outlook: Gesture and Speech

As described in the previous sections, one core issue in our work is the production of synthetic lifelike gesture from symbolic descriptions where natural motion and timing are central aspects. Particular emphasis lies on how to achieve temporal coordination with external information such as the signal generated by a text-to-speech system. As our model is particularly conceived to enable natural cross-modal integration by taking into account synchrony constraints, further work includes the integration of speech-synthesis techniques as well as run-time extraction of timing constraints for the coordination of gesture and speech. In this outlook, some remarks on ongoing work are given. In particular, we have managed to coordinate the gesture stroke of any formally described gesture with synthetic speech output. For instance, we can have the MAX agent say (in German) "now take this bolt and place it in this hole" and, at the times of peak prosodic emphasis, have MAX issue pointing gestures to the according locations. Thereby, the shape and specific appearance of the gesture is automatically derived from the gestuary and the motor system, while the gesture peak timing is derived from the emphasis (EMPH) parameters of the synthetized speech signal.

For text-to-speech (TTS) we currently use a combination of a module for orthographic-phonetic transcription and prosody generation, TXT2PHO, and MBROLA for speech synthesis. TXT2PHO was developed at the University of Bonn [12]. Its core part consists of a lexicon with roughly 50,000 entries and flexion tables which are used to convert German text to phonemes. Each word is marked with prominence values which support the subsequent generation of prosodic parameters (phoneme length; intonation) to produce a linguistic representation of text input. From this representation, speech output is generated by the use of MBROLA and the German diphone database provided for it [5]. MBROLA is a real-time concatenative speech synthesizer which is based on the multi-band resynthesis, pitch-synchronous overlap-add procedure (MBR-PSOLA). To achieve a variety of alterations in intonation and speech timing,

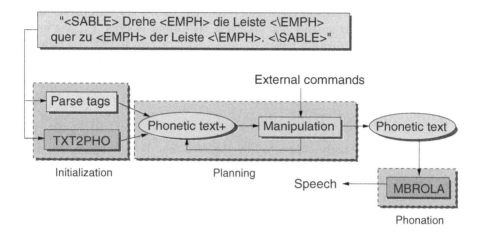

Fig. 10. "Turn *this bar* crosswise to *that bar*": Outline of our text-to-speech method allowing pitch scaling and time scaling; SABLE tags used for additional intonation commands

pitch can be varied by a factor within the range of 0.5 to 2.0, and phone duration can be varied within the range of 0.25 to 2.0.

Building on these features, a method was developed in our lab which allows to control a variety of prosodic functions in the TTS system by pitch scaling and time scaling. We use a markup language, SABLE [15] which is based on the extensible markup language (XML), to tag words or syllables to be emphasized in speech. Upon parsing such tags in the text input, phonetic text produced by TXT2PHO is altered accordingly and can further be manipulated to meet timing constraints from external commands generated in conjunction with the gesture planning procedure. Thus it is possible to preplan the timing of stressed syllables in the phonation for MBROLA and to synchronize accented behaviors in speech and gesture synthesis. Figure 10 roughly outlines the extended TTS method used in our lab. Not further explained in this paper, the face of the MAX agent is animated concurrently to exhibit lip-synchronous speech.

In conclusion, we have presented an approach for lifelike gesture synthesis and timing that should work well for a conversational agent and which is demonstrable by the MAX system. Our model is conceived to enable cross-modal synchrony with respect to the coordination of gestures with the signal generated by a text-to-speech system. In particular, the methods desribed can achieve precise timing for accented parts in the gesture stroke as a basis to synchronize them with stressed syllables in accompanying speech. Ongoing work is directed to have the system link multimodal discourse segments (chunks) in a incremental and smooth way. In multimodal communication, by which we mean the concurrent formation of utterances that include gesture and speech, a rhythmic alternation of phases of tension and relaxation can be observed. The issue of rhythm in communication has been addressed widely and has been a key idea in our earlier work on synchronizing gesture and speech in HCI input devices [17]. We intend to use the idea of production pulses to mark a grid on which accented elements (e.g., stressed syllables) are likely to occur, together with a low-frequency (2-3s) chunking mechanism to achieve natural tempo in multimodal discourse output.

Acknowledgment

This research is partially supported by the Deutsche Forschungsgemeinschaft (DFG) in the Collaborative Research Center "Situated Artificial Communicators" (SFB 360). The authors are indebted to Dirk Stößel who developed a scalable TTS system in his master thesis, and to the members in the Bielefeld AI group who provided general support in many ways.

References

1. P. Becheiraz and D. Thalmann. A behavioral animation system for autonomous actors personified by emotions. In *Proc. First Workshop on Embodied Conversational Characters (WECC '98)*, Lake Tahoe, CA, 1998. 123
2. J. Beskow and S. McGlashan. Olga - a conversational agent with gestures. In E. André, editor, *Proc. of IJCAI-97 Workshop on Animated Interface Agents: Making Them Intelligent*, 1997. 123
3. J. Cassell, J. Sullivan, S. Prevost, and E. Churchill, editors. *Embodied Conversatinal Agents*. The MIT Press, Cambridge (MA), 2000. 120, 122, 123
4. J. deRuiter. The production of gesture and speech. In D. McNeill, editor, *Language and gesture*, chapter 14, pages 284–311. Cambridge University Press, 2000. 124, 126
5. T. Dutoit. *An Introdutcion to Text-To-Speech Synthesis*. Kluwer Academic Publishers, Dordrecht, 1997. 130
6. S. Gibet, T. Lebourque, and P. Marteau. High-level specification and animation of communicative gestures. *Journal of Visual Languages and Computing*, 12(6):657–687, 2001. 123
7. S. Kopp and I. Wachsmuth. A knowledge-based approach for lifelike gesture animation. In W. Horn, editor, *ECAI 2000 Proceedings of the 14th European Conference on Artificial Intelligence*, pages 661–667, Amsterdam, 2000. IOS Press. 124
8. M. Latoschik, M. Fröhlich, B. Jung, and I. Wachsmuth. Utilize speech and gestures to realize natural interaction in a virtual environment. In *IECON'98 - Proceedings of the 24th Annual Conference of the IEEE Industrial Electronics Society*, volume 4, pages 2028–2033. IEEE, 1998. 121
9. D. McNeill. *Hand and Mind: What Gestures Reveal about Thought*. University of Chicago Press, Chicago, 1992. 126
10. C. Pelachaud and S. Prevost. Talking heads: Physical, linguistic and cognitive issues in facial animation. Course Notes, Computer Graphics International, June 1995. 123
11. K. Perlin and A. Goldberg. Improv: A system for scripting interactive actors in virtual worlds. In *SIGGRAPH 96, Proceedings of the 23rd annual conference on Computer Graphics*, pages 205–216, 1996. 123

12. T. Portele, F. Höfer, and W. Hess. A mixed inventory structure for german concatenative synthesis. In *Proceedings of the 2nd ESCA/IEEE Workshop on Speech Synthesis*, pages 115–118, 1994. 130

13. S. Prillwitz, R. Leven, H. Zienert, T. Hamke, and J. Henning. *HamNoSys Version 2.0: Hamburg Notation System for Sign Languages: An Introductory Guide*, volume 5 of *International Studies on Sign Language and Communication of the Deaf.* Signum Press, Hamburg, Germany, 1989. 121, 125

14. J. Rickel and W. Johnson. Animated agents for procedural training in virtual reality: Perception, cognition, and motor control. *Applied Artificial Intelligence*, 13:343–382, 1999. 123

15. Sable Consortium. Sable: A synthesis markup language (version 10.). <http://www.bell-labs.com/project/tts/sable.html>, 2000. Bell Laboratories, 25.2.2001. 131

16. T. Sowa and I. Wachsmuth. Interpretation of shape-related iconic gestures in virtual environments. This volume. 121

17. I. Wachsmuth. Communicative rhythm in gesture and speech. In A. Braffort et al., editor, *Gesture-Based Communication in Human-Computer Interaction - Proceedings International Gesture Workshop*, pages 277–289, Berlin, March 1999. Springer (LNAI 1739). 132

SignSynth:
A Sign Language Synthesis Application
Using Web3D and Perl

Angus B. Grieve-Smith

Linguistics Department, Humanities 526, The University of New Mexico
Albuquerque, NM 87131 USA
grvsmth@unm.edu
http://www.unm.edu/~grvsmth/signsynth

Abstract. Sign synthesis (also known as text-to-sign) has recently seen
a large increase in the number of projects under development. Many of
these focus on translation from spoken languages, but other applications
include dictionaries and language learning. I will discuss the architecture
of typical sign synthesis applications and mention some of the applications and prototypes currently available. I will focus on SignSynth, a
CGI-based articulatory sign synthesis prototype I am developing at the
University of New Mexico. SignSynth takes as its input a sign language
text in ASCII-Stokoe notation (chosen as a simple starting point) and
converts it to an internal feature tree. This underlying linguistic representation is then converted into a three-dimensional animation sequence
in Virtual Reality Modeling Language (VRML or Web3D), which is automatically rendered by a Web3D browser.

1 Introduction – What Is Sign Synthesis?

In recent years, there has been an increase in interest in the development of computer applications that deal with sign languages.[1] Many of these applications
involve some form of sign synthesis, where a textual representation is converted
into fluent signing. I will discuss some possible applications of sign synthesis
technology, describe the general architecture of sign synthesis, and give a summary of the existing sign synthesis applications and prototypes that I am aware
of. I will finish by describing the current state of SignSynth, the sign synthesis
prototype that I am developing at the University of New Mexico.

Despite the difference in the modes of communication, there is a great deal
of similarity between sign synthesis and its older cousin, speech synthesis. Over

---

[1] This project was supported by Grant Number 1R03DC03865-01 from the National
Institute on Deafness and Communication Disorders of the United States Govern-
ment (Jill P. Morford, Principal Investigator). This report is solely the responsibility
of the author, and does not necessarily represent the official views of the NIDCD
or the National Institutes of Health. I am also grateful to Sean Burke, Jill Morford,
Benjamin Jones and Sharon Utakis.

I. Wachsmuth and T. Sowa (Eds.): GW 2001, LNAI 2298, pp. 134–145, 2002.
© Springer-Verlag Berlin Heidelberg 2002

the past thirty years, linguists have shown that sign languages are full-fledged languages, with a level of complexity and creativity equal to that found in spoken languages. There are significant typological differences, but the best approach is to start from the assumption that sign language and spoken language are equivalent, in the absence of any reason to believe otherwise. It is important to note that this assumption does not imply that any given sign language is "the same" as any spoken language. The definition of sign synthesis that I will use in this paper thus parallels its spoken-language equivalent, speech synthesis:

> **Sign synthesis** is a way to convert sign language from a stored, textual medium to a fluid medium used for conversation.

2 Applications of Sign Synthesis Technology

There are many applications of sign synthesis technology, aimed at solving various problems. Some of the most popular are ideas that attempt to address communication problems, but are prohibitively complicated in reality. Others address problems that signers encounter when dealing with technology. A small number of applications do not fit in either category.

2.1 Popular Applications: The Communication Problem

One group of computational sign linguistics applications deals with the communication gap that exists between the deaf and hearing in every country. To begin with the obvious, almost all Deaf people are unable to hear or read lips well enough to understand a spoken language as it is spoken. Beyond this, most Deaf people have difficulty reading the written form of any spoken language. The Gallaudet Research Institute found that young Deaf and hard-of-hearing adults aged 17 and 18 had an average reading score in English that was the same as that of the average ten-year-old hearing child [1]. This means that Deaf people have difficulty conversing with non-signers, watching movies and television (even captioned), and reading books and the World Wide Web.

Sign synthesis is often seen as a solution to this Communication Problem. The assumption is that all that is necessary is to convert speech or writing into sign, and the Communication Problem is solved. The difficulty with this approach is that any sign language is vastly different from all spoken languages, even languages from the same country. For example, American Sign Language (ASL), the most widely used sign language in the United States, bears very little resemblance to English, Spanish, French or German, the most widely used spoken languages in the country. So translating from a spoken language to a sign language is a complex undertaking. Some sign synthesis applications try to avoid this problem by using "signed" versions of various spoken languages, but those systems are not understood by Deaf people much better than the written versions of those languages.

The only true way to convert between a spoken language and a signed language is with machine translation. Unfortunately, linguists have been trying to achieve machine translation between closely related Western European languages for decades with only limited success. The outlook is much less promising for a pair of languages that are as divergent as (for example) English and American Sign Language.

Machine translation has shown significant promise when it deals with highly conventional texts in specialized domains, such as weather reports, as opposed to the less standard, less predictable language used in face-to-face transactions. There are a few prototype spoken-language-to-sign-language machine translation applications that focus on weather ([2], [3]), and these are the most promising of the applications that target the Communication Problem. Of course, in these cases the focus is on machine translation; sign synthesis is only needed for other problems.

2.2   The Record Problem and Its Correlates, the Storage and Bandwidth Problems

The areas where synthesis shows the most promise deal with how we record and represent sign languages. At the core is the Record Problem. The most common way of recording spoken languages is to write them, but sign languages are almost never written. There have been several attempts to create writing systems for sign languages, but none have been widely adopted by any Deaf community.

One often-advocated solution to the Record Problem is video, but that raises two other problems, the Storage Problem and the Bandwidth Problem. The amount of disk space needed to store one average written sentence in a spoken language is less than 100 bytes. The amount of disk space needed to store one average written sentence in a sign language is well under 1KB. The amount of disk space needed to store an audio recording of one average sentence in a spoken language is less than 200KB. The amount of disk space needed to store a video recording of one average sentence in a sign language is more than 1MB. The Storage Problem is that video takes up more storage media (disk, tape, etc.), and the Bandwidth Problem is that it takes longer to transmit from one place to another.

Sign Synthesis technology can solve these problems. For a given passage (public address, poem, love letter, contract) to be stored, all that is necessary is for it to be written down, by the author or by a third party, using one of the existing writing systems. It can then be stored and transmitted in written form, and synthesized into fluent sign whenever it needs to be accessed.

One specific application of this solution is electronic dictionaries. For the first sign language dictionaries, the Record Problem was solved by using pictures of each sign, often hand drawings. These pictures take up a lot of space, and were usually used for just a gloss of a spoken-language term. Significantly, they do not allow for the sign language equivalent of a Webster's dictionary: a monolingual dictionary where signs are defined in terms of other signs.

More recent multimedia sign dictionaries are able to overcome the limitations of hand drawings by using video, but they run into the Storage Problem. By overcoming the Record Problem, sign synthesis can allow sign language dictionaries to take up no more space than similar spoken language dictionaries.

There are also many computer learning applications that use prerecorded and synthesized speech to teach a wide variety of subjects and skills to children and adults. The analogous applications for sign languages run into the Storage and Bandwidth problems, which can be solved using sign synthesis. VCom3D [4] has produced several educational applications using their sign synthesis technology, but unfortunately they only produce a North American variety of Signed English, which is not the kind of true sign language that can be understood by most Deaf people.

2.3   New Ideas

There are two other applications for sign synthesis that have been discussed in the literature. One is cartoons: there are currently very few cartoons with sign language. Sign synthesis can make it easy to generate cartoon characters that sign fluently. VCom3D has taken steps in this direction with a cartoon frog that can produce Signed English.

One of the first applications of speech synthesis was in experiments investigating categorical perception. Researchers used synthetic speech to test the way that hearers perceive category boundaries [5]. Attempts to replicate these tests with sign language have produced mixed results ([6], [7]), but none have used fully synthesized signs.

3   Sign Synthesis Architecture

The process of sign synthesis is essentially the same as that of speech synthesis; what is different is the form of the output. The architecture of a sign synthesis application is thus almost identical to that of a speech synthesis application.

3.1   Basic Architecture

Klatt [8] describes the basic architecture of a text-to-speech application: input text is acted on by a set of analysis routines that produce an abstract underlying linguistic representation. This representation is fed into synthesis routines that then produce output speech. The architecture is summarized in Figure 1.

In the next few sections, I will discuss the analogs of these three stages in sign synthesis.

3.2   Input Text

In the early stages of development, speech synthesis really had no "input text"; the input was all in the underlying linguistic representation. But in order to be accessible to a larger base of users, the application had to be able to analyze

Fig. 1. Diagram from Klatt [8], showing the architecture of a speech synthesis application

input text. The notion of "input text" is relatively straightforward for speech, but what do we really mean when we talk about "text-to-sign"?

The most common assumption among newcomers to sign language is that the input text is one of the national standards, for example Standard English for the United States. But as I discussed above, English and American Sign Language are two very different languages, and to take a spoken language as input requires a machine translation component. Another form that has been suggested for input is a "gloss" system: for example, in the United States, to represent American Sign Language as a gloss, the writer performs a mental word-for-word translation from ASL into English, keeping the words in ASL word order and using semi-conventionalized forms to represent ASL words that have no direct equivalent in English. An example of this can be found in some textbooks for teaching ASL [9]. Gloss systems are almost completely unused by the Deaf outside of language study and teaching.

Another possibility for input text is one of the writing systems that have been developed for sign languages. The most popular is SignWriting, a phonetically-based writing system with iconic symbols for handshapes, movements and non-manuals arranged iconically. SignWriting is used in the education of the Deaf in several pilot projects around the world [10]. Another, SignFont [11], is somewhat iconically-based, but has a few technological advantages over SignWriting, including a linear alignment and a much smaller symbol set, but it has found limited acceptance.

Since knowledge of writing systems is relatively limited, a third possibility is to avoid the use of text, and instead use a more interactive interface, similar to the "wizards" used in popular office applications. This would take the author of a text step-by-step through the process of specifying the signs he or she wanted to record. Such a system could also train authors to use one of the writing systems.

3.3   Underlying Linguistic Representation

Speech synthesis systems use a form of phonetic notation for their underlying linguistic representation. Analogously, numerous notation systems for sign languages have been invented, and thus there are several candidates for the form of the underlying linguistic representation.

One of the earliest notation systems is Stokoe notation [12], which uses a combination of arbitrary and iconic symbols, arranged in almost-linear order. ASCII-Stokoe notation [13] is an adaptation of Stokoe notation for the ASCII character set and a strictly linear arrangement. Unfortunately, Stokoe (and ASCII-Stokoe) do not cover the full range of expression that is possible in sign languages, and notably have no way of notating nonmanual gestures.

SignFont is phonetically-based enough to serve as a possible form for the underlying linguistic representation. The Literal Orthography [14] is a similar, but ASCII-based, forerunner to SignFont. Both systems are being more complete than Stokoe, but still easily machine-readable.

HamNoSys [15] is currently the most extensive gesture notation system. It is composed of a large set of mostly iconic symbols, arranged linearly. Development is underway on SiGML, an XML-based variant of HamNoSys [16].

QualGest [17] is a representation system specifically designed to serve as the underlying linguistic representation for the GesSyCa system. It is based on the Stokoe/Battison parameters of handshape, movement, location and orientation, with facial expression added.

SignWriting is also complete enough and phonologically-based enough to be used for the underlying linguistic representation, but the SignWriter binary file format is difficult to deal with. SWML, an XML-based variant of SignWriting, is available instead [18]. Another movement notation system originally developed for dance is Labanotation. Some work is being done to adapt the Laban system for use as an underlying linguistic description [19].

3.4   Output Gesture

Speech synthesis systems usually have as their output speech a synthetic sound wave. For sign languages, there are a few possibilities for the form of the output gesture. Deaf-blind users may require a tactile output, such as a robotic hand, but for sighted Deaf users the easiest is some form of animated video.

Concatenative vs. Articulatory Synthesis Klatt broadly groups speech synthesis applications into two types: articulatory, where the output speech is synthesized by rules that are intended to correspond to the process by which humans articulate speech, and concatenative, where prerecorded speech is broken down into small segments that are recombined (with some smoothing) to create new words. Almost all currently available speech synthesis systems are concatenative, but there is at least one articulatory synthesizer, CASY [20]. The same two types exist in sign synthesis.

A few concatenative sign synthesis systems aim to simply recombine videos of individual signs. Most take gesture recognition data (often from a glove-type gesture recording device) for individual signs and recombine them. Because each sign is usually a single word, this amounts to storing each word. Klatt discusses the problems for speech synthesis inherent in this strategy:

> However, such an approach is doomed to failure because a spoken sentence is very different from a sequence of words uttered in isolation. In a sentence, words are as short as half their duration when spoken in

isolation – making concatenated speech seem painfully slow. The sentence stress pattern, rhythm, and intonation, which depend on syntactic and semantic factors, are disruptively unnatural when words are simply strung together in a concatenation scheme. Finally, words blend together at an articulatory level in ways that are important to their perceived naturalness and intelligibility. The only satisfactory way to simulate these effects is to go through an intermediate syntactic, phonological and phonetic transformation.

The problems that Klatt describes all apply to sign synthesis, and are even more significant because in all sign languages that have been studied, the "stress pattern, rhythm and intonation" of sentences actually encodes grammatical features that in spoken languages are usually encoded by affixes and function words.

The way that concatenative speech synthesis systems typically overcome these problems is to use prerecorded single phonemes or diphones, but the internal organization of signs is more simultaneous than consecutive. To perform concatenative sign synthesis in a way analogous to contemporary speech synthesizers would require such a deep analysis of the sign data that it is more feasible to perform articulatory synthesis in most cases.

Humanoid Animation All of the major articulatory sign synthesis systems, and some of the concatenative systems, use as output some form of three-dimensional humanoid animation. Many conform to the open standards of the Web3D Humanoid Animation Working Group (a.k.a. H-Anim) [21]. Some systems use the keyframe animation and interpolation provided by Web3D (also known as Virtual Reality Modeling Language or VRML) browsers, but others, like the latest ViSiCAST articulatory synthesis project [16], make naturalness of movement a priority and provide intermediate frames to control the movement more closely.

Real-World Priorities Any discussion of the output format should include a mention of some of the priorities set by real-world applications of the technology. A 1971 census of Deaf people in the United States found that they earned twenty-five percent less than the national average [22], and it is likely that similar situations exist today around the world. This means that although some institutions might be able to afford the latest in three-dimensional animation technology, the average Deaf individual cannot, and Deaf individuals in less-industrialized areas have even less access to technology. Whatever systems are used should ideally be able to work with older, less advanced computers.

The highest priority, of course, is the ability of Deaf people to understand the signing. However, realism in articulation may not always contribute to understanding, and it is possible that realism may detract from understanding. In comic books, for example, the more detail in the characters, the more they are perceived as individuals, and the less their words are perceived as generic [23]. If the synthesis is meant to deliver an official message, or if the source of the input is unknown, then it is important to avoid introducing too much individual detail.

4   Current Sign Synthesis Systems

There are several groups around the world that are working on sign synthesis applications, and two groups (VCom3D and SignTel) have released applications to the public.

There are several concatenative synthesis applications. In Japan, Hitachi's Central Research Laboratory [24] and the Communications Research Laboratory [25] are both developing concatenative synthesis systems as part of larger speech-to-sign interpretation prototypes. Many of the concatenative systems produce output that is not in a true sign language, but in a signed version of the spoken input. Two American companies, VCom3D [4] and SignTel [26] have released concatenative synthesis applications that produce a North American variety of Signed English, with VCom3D using motion capture data and SignTel using videos. The English company Televirtual developed a prototype captioning system known as Simon the Signer [16], now incorporated into the ViSiCAST system, which also produced signed English (a British variety). Some ViSiCAST applications still use concatenations of DataGlove data [3].

There are also a number of articulatory synthesis prototypes, including Sister Mary O'Net [27], developed at the University of Delaware, and TEAM [19], developed by the University of Pennsylvania's Human Modeling and Simulation group. In France, LIMSI [17] and I3D Labs [28] are working on articulatory synthesis prototypes. As mentioned above, ViSiCAST is working towards an articulatory model as well.

The next section of this paper will focus on SignSynth, a prototype that I am developing at the University of New Mexico. SignSynth currently uses ASCII-Stokoe for the underlying representation and produces H-Anim-compliant Web3D.

5   Description of SignSynth

SignSynth is a prototype sign synthesis application currently under development at the University of New Mexico. It uses Perl scripts through the Common Gateway Interface (CGI) to convert input text in American Sign Language into Web3D animations of the text as signed.

5.1   Features

SignSynth has a number of features that set it apart from other sign synthesis prototypes and applications. It is free, non-profit and open source (except for the third-party inverse kinematics library). It conforms to open standards, so the output can be viewed on any Web3D browser. It has a simple humanoid, which is quick to download and render even on older equipment, and does not distract the viewer with too much detail.

The Perl CGI itself runs on any Web server. It can be accessed through a low-level underlying representation interface, and a higher-level input text interface will eventually be available. It performs articulatory synthesis and can produce true sign language, including the full range of nonmanual gestures used with American Sign Language. Its modular architecture makes it flexible and easy to expand.

5.2   Architecture

Figure 2 shows the architecture of SignSynth:

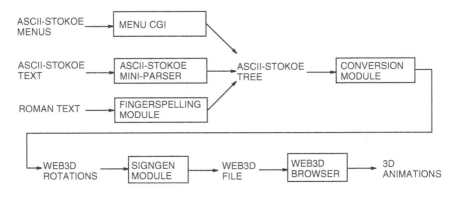

Fig. 2. Architecture of SignSynth

SignSynth currently has three interfaces. The Menu CGI offers the user a set of menus for each point in time, by which he or she can specify the phonological parameters for that time. By using the menus, a user who does not know ASCII-Stokoe can select the handshape, location and orientation for each hold. Additional menus allow the user to specify the timing of the gesture and accompanying nonmanual gestures.

The ASCII-Stokoe Mini-Parser allows a more advanced user to type in an underlying linguistic representation of any length in ASCII-Stokoe, with additions for timing and nonmanuals.

The Fingerspelling Module allows the user to type in any text in the Roman alphabet and converts it into the fingerspelling signs of the American Manual Alphabet.

Each of these modules outputs an ASCII-Stokoe Tree, stored in a nested hash table. This Tree then serves as input to the Conversion Module, which produces a set of Web3D rotation values for each of the joints used, plus Coord displacement values for non-jointed articulators such as the lips.

The ASCII-Stokoe Tree does not specify the Movement values that were part of the original Stokoe representation. Instead, it has a set of features for a given point in time, similar to the Movement-Hold Model of Liddell and Johnson [29]. These points in time are used primarily because they work well with Web3D's keyframe animation.

Once these Web3D rotations have been created and are stored in their own hash table, they are fed into the SignGen Module, which integrates them with the chosen Web3D humanoid to create a full Web3D "world" file including animation data. This Web3D file is returned to the user's Web browser by the CGI and then rendered by a Web3D browser plugin into three-dimensional animated American Sign Language.

5.3   Current Application – Categorical Perception

As described in Section 2.3, sign synthesis can be very valuable in the investigation of categorical perception phenomena in sign language. SignSynth is currently being used in categorical perception experiments funded by the United States National Institute for Deafness and Communication Disorders.

Categorical perception requires a set of stimuli taken at intervals from a continuous range between a pair of categories. A custom interface was designed that allows the user to specify the phonological parameters of each member of the chosen pair. The interface uses interpolation to produce a Web3D world for each interval. Each of these synthetic signs is converted to a video file using a screen capture utility. The stimuli are then integrated into an interactive test program and presented to subjects. The experiment is currently being performed.

5.4   Future Developments

SignSynth is still in the early stages of development. It can produce signs that are understandable, as well as fingerspelling, but there is a significant amount of work to be done.

One of the main challenges for articulatory sign synthesis is the inverse kinematics problem [30]. In general, it is very difficult to determine the necessary rotations of the shoulder, elbow and wrist to place the hand at a desired position and angle. This is partly because the human anatomy allows for multiple solutions. For SignSynth, the strategy I have chosen is to use the IKAN (Inverse Kinematics with Analytical Methods) library, developed by the University of Pennsylvania [31]. This library was created in C++ for the Jack system, so there are difficulties involved in adapting it to work with Perl and with the H-Anim standard.

Once the inverse kinematics problem is resolved, the next step is to extend the underlying linguistic representation. ASCII-Stokoe was chosen as an initial representation because of its simplicity, but it does not cover anywhere near the number of handshapes, locations and movements used in American Sign Language, let alone other sign languages. It also does not provide a way of representing nonmanual gestures or timing, so one had to be developed specially for SignSynth. A more extensive and flexible system will eventually be adopted, such as the Literal Orthography [14] or HamNoSys/SiGML [16].

After a more complete underlying representation is in place, the input will be extended to allow complete texts in a true writing system, such as SignFont or SignWriting. More beginner-oriented systems will also be created, to give the novice user an easier way to input signs.

Another limitation is Web3D, which currently requires any user to download and install a separate browser plugin (although some current versions of Internet Explorer include a Web3D plugin). Third-party products like Shout3D allow the system to display Web3D-like files using Java applets, rather than plugins. Eventually, SignSynth will be modified to use this kind of technology.

References

1. Gallaudet Research Institute: Stanford Achievement Test, 9th Edition, Form S, Norms Booklet for Deaf and Hard-of-Hearing Students. Gallaudet University, Washington (1996) 135
2. Grieve-Smith, A.: English to American Sign Language Translation of Weather Reports. In: Nordquist, D. (ed): Proceedings of the Second High Desert Student Conference in Linguistics. High Desert Linguistics Society, Albuquerque. (to appear) 136
3. Verlinden, M., Tijsseling, C., Frowein, H: A signing Avatar on the WWW. This volume (2002) 136, 141
4. VCom3d: SigningAvatar   Frequently   Asked   Questions.   Available   at http://www.signingavatar.com/faq/faq.html (2000) 137, 141
5. Liberman, A., Harris, K., Hoffman, H., Griffith, B.: The Discrimination of Speech Sounds Within and Across Phoneme Boundaries. Journal of Experimental Psychology **34** (1957) 358–368 137
6. Newport, E.: Task Specificity in Language Learning? Evidence from Speech Perception and American Sign Language. In: Wanner, E. and Gleitman, L. (eds): Language Acquisition: The State of the Art Cambridge University Press, Cambridge (1982) 137
7. Emmorey, K., McCullough, S., Brentari, D.: Categorical Perception in American Sign Language (to appear) 137
8. Klatt, D.: Review of Text-to-Speech Conversion for English. Journal of the Acoustic Society of America **82** (1987) 737–793 137, 138
9. Baker-Shenk, C., Cokely, D.: American Sign Language Gallaudet University Press, Washington (1991) 138
10. Sutton, V.: Lessons in SignWriting. SignWriting, La Jolla (1985) 138
11. Newkirk, D.: SignFont Handboook. Emerson and Associates, San Diego (1987) 138
12. Stokoe, W., Casterline, D., Croneberg, C.: A Dictionary of American Sign Language on Linguistic Principles. Linstok Press, Silver Spring (1965) 138
13. Mandel, M.: ASCII-Stokoe Notation: A Computer-Writeable Transliteration System for Stokoe Notation of American Sign Language. Available at http://world.std.com/~mam/ASCII-Stokoe.html (1983) 138
14. Newkirk, D.: Outline of a Proposed Orthography of American Sign Language. Available at http://members.home.net/dnewkirk/signfont/orthog.htm (1986) 139, 144

15. Prilliwitz, S., Leven, R., Zienert, H., Hanke, T., Henning, J.: HamNoSys, Version 2.0: Hamburg Notation System for Sign Languages – An Introductory Guide. Signum Press, Hamburg (1989) 139
16. Kennaway, R.: Synthetic Animation of Deaf Signing Gestures. This volume (2002) 139, 140, 141, 144
17. Lebourque, T., Gibet, S.: A Complete System for the Specification and the Generation of Sign Language Gestures. In: Braffort, A. et al. (eds): Gesture-based Communication in Human Computer Interaction: International Gesture Workshop, GW '99. Lecture Notes in Artificial Intelligence **1739**. Springer-Verlag, Berlin Heidelberg New York (1999) 227–238 139, 141
18. da Rocha Costa, A. C., Pereira Dimuro, G.: A SignWriting-Based Approach to Sign Language Processing This volume. (2002) 139
19. Zhao, L., Kipper, K., Schuler, W., Vogler, C., Badler, N., Palmer, M.: A Machine Translation System from English to American Sign Language. Presented at the Association for Machine Translation in the Americas conference (2000) 139, 141
20. Browman, C., Goldstein, L., Kelso, J. A. S., Rubin, P., Saltzman, E.: Articulatory Synthesis from Underlying Dynamics. Journal of the Acoustic Society of America **75** (1984) S22–S23 139
21. Roehl, B. (ed): Specification for a Standard VRML Humanoid. Available at http://www.H-anim.org (1998) 140
22. Schein, J. D., Delk, M. T.: The Deaf Population of the United States. National Association of the Deaf, Silver Spring (1974) 140
23. McCloud, S.: Understanding Comics: The Invisible Art. Harper Perennial, New York (1994) 140
24. Ohki, M., Sagawa, H., Sakiyama, T., Oohira, E., Ikeda, H., Fujisawa, H.: Pattern Recognition and Synthesis for Sign Language Transcription System. ASSETS **10** (1994) 1–8 141
25. Lu, S., Igi, S., Matsuo, H., Nagashima, Y.: Towards a Dialogue System Based on Recognition and Synthesis of Japanese Sign Language. In: Wachsmuth, I., Fröhlich, M. (eds): International Gesture Workshop, GW'97. Lecture Notes in Artificial Intelligence. Springer-Verlag, Berlin Heidelberg New York (1997) 259–271 141
26. SignTel: How It Works Available at http://www.signtelinc.com/products-howitworks.htm (2001) 141
27. Messing, L., Stern, G.: Sister Mary Article. Unpublished manuscript (1997) 141
28. Losson, O., Vannobel, J.-M.: Sign Specification and Synthesis. In: Braffort, A. et al. (eds): Gesture-based Communication in Human Computer Interaction: International Gesture Workshop, GW '99. Lecture Notes in Artificial Intelligence **1739**. Springer-Verlag, Berlin Heidelberg New York (1999) 227–238 141
29. Liddell, S., Johnson, R.: American Sign Language: The Phonological Base. In: Valli, C., Lucas, C. (eds): Linguistics of American Sign Language: A Resource Text for ASL Users. Gallaudet University Press, Washington (1992) 143
30. Gibet, S., Marteau, F., Julliard, F.: Internal Models for Motion Control. This volume. (2002) 143
31. Tolani, D., Goswami, A., Badler, N.: Real-time Inverse Kinematics Techniques for Anthropomorphic Limbs. Available at http://hms.upenn.edu/software/ik/ikan_gm.pdf (2000) 143

Synthetic Animation of Deaf Signing Gestures

Richard Kennaway

School of Information Systems, University of East Anglia
Norwich, NR4 7TJ, U.K.
jrk@sys.uea.ac.uk

Abstract. We describe a method for automatically synthesizing deaf signing animations from a high-level description of signs in terms of the HamNoSys transcription system. Lifelike movement is achieved by combining a simple control model of hand movement with inverse kinematic calculations for placement of the arms. The realism can be further enhanced by mixing the synthesized animation with motion capture data for the spine and neck, to add natural "ambient motion".

1 Introduction

1.1 Background

The object of the ViSiCAST project is to facilitate access by deaf citizens to information and services expressed in their preferred medium of sign language, using computer-generated virtual humans, or avatars. ViSiCAST addresses three distinct application areas: broadcasting, face-to-face transactions, and the World-Wide Web (WWW). For an account of the whole project see [2].

The task of signing textual content divides into a sequence of transformations:

1. From text to semantic representation.
2. From semantic representation to a sign-language specific morphological representation.
3. From morphology to a signing gesture notation.
4. From signing gesture notation to avatar animation.

This paper deals with the last of these steps, and describes an initial attempt to create synthetic animations from a gesture notation, to supplement or replace our current use of motion capture.

Related work by others in this area includes the GeSsyCa system [3,10] and the commercial SigningAvatar [17].

1.2 Motion Capture vs. Synthetic Animation

ViSiCAST has developed from two previous projects, one in broadcasting, *Sign-Anim* (also known as *Simon-the-Signer*) [12,13,19], and one in Post Office transactions, *Tessa* [11]. Both these applications use a signing avatar system based on *motion capture*: before a text can be signed by the avatar, the appropriate

I. Wachsmuth and T. Sowa (Eds.): GW 2001, LNAI 2298, pp. 146–157, 2002.
© Springer-Verlag Berlin Heidelberg 2002

lexicon of signs must be constructed in advance, each sign being represented by a data file recording motion parameters for the body, arms, hands, and face of a real human signer. For a given text, the corresponding sequence of such motion data files can be used to animate the skeleton of the computer-generated avatar. The great merit of this system is its almost uncanny authenticity: even when captured motion data is "played back" through an avatar with physical characteristics very different from those of the original human signer, the original signer (if already known to the audience) is nevertheless immediately recognizable.

On the other hand, motion capture is not without drawbacks.

- There is a substantial amount of work involved in setting up and calibrating the equipment, and in recording the large number of signs required for a complete lexicon.
- It is a non-trivial task to modify captured motions.

There are several ways in which we would wish to modify the raw motion capture data. Data captured from one signer might be played back through an avatar of different body proportions. One might wish to change the point in signing space at which a sign is performed, rather than recording a separate version for every place at which it might be performed. Signs recorded separately, and perhaps by different signers, need to be blended into a continuous animation. Much research exists on algorithmically modifying captured data, though to our knowledge none is concerned specifically with signing, a typical application being modification of a walking character's gait to conform to an uneven terrain. An example is Witkin and Popović's "motion warping" [14,20]. We are therefore also interested in synthetic animation: generation of the required movements of an avatar from a more abstract description of the gestures that it is to perform, together with the geometry of the avatar in use. The animation must be feasible to generate in real time, within the processing power available in the computers or set-top boxes that will display signing, and the transmitted data must fit within the available bandwidth.

Traditional animation (i.e. without computers) is a highly laborious process, to which computers were first introduced to perform in-betweening, creating all the frames between the hand-drawn keyframes. Even when animations are entirely produced on computer, keyframes are still typically designed by hand, but using 3D modelling software to construct and pose the characters. In recent years, physics-based modelling has come into use, primarily for animating inanimate objects such as racing cars or stacks of boxes acted on by gravity. Synthetic animation of living organisms poses a further set of problems, as it must deal not only with gravity and the forces exerted by muscles, but also with the biological control systems that operate the muscles. It is only in the last few years that implementation techniques and fast hardware have begun to make synthetic real-time animation of living movement practical. The tutorial courses [4,7] give an overview of this history and the current state of the art.

We present here a simplified biomechanical model dealing with the particular application of signing and the constraints of real-time animation.

2   Description of Signs

We start from the HamNoSys [16] notation for transcribing signing gestures. This
was developed by our partners at IDGS, University of Hamburg, for the purpose
of providing a way of recording signs in any sign language. It is not specialised
to any individual sign language. This makes it well suited to the ViSiCAST
project, which includes work on British, Dutch, and German Sign Languages.
The IDGS already has a substantial corpus of HamNoSys transcriptions of over
3000 German signs.

HamNoSys describes the physical action required to produce a sign, not the
sign's meaning. It breaks each sign down into components such as hand position,
hand orientation, hand shape, motion, etc. HamNoSys up to version 3 only
records manual components of signs. The current version 4 also describes facial
components, and we are beginning to extend the work reported here to include
them.

We will not attempt to describe HamNoSys in detail (see the cited manual),
but indicate its main features and the issues they raise for synthetic animation.
A typical HamNoSys transcription of a single sign is displayed in Figure 1. This
is the DGS (German Sign Language) sign for "GOING-TO". The colon sign
specifies that the two hands mirror each other. The next symbol specifies the
handshape: the finger and thumb are extended with the other fingers curled. The
next (the vertical line and the bat-wing shape) specifies that the hand points
upwards and forwards. The oval symbol indicates that the palm of the right hand
faces left (so by mirroring, the palm of the left hand faces right). The part in
square brackets indicates the motion: forwards, in an arc curved in the vertical
plane, while changing the orientation of the hand so as to point forwards and
down.

Note that HamNoSys describes signs in terms of basic concepts which are not
themselves given a precise meaning: "upwards and forwards", "curved", "fast",
"slow", etc. Other aspects are not recorded at all, and are assumed to take
on some "default" value. For example, HamNoSys records the positions of the
hands, but usually not of the rest of the arms. For signs which do not, as part of
their meaning, require any particular arm movements beyond those physically
necessary to place the hands, no notation of the arms is made. It is a basic
principle of HamNoSys that it records only those parts of the action which
are required to correctly form the sign, and only with enough precision as is
necessary. People learning to sign learn from example which parts of the action
are significant, and how much precision is required for good signing. To synthesise

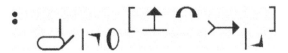

Fig. 1. HamNoSys transcription of the DGS sign for GOING-TO

an animation from a HamNoSys description requires these choices to be made by the animation algorithms.

HamNoSys was originally developed to meet the needs of humans recording signs. As can be seen from the example, it uses a specialised font; in addition, a grammar was only developed for it *post facto* to formalise the practice of Ham-NoSys transcribers. As these features make it awkward to process by computer, we have developed a version of HamNoSys encoded in XML [1], called SiGML (Signing Gesture Markup Language). Abstractly, the SiGML representation is identical to the HamNoSys, but represents it as ordinary text without special fonts, and like all XML languages, it can be parsed by standard and widely available XML software. In SiGML, the GOING-TO sign is represented thus:

```
<sign_manual both_hands="true" lr_symm="true">
        <handconfig extfidir="uo" palmor="l"
                handshape="finger2" thumbpos="out"/>
        <tgt_motion>
                <directed_motion direction="o" curve="u"/>
                <handconfig extfidir="do"/>
        </tgt_motion>
</sign_manual>
```

3   Synthesis of Static Gestural Elements

3.1   Disambiguation

We describe by examples how we have approached the task of making precise the fuzzy definitions of HamNoSys concepts.

HamNoSys defines a set of 60 positions in the space in front of the signer. These are arranged in a grid of four levels from top to bottom, five from left to right, and three from near to far. The four vertical levels are indicated by the glyphs ⊔ (shoulder level), ⊟ (chest level), ⊔ (abdomen level), and ⊟ (below abdomen level). For any given avatar, we define these to be respectively at heights s, $(s+e)/2$, e, and $(e+w)/2$, where s, e, and w are the heights of the shoulder, elbow, and wrist joints of the standing avatar when both arms hang vertically. We have defined these heights in terms of the arms rather than the torso, because these measurements are guaranteed to be present for any avatar used for signing.

HamNoSys indicates left-to-right location by a modification to these glyphs. The five locations at chest level are represented by the glyphs ▫ ⊟, ▪ ⊟, ⊟, ⊟ ▪, and ⊟ ▫. We define their left-to-right coordinates in terms of the positions of the shoulders: centre is midway between the shoulders, and the points left and right are regularly spaced with gaps of 0.4 times the distance between the shoulders.

The three distances from the avatar are notated in HamNoSys by)((near), no explicit notation for neutral, and ⤳ (far); we define these in terms of the shoulder coordinates and the length of the forearm.

An important feature of this method of determining numerical values is that it is done in terms of measurements of the avatar's body, and can be applied automatically to any humanoid avatar. The precise choice of coordinates for these and other points must be judged by the quality of the signing that results. The animation system is currently still under development, and we have not yet brought it to the point of testing.

Hand orientations are described by HamNoSys in terms which are already precise: the possibilities are all of the 26 non-zero vectors (a, b, c), where each of a, b, and c is -1, 0, or 1. When more precision is required, HamNoSys also allows the representation of any direction midway between two such vectors. These directions are the directions the fingers point (or would point if the fingers were extended straight); the orientation of the palm around this axis takes 8 possible values at 45 degree intervals.

To complete the specification of HamNoSys positions requires definitions along similar lines of all the positions that are significant to HamNoSys. In addition to these points in space, HamNoSys defines many contact points on the body, such as positions at, above, below, left, or right of each facial element (eyes, nose, cheeks, etc.), and several positions on each finger and along the arms and torso. The total number of positions nameable in HamNoSys comes to some hundreds. Any avatar for use in Hamnosys-based signing must include, as part of its definition, not only the geometry of its skeleton and surface, and its visual appearance, but also the locations of all of these "significant sites".

3.2   Inverse Kinematics

Given a definition of the numerical coordinates of all the hand positions described by HamNoSys, we must determine angles of the arm joints which will place the hand in the desired position and orientation. This is a problem in "inverse kinematics" (forward kinematics being the opposite and easier problem, of computing hand position and orientation from the arm joint angles).

The problem can mostly be solved by direct application of trigonometric equations. Two factors complicate the computation: firstly, the arm joint angles are not fully determined by the hand, and secondly, care must be taken to ensure that physiologically impossible solutions are avoided.

If the shoulder joint remains fixed in space, and the hand position and orientation are known, then one degree of freedom remains undetermined: the arm can be rotated about the line from the shoulder to the wrist joint. If the sign being performed requires the elbow to be unusually elevated, this will be notated in the HamNoSys description; otherwise, the animator must choose a natural position for the elbow. The solution we adopted requires the plane containing the upper arm and forearm to have a certain small angle to the vertical, pointing the elbows slightly outwards. The angle is increased when the arm reaches into signing space on the opposite side of the body, to prevent the upper arm passing into the torso. In addition, for such reaches, and for reaches into the "far" part of signing space, greater realism is obtained by using the sternoclavicular joint to let the shoulder move some distance towards the target point. A repertoire of inverse kinematic constraints and formulae for arm positioning can be found in [18].

For positions around the head, care must be taken to avoid the hand penetrating the head. It should be noted that HamNoSys itself does not attempt to syntactically exclude the description of physiologically impossible signs. One can, for example, specify a hand position midway between the ears. This is not a problem; the real signs that we must animate are by definition possible to perform. This implies that a synthetic animation system for signing does not have to solve general collision problems (which are computationally expensive), but only a few special cases, such as the elbow positioning described above.

3.3   Contacts

Besides specifying a hand position as a point in space or on the body, HamNoSys can also notate contacts between parts of both hands. The BSL two-handed spelling signs are an example of these. The inverse kinematic problem of calculating the arm angles necessary to bring the two hand-parts into contact can be reduced to two one-arm problems, by determining for each arm separately the joint angles required to bring the specified part of the hand to the location in space at which the contact is required to happen.

This is an area in which motion capture has difficulty, due to the accuracy with which some contacts must be made. A contact of two fingertips, for example, may appear on playback of the raw data to pass the fingers through each other, or to miss the contact altogether. This is due to basic limitations on the accuracy of capture equipment. When using motion capture data we deal with this problem by editing the calibration data for the equipment in order to generate the correct motion.

3.4   Handshapes

HamNoSys distinguishes around 200 handshapes. Each of these must be represented ultimately as the set of rotations required in each finger and thumb joint to form the handshape. We split this task into two parts. First, we specify the required joint angles as a proportion of the maximum values, independently of the avatar. Secondly, we specify for each avatar the maximum bendings of its finger and thumb joints. For those handshapes which require a contact to be made (e.g. the "pinch" shape requiring the tips of the thumb and index finger to meet), the precise angles that will produce that contact should be calculated from the geometry of the avatar's hands.

4   Motion Synthesis

We have so far discussed how to synthesise a static gesture. Most signs in BSL include motion as a semantic component, and even when a sign does not, the signer must still move to that sign from the preceding sign, and then move to the next.

If we calculate the joint angles required for each static gesture, and then linearly interpolate over time, the effect is robotic and unnatural. Artificial smooth trajectories can be synthesised (e.g. sine curve, polynomial, Bezier, etc.), but we take a more biologically based approach and model each joint as a control system.

For the particular application of signing, the modelling problem is somewhat easier than for general body motion, in that accurate physics-based modelling is largely unnecessary. Physics plays a major role in motions of the lower body, which are primarily concerned with balancing and locomotion against gravity. This is also true of those motions of the upper body which involve exertions such as grasping or pushing. Signing only requires movement of upper body parts in space, without interaction with external objects or forces. The effect of gravity in shaping the motion is negligible, as the muscles automatically compensate.

We therefore adopt a simplified model for each joint. The distal side of the joint is represented as a virtual mass or moment of inertia. The muscles are assumed to exert a force or torque that depends on the difference between the current joint angle and the joint angle required to perform the current sign, the computation being described in detail below.

Simplifications such as these are essential if the avatar is to be animated in real time.

4.1   A Brief Introduction to Control Systems

This brief summary of control theory is based on [15]. We do not require any mathematics.

A *control system* is any arrangement designed to maintain some variable at or near a desired value, independently of other forces that may be acting on it.

In general, a control system consists of the following parts:

1. The *controlled variable*, the property which the controller is intended to control.
2. A *perception* of the current value of the controlled variable.
3. A *reference signal* specifying the desired value of the controlled variable.
4. An *error signal*, the difference between reference and perception.
5. The *output function*, which computes from the error signal (and possibly its derivatives or integrals) the *output signal*, which has some physical effect.

The effect of the output signal in a functioning control system is to bring the value of the controlled variable closer to the reference value. The designer of a real (i.e. non-virtual) control system must choose an output function which will have this effect. For the present application, our task is to choose an output function such that, when the reference angle for a joint is set to a new value, the joint angle changes to reach that value in a realistic manner.

4.2   Hinge Joints

Our application of this to avatar animation places a controller in each joint. For a hinge joint such as the elbow, the controlled variable is the angle of the joint. The reference value is the angle which is required in order to produce some gesture. The perception is the current angle. The output is a virtual force acting on the distal side of the joint. The latter is modelled as a mass, whose acceleration is proportional to the force. We also assume a certain amount of damping, that is, a force on the mass proportional to and in the opposite direction to its velocity.

The mass, the force, and the damping are fictitious, and not intended as an accurate physical model; their values are tuned to provide a realistic-looking response to changes in the reference value.

Figure 2 illustrates the response of this system to a series of sudden changes in the reference value. For a sequence of static gestures, the reference value will change in this manner, taking on a new value at the time when the next gesture is to be begun. Parameters of the control system can be adjusted to give whatever speed of response is required. This can be different for different signs: Hamnosys can notate various tempos such as fast, slow, sudden stop, etc., and these will be implemented by translating them into properties of the controller.

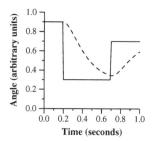

Fig. 2. Solid line: reference value. Dashed line: current value

4.3   Higher Degree Joints

A turret or universal joint has two degrees of freedom. An example is the joint at the base of each finger, which can move the finger up and down, or left and right, but cannot rotate the finger about its own axis. This can be modelled as a pair of hinge joints at right angles, and the method of the preceding section applied to each hinge. This will not be accurate if the rotation of either hinge approaches a right angle, but the universal joints in the upper body all have sufficiently limited mobility that this is not a problem.

A ball and socket joint such as the shoulder has three degrees of freedom. In principle, it can be modelled by three hinge joints in series.[1] However, there is no

---

[1] The angles of the hinges are known as an *Euler angle* representation of the rotation.

obvious way to choose axes for the hinges that corresponds to the actual articulation of the muscles. Mathematically, there are singularities in the representation, which correspond to the physical phenomenon of "gimbal lock", which does not occur in a real shoulder joint. Aesthetically, animation of the three hinge angles separately tends to give wild and unnatural movements of the arm.

Instead, we reduce it to a single one-dimensional system. If the current rotation of the shoulder joint is q, and the required rotation is q', we determine a one-dimensional trajectory between these two points in the space of three-dimensional rotations.[2] The trajectory is calibrated by real numbers from 0 to 1, and this one-dimensional calibration used as the controlled variable. When the reference value is next changed, a new trajectory is computed and calibrated. Thus we use interpolation to define the trajectory, and a controller to determine acceleration and velocity along that trajectory.

4.4   Moving Signs

Some moving signs are represented in HamNoSys as a succession of static postures. They may also be described as motion in a particular direction by a particular amount, with or without specifying a target location. These can be animated by considering them as successive static postures, in the latter case calculating a target location from the direction and size of the movement.

An explicit route from one posture to the next may be indicated (straight, curved, circular, zigzag, etc.). Since the method described above of animating the joints tends to produce straight movements of the hands, explicitly straight movements can be handled in the same way. More complex movements are handled by computing the desired trajectory of the hands, computing trajectories for the joint rotations by inverse kinematics, and using those as moving targets for the joint controllers.

HamNoSys can also specify the tempo of the motion. If the tempo is unmarked, it is performed in the "default" manner, which is what we have attempted to capture by our control model of motion. It may also be fast or slow (relative to the general speed of the signer), or modulated in various ways such as "sudden stop" or "tense" (as if performed with great effort). These concepts are easily understood from example by people learning to sign. Expressing them in terms of synthesised trajectories is a subject of our current work.

4.5   Sequences of Signs

Given a sequence of signs described in HamNoSys or SiGML, and the times at which they are to be performed, we can generate a continuous signing animation by determining the joint angles required by each sign, and setting the reference values for the joint controllers accordingly at the corresponding times (or slightly in advance of those times to account for the fixed lag introduced by

---

[2] We use spherical linear interpolation (also known as *slerp*), a standard method of interpolating between rotations which is free of the singularities of Euler angles.[6]

the controllers). The blending of motion from each sign to the next is performed automatically, without requiring the avatar to go to the neutral position between signs.

4.6   Ambient Motion

Our current avatar has the ability to blend signing animation data with "ambient" motion — small, random movements, mainly of the torso, head, and eyes — in order to make it appear more natural. This can be used even when the animation data come from motion capture. It is particularly important for synthetic animation, since we only synthesise movements of those joints which play a part in creating the sign. If the rest of the body does not move at all — and there are few signs which require any torso motion as part of their definition — the result will look unnaturally stiff. Motion capture data can be blended with synthesized data; a possible future approach is to generate these small random movements algorithmically.

5   Target Avatars

These ideas were initially prototyped in VRML 97, the Virtual Reality Modelling Language [8]. This is a textual description language for 3D worlds with animation and user interaction. Associated with VRML is a standard called H-Anim [5], which specifies a standard hierarchy of joints for a humanoid skeleton, and a standard method of representing them in VRML. We have written software to convert a description of the geometry of an avatar into an H-Anim-compliant ball-and-stick model, and to convert a sequence of gestures described in SiGML into animation data for the avatar.

A ball-and-stick avatar is not suitable for production-quality signing, but there are two reasons for using one in prototyping. It can be drawn faster, and thus allows more frames per second and smoother animation, which is important for closely judging the quality of motion. More importantly, a ball-and-stick model gives the animator a clearer view of the motion. (It is quite illuminating to play motion capture data back through such an avatar. The motion appears every bit as lifelike.) A fully modelled avatar can obscure the finer details of the movement, although it is of course more realistic for the end user.

The same software can generate animations for any H-anim compliant avatar, such as that shown in Figure 3(b), and with minor changes, for any other avatar based on a similar hierarchically structured skeleton. The avatar currently in use by ViSiCAST is called Visia, and was developed by Televirtual Ltd., one of the ViSiCAST partners. Besides having a body structure with fully articulated hands, it is also capable of performing facial animation. It has a seamless surface mesh, that is, its surface deforms continuously as the underlying bone skeleton is animated. Visia is illustrated in Figure 3(c). Under consideration as a possible target is BAPs (Body Animation Parameters), a part of the MPEG-4 standard concerned with animation of humanoid figures [9], and closely connected with H-anim.

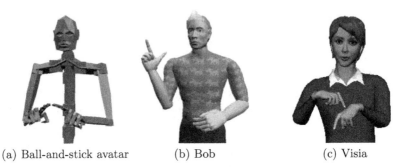

(a) Ball-and-stick avatar   (b) Bob   (c) Visia

Fig. 3. Signing avatars

6   Conclusions

We have described above the design and initial implementation of a method of automatic synthesis of manual signing gestures from their transcriptions in the HamNoSys/SiGML notations. We are confident that the approach described here will produce results that compare favourably with existing alternatives. Perhaps the most interesting comparison will be with the system based on motion capture, as already used in ViSiCAST. As our synthetic signing can target the same avatar model as the motion capture system, this provides the opportunity to undertake two kinds of comparison. Firstly, we will be able to do a meaningful comparison of user reaction to our synthetic signing with reaction to signing based on motion capture. In addition, as our synthesis process drives the avatar via a stream of data whose form is identical to that produced from motion-capture, we are also in a position to perform quantitative comparisons between the two methods.

Acknowledgements

We acknowledge funding from the EU under the Framework V IST Programme (Grant IST-1999-10500). The "Bob" avatar of Figure 3(b) is available at `http://ligwww.epfl.ch/~babski/StandardBody/`, and is due to Christian Babski and Daniel Thalmann of the Computer Graphics Lab at the Swiss Federal Institute of Technology. Body design by Mireille Clavien.

References

1. D. Connolly. *Extensible Markup Language (XML)*. World Wide Web Consortium, 2000. 149
2. R. Elliott, J. R. W. Glauert, J. R. Kennaway, and I. Marshall. The development of language processing support for the ViSiCAST project. In *ASSETS 2000 - Proc. 4th International ACM Conference on Assistive Technologies, November 2000, Arlington, Virginia*, pages 101–108, 2000. 146

3. S. Gibet and T. Lebourque. High-level specification and animation of communicative gestures. *J. Visual Languages and Computing*, 12:657–687, 2001. On-line at `http://www.idealibrary.com`. 146

4. A. Glassner. Introduction to animation. In *SIGGRAPH '2000 Course Notes*. Assoc. Comp. Mach., 2000. 147

5. Humanoid Animation Working Group. *Specification for a Standard Humanoid, version 1.1*. 1999. `http://h-anim.org/Specifications/H-Anim1.1/`. 155

6. A. J. Hanson. Visualizing quaternions. In *SIGGRAPH '2000 Course Notes*. Assoc. Comp. Mach., 2000. 154

7. J. Hodgins and Z. Popović. Animating humans by combining simulation and motion capture. In *SIGGRAPH '2000 Course Notes*. Assoc. Comp. Mach., 2000. 147

8. The VRML Consortium Incorporated. *The Virtual Reality Modeling Language: International Standard ISO/IEC 14772-1:1997*. 1997. `http://www.web3d.org/Specifications/VRML97/`. 155

9. R. Koenen. *Overview of the MPEG-4 Standard*. ISO/IEC JTC1/SC29/WG11 N2725, 1999. `http://www.cselt.it/mpeg/standards/mpeg-4/mpeg-4.htm`. 155

10. T. Lebourque and S. Gibet. A complete system for the specification and the generation of sign language gestures. In *Gesture-based Communication in Human-Computer Interaction*, Lecture Notes in Artificial Intelligence vol.1739. Springer, 1999. 146

11. M. Lincoln, S. J. Cox, and M. Nakisa. The development and evaluation of a speech to sign translation system to assist transactions. In *Int. Journal of Human-computer Studies*, 2001. In preparation. 146

12. I. Marshall, F. Pezeshkpour, J. A. Bangham, M. Wells, and R. Hughes. On the real time elision of text. In *RIFRA 98 - Proc. Int. Workshop on Extraction, Filtering and Automatic Summarization, Tunisia*. CNRS, November 1998. 146

13. F. Pezeshkpour, I. Marshall, R. Elliott, and J. A. Bangham. Development of a legible deaf-signing virtual human. In *Proc. IEEE Conf. on Multi-Media, Florence*, volume 1, pages 333–338, 1999. 146

14. Z. Popović and A. Witkin. Physically based motion transformation. In *Proc. SIGGRAPH '99*, pages 11–20. Assoc. Comp. Mach., 1999. 147

15. W. T. Powers. *Behavior: The Control of Perception*. Aldine de Gruyter, 1973. 152

16. S. Prillwitz, R. Leven, H. Zienert, T. Hanke, J. Henning, et al. *HamNoSys Version 2.0: Hamburg Notation System for Sign Languages — An Introductory Guide*. International Studies on Sign Language and the Communication of the Deaf, Volume 5. University of Hamburg, 1989. Version 3.0 is documented on the Web at `http://www.sign-lang.uni-hamburg.de/Projects/HamNoSys.html`. 148

17. SigningAvatar. `http://www.signingavatar.com`. 146

18. D. Tolani and N. I. Badler. Real-time inverse kinematics of the human arm. *Presence*, 5(4):393–401, 1996. 151

19. M. Wells, F. Pezeshkpour, I. Marshall, M. Tutt, and J. A. Bangham. Simon: an innovative approach to signing on television. In *Proc. Int. Broadcasting Convention*, 1999. 146

20. A. Witkin and Z. Popović. Motion warping. In *Proc. SIGGRAPH '95*, 1995. 147

From a Typology of Gestures
to a Procedure for Gesture Production

Isabella Poggi

Dipartimento di Scienze dell'Educazione - Università Roma Tre
Via del Castro Pretorio 20 - 00185 Roma - Italy
poggi@uniroma3.it

Abstract. A typology of gesture is presented based on four parameters: whether the gesture necessarily occurs with the verbal signal or not, whether it is represented in memory or created anew, how arbitrary or motivated it is, and what type of meaning it conveys. According to the second parameter, gestures are distinguished into codified gestures, ones represented in memory, and creative gestures, ones created on the spot by applying a set of generative rules. On the basis of this typology, a procedure is presented to generate the different types of gestures in a Multimodal Embodied Agent.

1 Introduction

As all gesture scholars have shown, we make gestures of different kinds. They differ in their occurrence (some are necessarily performed during speech, others are not) and their physical structure (some are bi-phasic, other ones are tri-phasic, [1]). But they also bear different kinds of meanings and fulfill different communicative functions.

A large number of gesture taxonomies have been proposed in literature. Sometimes, though, we feel that a gesture does not belong to a single type but is better carachterized against several parameters. In Sections 2 – 4, I propose a parametric typology of gestures, while in Section 5, on the basis of this typology I outline a model of gesture production: I present a hypotheses on the procedure through which we decide which type of gesture to make during discourse, according to our communicative goals and to our knowledge of the context.

2 A Parametric Typology of Gestures

The parameters I propose to classify a gesture are the following: its relationship to other concomitant signals; its cognitive construction; the relationship between the gesture and its meaning.

I. Wachsmuth and T. Sowa (Eds.): GW 2001, LNAI 2298, pp. 158-168, 2002.
© Springer-Verlag Berlin Heidelberg 2002

2.1 Relationship to Other Signals

The first relevant parameter to distinguish gestures is its *relationship to other signals* in other modalities: that is, whether the gesture co-occurs with other signals or it can be produced by itself. From this point of view, we can distinguish *autonomous* and *coverbal* gestures. An *autonomous* gesture may or may not be produced during speech, while a *coverbal* gesture is awkward if produced in absence of speech. Emblems are generally of the former type, beats of the latter [2],[3].

2.2 Cognitive Construction

Another relevant parameter to distinguish gestures is their *cognitive construction*, that is, whether and how they are represented in the Speaker's mind. From this point of view, we may distinguish *codified* and *creative* gestures [4].

A *codified* gesture is one steadily represented in the Speaker's mind somehow as a lexical item of a gestural lexicon. On the signal side the motor and perceptual features of the gesture are represented, on the meaning side the semantic information it bears. The signal may be represented in terms of mental images or motor programs, and the meaning in the same format and/or in a propositional format. Codified gestures then form a lexicon in our minds, that is a list of gesture-meaning pairs, where the speaker "knows" how is its standard form and which is its precise meaning, and believes that such gesture means such meaning to the interlocutor as well as to oneself: that is, the representation of the gesture-meaning pair is also believed to be shared. «Emblems» [2] are a typical case of codified gestures.

A *creative* gesture is not steadily represented in our mind, but one we invent on the spot as we want to illustrate our speech more vividly. It is created on the basis of a small set of generative rules that state how to create a gesture which is new (never produced and never seen before), but nonetheless comprehensible [4]. McNeill's [3] iconics, but also deictics, as I will argue below, are generally creative gestures.

2.3 Gesture-Meaning Relationship

The third parameter is the *gesture - meaning relationship*. A gesture may be either motivated or arbitrary. It is *motivated* when the meaning can be inferred from the signal even by someone who has never perceived it before; that is, when the meaning is linked to the signal in a non-random way.

A signal and a meaning may be linked to each other non-randomly in two ways: either by *similarity* or by *mechanical determinism*. Gestures linked to their meaning by *similarity* (resemblance, imitation) are *iconic* signals. A signal is iconic when some perceptual part or aspect of the signal in some way resembles some perceivable part or aspect linked to the meaning. Drawing the outline of a cello in the air to mean "cello" is an iconic gesture. Gestures linked to their meaning by *mechanical determinism* are *biological* or *natural* signals. Take the gesture of shaking fists up, that expresses joy or triumph for succeding in something (say, at the end of a race). This is not an iconic gesture, since it does not "imitate" joy; but it is determined by the physiological arousal produced by the emotion of joy, that necessarily causes outward or upward movements. A gesture is a natural or biological signal when its perceptual motor aspects are the same as those produced by a biological event linked to the meaning of the signal itself.

Finally, a gesture is *arbitrary* when signal and meaning are linked neither by a relationship of similarity nor by any other relationship that allows one to infer the meaning from the signal even without knowing it.

2.4 Semantic Content

The fourth parameter to distinguish gestures is their semantic content. The semantic contents of our communicative acts may concern either Information on the World or Information on the Speaker's Mind [5]: we communicate both about abstract and concrete objects, persons, animals, events - the world outside ourselves - and about our mental states, namely our beliefs, goals, and emotions. Among the gestures that inform about objects and events of the outside world, for instance, Italian emblems mention persons ("indian", "communist"), animals ("horse"), objects ("scissors", "cigarette"), actions ("to cut", "to smoke", "to walk"), properties ("thin", "stubborn", "stupid"), relations ("link between two things"), times ("yesterday"), quantifiers ("two"). Other gestures, among both Emblems and other types of gestures, are Gestural Mind Markers, that is, hand movements devoted to convey Information about the Speaker's Mind. Among Gestural Belief Markers some provide information about the **degree of certainty** of the beliefs we are mentioning: the *"palm up open hand"* [6] means that what we are saying is obvious, self-evident, while *"showing empty hands while lowering forearms"* means we are quite uncertain about something. Instead, we provide **metacognitive information** as we inform about the source of what we are saying: we may be trying to retrieve information from our long-term memory, as we imply when we *"snap thumb and middle finger"* (= "I am trying to remember"); or we may try to concentrate, that is, to draw particular attention to our own thinking, as we imply by *"leaning chin on fist"* (Rodin's "Thinker" posture).

Within Gestural Goal Markers, some express a **performative**, that is, the goal of a single sentence: *raising the flat hand or the index finger near the shoulder* is like saying: "attention please"; the Italian *purse hand* [7] means "I ask you a question". Other gestures distinguish **topic** and **comment** in a sentence, thus marking the Speaker's goal in the sentence planning - what s/he wants to stress vs. what s/he takes for granted: this is the function of the up and down movement of beats in general [8] and of specific gestures such as Kendon's [8] ring and finger bunch. Then we have **metadiscursive** gestures, that inform on the hierarchical structure and on the logical relations among sentences in a discourse. By *"bending index and middle fingers of both hands"* (= "quotes"), the Speaker is distancing from what s/he's saying; in Italy, *"fist rotating on wrist with curved thumb and index finger"* states a link, say, of cause-effect or proximity, between two things. But also locating discourse characters or topics in the space and then pointing at them, thus meaning "Now I come back to this" is a metadiscursive gesture. Again, *"raising a hand"* is a **turn-taking** device. Finally, we have Gestural Emotion Markers. *"Raising fists"* to express elation, or *"pulling one's hair"* to express despair inform about the Speaker's **emotion** (Table 1).

Table 1

GESTURAL MIND MARKERS		
BELIEFS	Degree of certainty	Palm open hand = this is evident Forearms down with empty hands = I am uncertain
	Meta-cognitive Information	*Index on nose* = I am concentrating Snap fingers = I'm trying to remember
GOALS	Performative	Raised index finger = attention
	Sentence plan	*Hand up* = this is the topic *Hand down* = this is the comment
	Discourse plan	Count on fingers = n.1... n.2... Bent index and middle fingers = I don't really mean that Rotating fist with curved thumb and index = there is a link
	Conversation plan	*Hand raised* = I want to speak
EMOTIONS		Fists raised =I am exulting *Pulling hair* =I am in despair

3   Crossing Parameters

Parameters cut across each other. The parameter "Cognitive construction" of the gesture (creative vs. codified) crosses with   "Relationship to other signals" (autonomous vs. coverbal): when communicating through a glass window, with the acoustic modality ruled out, the autonomous gestures I use may either be already represented in the mind or created on the spot. Again, "Semantic content" (World vs. Speaker's Mind) cuts across with "Cognitive construction", since not only objects, persons, actions but also beliefs, goals and emotions can be conveyed both by codified and by creative gestures. To refer to scissors or cigarettes, in Italy you can resort to emblems; but to mention an acid material an Aphasic patient [4] scratches his hand, inventing a creative gesture (acid = something that scratches); to assert something as self evident you can use the ready-made Belief Marker "*palm up open hands*", but a politician opens hands as something that opens up, thus giving the idea of something visible to everybody [7]: a metaphorical creative gesture.

Finally, as Fig. 1 shows, "Cognitive Construction" can cut across "Gesture – Meaning Relationship". Both codified and creative gestures can be classified as to their motivatedness. Raising hand or index finger to ask for speaking turn is a *codified* gesture, in that we all know what it means; and it is *natural* (biologically codified) since raising a hand naturally captures attention by enhancing visibility. Sliding one's hand under one's chin (= "I couldn't care less") is a *culturally codified arbitrary* neapolitan gesture, while the extended index and middle finger near the mouth (= "to smoke") is *cultural codified iconic*. A *creative iconic* gesture is, for instance, to draw the outline of a cello to mean "cello" (Fig.1).

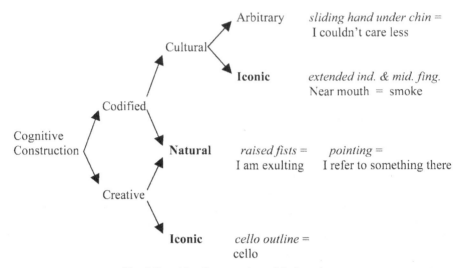

Fig. 1 Cognitive Construction x Motivatedness

4   Creative Gestures

Suppose you want to convey some Meaning (say, mention some referent) by a gesture, and no gesture is codified for that Meaning. If the referent is present in the surrounding context, you can simply point at it (that is, use a deictic); but if it is not there, you have to invent a creative gesture for it. In fact, both pointing and creating a new gesture share one and the same cognitive task: both characterize the referent by selecting one or a few relevant features of it.

4.1 Deictics

When the referent is in the physical contexts, its most salient feature is its present location, so that the easiest thing to do is to point at it, using hands or fingers (or nose tip, chin, lip, gaze) to mark in which direction the Addressee can find that referent. This is the origin of deictic gestures, and it is so much biological that any child starts pointing at the right age. This is why I consider deictics creative natural gestures. We may also have abstract deixis [3] in that one may not point at the concrete referent itself, but to something, in the surrounding space, which is in some way linked to that referent.

4.2 Iconics

Creative Iconic Gestures are cognitively more difficult to create than Deictics are. This is the difference between a noun and a deictic [9]: when the referent is not present, you must invent a noun for it, which implies constructing the concept of it, that is, abstracting all its defining features. This is what happens when any (verbal or nonverbal) noun is created for the first time: also for a "gestural noun", you have to abstract all the defining features of a referent, and then depict one or some by hands [4], [7]. An iconic gesture is in fact a gesture that, by using the shape, location and movement of your hand, imitates some distinctive features of the referent: its form, its typical location, the actions that the referent itself typically performs and those others perform with it. This is why an iconic gesture is cognitively more complex than a deictic one, which only requires selecting the feature of location to mention a referent.

As shown in [7], to create a gesture anew for naming a referent, we select the most distinctive visible feature of the referent and represent it mimically: to represent "guitar" we may depict its shape, for "hat" its location, for "bird" its action of flying, for "salt" our action of spreading it. If the referent is not visible, by our hands we'll represent another visible referent inferentially linked to it, (waving hair for "wind", talking people for "democracy"). In such cases, in order to create a gestural noun, we have to find a "bridge" between that non visible referent and the handshapes or hand movements we can perform: the "bridge" is another referent which, on the one side, can be represented by handshapes and movements, but on the other side is linked to the target referent by a small number of logical inferences, like cause-effect, object-function, class-example, opposite, negation and so on. For example, to mean "noise" I imitate a person closing one's ears with hands (*cause-effect*), to mean "noun" I depict a label (*object-function*), to mean "freedom" I pretend to be behind the bars of a prison and then deny this (*negation of the opposite*) [7].

4.3 Deictics and Iconics as Creative Gestures

Deictics and Iconics are then both creative gestures in that it is implausible for them to be represented in memory: both are produced out of a set of generation rules. I argue for this because I think It would not be very economic, say, for us to memorize a specific deictic gesture for each angle made by our hand with our body; it is more likely that we memorise a single rule like: "position your hand so as to describe an imaginary line going from your finger to the referent".

5   Codified Gestures

A codified gesture is in some way simpler than a creative gesture, from a cognitive point a view, since in it the signal (the particular handshape, movement, location and orientation you impress to the hand), the meaning, and their link to each other, are coded in the Speakers long term memory: when the Speaker wants to communicate that meaning s/he only has to retrieve it in memory, find out the corresponding signal, and produce it; while the Interlocutor, when seeing the signal, simply has to search it in memory and retrieve the corresponding meaning. Whatever the origin of a codified gesture, whether cultural or natural, iconic or arbitrary, retrieval in memory is the only cognitive task required to produce a codified gesture: no selection of features is needed, and hence no work of abstraction, no search for similarities is required.

6   A Procedure for the Generation of Gesture in Speech

So far I have outlined a static typology of gestures based on some relevant parameters that allow us to distinguish the gestures we see. But what happens as we produce gestures ourselves? While we are talking, what determines whether we also resort to gesture, beside speech? And, provided we use also the gestural modality to be more clear or effective, how do we decide which gesture to use at each step of our discourse?

6.1 When to Produce a Gesture

Gesture in bimodal communication is triggered by context. Any time we engage in a communicative process, to fulfill our communicative goals we have to take into account not only the meanings we want to convey (say, what are the "surprising features" of something we are describing [10]), but also the communicative resources at hand [11], [7], [10]. These resources encompass, on the one hand, our *internal capacities*: both transitory *pathological conditions* such as slips of the tongue or aphasia, and general *linguistic competence* conditions, like, say, the fact that as a foreigner we do not master a language completely, or that our own language is not rich in motion verbs [3]. On the other hand, these resources encompass *external conditions* (context proper), that include our assumptions about the Addressee and the physical and social situation at hand: the *Addressee's cognitive features* like *linguistic competence*, *knowledge base* and *inference capacity*, as well as its *personality traits*; the availability of *only visual* or both *visual and acoustic* modality and, finally, our *social relationship* to the Addressee and the *type of social encounter* we are engaged in, whether formal or informal, and whether aimed at affective goals - like with a friend or acquaintance - or at goals typical of a service encounter [12].

Now, the Speaker may decide, at a higher or lower level of awareness, to use a gesture instead of a word to convey some specific meaning, and can do this for whatever reason: s/he may be an aphasic, or have a slip of the tongue, or his/her language may not provide a word specific enough for that concept or visual image (*internal capacities*), or there is a lot of noise in the room, or to utter the right word would be impolite, or too difficult for the Listener's low linguistic competence (*external conditions*). However, to state which gesture to make, context determines only the choice between autonomous and coverbal gestures: if only visual modality is available, then autonomous gestures have to be chosen, while if speech cooccurs, both autonomous and coverbal gestures may be produced. But whatever kind of gesture we produce, the question is: how do we choose which gesture to make, what is the mental route from the goal of conveying some meaning via gesture to a specific gestural output?

6.2 What Gesture to Produce

In order to answer this questions, a simulation approach can be usefully adopted. Presently, a whole research area is concerned with the construction of Believable Interactive Embodied Agents, Artificial Agents capable of interactive with the User not only by written text but also with voice, facial expression, body posture and gestures [13]. In this area, several systems that make gestures have been produced (just to give some examples, [10], [14] for iconics, [15] for deictics). With the aim of producing, in future work, an Embodied Agent that can generate any type of gesture [16], in this Section I present a hypothesis about the generation of gesture during speech that stems from the gesture typology presented above. I propose a procedure for the generation of gestures that leads from the "decision" to make a gesture, triggered by consideration of the context, to the production of different gestures, by taking into account not only the meaning to convey but also the cognitive construction of the gestures to make (Figure 2).

The hypothesis underlying this procedure is that the choice of the Gesture to make in order to communicate some Meaning is in the first place determined by requirements of cognitive economy (whatever its meaning, on the World or on the Speaker's Mind): for the law of the least effort, a codified gesture, when available, should in principle be preferred to a creative gesture, since simply retrieving a ready-made gesture from memory looks easier than engaging in the creation of a gesture anew. In the second place, it is the type of meaning to convey, either on the World or on the Mind, that leads our choice of which gesture to make.

The first question in the procedure, then (see Figure 2, Question 1), aims at finding out a Codified Gesture for the Meaning to convey. If such a ready-made gesture is not found in memory, the second easier step is to check if a deictic may be used (Questions 2 and 3): again, since to select only the feature of location is easier than to sort out a whole bunch of defining features, concrete and abstract deixis will be the second and third choice.

If neither codified nor deictic gestures can be used, for Information on the World (Question 4) we resort to creative iconic gestures: that is, we look for the most distinctive feature of the referent (shape, location, action we perform on it or action performed by it) and we represent it by our hands. (This quite complex subprocedure, that I cannot illustrate in detail here, is the one sketched in Sect.4.2. and described at lenght in [7]).

If the Meaning to convey is Information on the Speaker's Mind, for topic and comment markers (Question 5) we use beats (hands, however shaped, that generally move up during the topic and down during the comment); to provide discourse reference, for instance, to mention a new character or refer to one already mentioned, which I classify as metadiscursive information (Question 6) we point in the space around (again an abstract deictic gesture).

Finally, if information on the Speaker's Mind is neither about topic-comment distinction, nor about discourse planning, in order to perform a creative gestural Mind Marker the Speaker will necessarily produce a metaphorical gesture, that is, mimically represent a visible referent that is inferentially linked to the Meaning: like the politician who opens hands to mean "self-evident".

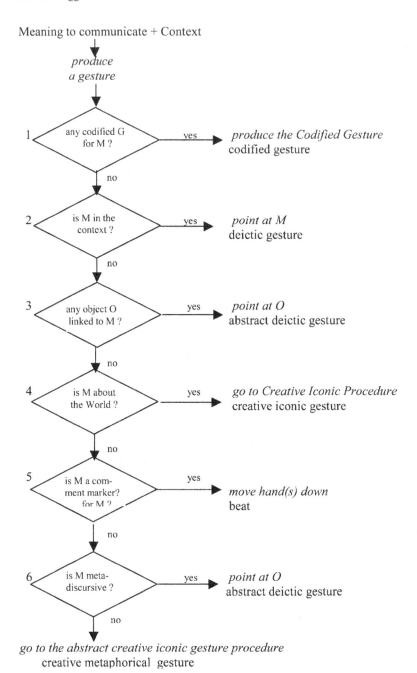

Meaning to communicate + Context

*produce
a gesture*

1 | any codified G for M ?
yes → *produce the Codified Gesture*
codified gesture

no

2 | is M in the context ?
yes → *point at M*
deictic gesture

no

3 | any object O linked to M ?
yes → *point at O*
abstract deictic gesture

no

4 | is M about the World ?
yes → *go to Creative Iconic Procedure*
creative iconic gesture

no

5 | is M a comment marker? for M ?
yes → *move hand(s) down*
beat

no

6 | is M meta-discursive ?
yes → *point at O*
abstract deictic gesture

no

go to the abstract creative iconic gesture procedure
creative metaphorical gesture

Fig. 2 A procedure for the generation of gestures

QUESTION N.1:
for the meaning M you want to convey, is there a ready-made gesture-meaning pair CG (Codified Gesture) already represented in your lexicon of gestures?
If **yes**, --> produce that gesture (***CODIFIED GESTURE***).
If **no**, --> resort to a creative gesture

QUESTION N.2:
is your referent present in the surrounding context?
If **yes**, --> point at it (***DEICTIC GESTURE***)
If **no**, --> go to Question n.3

QUESTION N.3:
is there in the context some place or object linked to the intended referent?
If **yes**, --> point at it (***ABSTRACT DEICTIC GESTURE***)
If **no**, --> go to Question n.4

QUESTION N.4:
is the meaning to convey about the World?
If **yes**, --> invent a new gesture (***CREATIVE ICONIC GESTURE***)
Sort out and imitate the most distinctive visible feature of the referent: shape, action by the referent, action by an Agent, location. With no visible features, find a visible referent inferentially linked to M, and represent it mimically.
If **no**, --> go to Question n.5

QUESTION N.5:
does the meaning to convey concern the topic-comment planning of a sentence?
If **yes**, --> go up while uttering the topic and down while uttering the comment (***BEAT***).
If **no**, --> go to Question n.6

QUESTION N.6:
do you want to locate a character or topic you will refer to later?
If **yes**, --> point at some place in context (***METADISCURSIVE GESTURE***).
or
do you want to locate a character or topic you located before?
If **yes**, --> point at that place in context (***METADISCURSIVE GESTURE***).
If **no**, --> produce a ***METAPHORICAL GESTURE***

7   Conclusion

I have presented a parametric typology that distinguishes gestures as to their relationship to other signals, cognitive construction, gesture-meaning relationship and semantic content. Based on this typology, I have proposed a procedure for the generation of gesture during speech, that generates different kinds of gestures according to the different meanings they convey and to their different cognitive representation in the Agent's mind.

By starting from a theoretical typology of gestures and from this drawing a procedure for the construction of an Artificial Gestural Agent, I hope I have shown how the production of Artificial Agents that communicate multimodally can (and should) take advantage of theoretical research in order to render all the richness and subtleties of Human multimodal communication.

References

1. Kendon, A.: Some relationships between body motion and speech. In: Siegman, A., Pope, B. (eds.): Studies in dyadic communication. Pergamon Press, New York (1972).
2. Ekman, P., Friesen, W.: The repertoire of nonverbal Behavior. Semiotica 1 (1969) 49-98.
3. Mc Neill, D.: Hand and Mind. Chicago University Press, Chicago (1992).
4. Magno Caldognetto, E., Poggi, I.: Creative iconic gestures. In: Simone, R. (ed.): Iconicity in Language. John Benjamins, Amsterdam (1995).
5. Poggi, I.: Mind Markers. In: Rector, M., Poggi, I., Trigo, N. (eds.): Gestures, Meaning and Use. Universidad Fernando Pessoa, Porto (2002).
6. Mueller, C.: Conventional Gestures in Speech Pauses. In: Mueller,C., Posner, R. (eds.): The Semantics and Pragmatics of everyday Gestures. Berlin Verlag Arno Spitz, Berlin (2001).
7. Poggi, I., Magno Caldognetto, E.: Mani che parlano. Gesti e psicologia della comunicazione. Padova, Unipress (1997).
8. Kendon, A.:. Gestures as Illocutionary and Discourse Markers in Southern Italian Conversation. Journal of Pragmatics 23 (1995) 247-279.
9. Antinucci, F.: I presupposti teorici della linguistica di Franz Bopp. In: Vignuzzi, U., Ruggiero, G., Simone, R. (eds.): Teoria e Storia degli studi linguistici. Bulzoni, Roma (1975).
10. Yan, H.: Paired speech and gesture generation in embodied conversational agents. Master dissertation, MIT, Cambridge, Mass., 2000.
11. Cassell, J., Stone, M., Yan, H.: Coordination and context-dependence in the generation of embodied conversation. INLG 2000, Mitzpe Ramon, Israel, 2000.
12. Poggi, I., Pelachaud, C.: Performative Facial Expressions in Animated Faces. In: Cassell, J., Sullivan, J., Prevost, S., Churchill, E. (eds.): Embodied Conversational Agents. MIT Press, Cambridge, Mass. (2000).
13. Pelachaud, C., Poggi, I. (eds.): Multimodal Communication and Context in Embodied Agents. 5th International Conference on Autonomous Agents. Montreal, Canada, May 29, 2001.
14. Sowa, T., Wachsmuth, I.: Coverbal iconic gestures for object descriptions in virtual environments: an empirical study. In: Rector, M., Poggi, I., Trigo, N. (eds.): Gestures, Meaning and Use. Universidad Fernando Pessoa, Porto (2001).
15. Lester, J., Stuart, S. Callaway, C, Voerman, J., Fitzgerald, P.: Deictic ad emotive communication in animated pedagogical agents. In: Cassell, J., Sullivan, J., Prevost, S., Churchill, E. (eds.): Embodied Conversational Agents. MIT Press, Cambridge, Mass (2000).
16. Pelachaud, C., Hartmann, B., Poggi, I.: Gesture in a Conversational Agent. In prep..

A Signing Avatar on the WWW

Margriet Verlinden, Corrie Tijsseling, and Han Frowein

Instituut voor Doven, Research/Development & Support,
Theerestraat 42, 5271 GD Sint-Michielsgestel, the Netherlands
{m.verlinden,c.tijsseling,h.frowein}@ivd.nl

Abstract. The work described is part of the European project ViSiCAST, which develops an avatar presented in a web page, signing the current weather forecast. The daily content of the forecast is created by semi-automatic conversion of a written weather forecast into sign language. All signs needed for weather forecasts were separately recorded by way of motion capturing. The sign language forecast is displayed on a computer by and animating fluently blended sequences of pre-recorded signs. Through a browser plug-in, this application will work on the world-wide-web.

Keywords: animation, avatars, sign language, translation, world-wide-web

1   Introduction

This paper describes the work involved in the realisation of a Virtual Human (i.e. an "avatar") signing the weather forecast on an Internet page. This application is part of the EU-subsidised ViSiCAST project. The project aims to achieve semi-automatic translation from text to sign language. With the use of a computer, sentences in a written language are analysed and transposed to a sequence of signs. Then the signs are displayed using computer animation of an avatar.

Several applications of the avatar are being developed in the ViSiCAST project, some using synthetic signing ([1,2]), others using recorded signing (i.e. "captured motions", [3]). This paper describes the use of the avatar presenting the weather forecast daily in a web page, using captured motions. In the following sections, we explain the conversion from text to sign language, the capturing and animation of signs, and embedding of the avatar in a web page.

2   Conversion from Text to Sign Language

The signed weather forecast that is presented on the Internet-page is based on the forecast written by a Dutch meteorological institute. The text in these forecasts consists of free-form sentences. To make this input suitable for automatic conversion to sign language, the structure of the sentences is simplified while maintaining the

I. Wachsmuth and T. Sowa (Eds.): GW 2001, LNAI 2298, pp. 169-172, 2002.
© Springer-Verlag Berlin Heidelberg 2002

content. Based on previous forecasts, we determined 20 possible sentence structures, henceforth "templates". Two examples of such templates are given below.

1. *Summary forecasts* [weekday] [month] [number].
2. [time] [weather condition].

As you can see, the templates have slots for varying information. Template 1 has three slots, respectively for the day of the week, the month and the day of the month. Template 2 consists entirely of slots. For each slot, a fixed set of phrases can be filled in. For 'TIME' phrases can be filled in such as "*in the morning*", "*during the whole day*", "*now and then*", etc., and for WEATHER CONDITION' phrases as, "*sunny*", "*fog*", "*heavily clouded*", "*veil clouds*" etc. Altogether there are 350 phrases for the templates and the slots. We based the templates and sets of slot filling phrases on 100 actual weather forecasts from the Netherlands. Therefore most types of Dutch weather can be expressed with it.

Subsequently, for each template it has been described how it corresponds to sign language. Since the grammar of English differs from the grammar of Sign Language of the Netherlands, the mapping is not one-on-one. Template 2 for example can be used for "*In the morning rain*", or "*Now and then rain*". In Sign Language of the Netherlands, the first sentence is signed as: the sign for morning followed by the sign for rain. However, the second sentence is signed as: first the sign for rain and then the sign for now and then. Therefore, the rule for the expressing [TIME] [WEATHER CONDITION] in Sign Language of the Netherlands is:

```
if       TIME= "now and then"
then   sign([WEATHER CONDITION])    sign("now and then")
else    sign([TIME])    sign([WEATHER CONDITION])
```

In this rule 'sign(...)' refers to the sign that has been recorded. Rules like this one have been described for all templates, for three sign languages. Following these rules (implemented in a software tool), a weather forecast in text is converted to a sequence of signs, in fact a playlist of recorded motions.

3   Recording and Visualisation of the Motion Data of Signs

All signs possibly occurring in weather forecasts were recorded (henceforth "captured"). Capturing a sign means that the motions of a signer are tracked with the following input devices: an optical system for tracking facial expression, a magnetic body suit for posture, and data gloves for hand and finger shapes (see Figure 1).

With these devices on the body, the signer makes the signs for weekdays, months, numbers, cloud types, parts of the day etc. This has been done for three sign languages: Sign Language of the Netherlands, German Sign Language and British Sign Language, each with a native signer. The equipment was originally developed for capturing global movements like running or dancing. For signing, the quality requirements of the capturing are much higher than for previous applications of these techniques. In sign language, even the slightest difference in speed, direction or gesture has a bearing upon meaning. But with the latest version, motions, posture, handshapes and facial expressions are captured at very high resolution. The signs can be easily recognised when the recorded motion data are visualised on the computer.

For visualisation an avatar program is fed with the recorded data from the three tracking systems. The avatar then makes the same movements as the live person made with the equipment on. Visualisation of the motion data is of course needed in order to see the signed weather forecast. Moreover, it is needed for manual post-processing. Several signs were recorded in one file, so the data had to be cut to obtain separate signs. Furthermore, the data needed to be reviewed and edited where needed. Sometimes the signer herself judged that she did not make the sign perfectly or the glove system did not register enough bending in a certain finger because the gloves did not fit perfectly. In ViSiCAST software has been developed that makes post-processing possible.

The avatar designed for ViSiCAST is called 'Visia'. The captured (and possibly edited) signs can be re-performed by Visia one by one separately, but also in a fluent sequence. Real-time visualisation software blends one captured motion into another, resulting in a smooth signed performance. This allows new signing sequences every day, without new recordings. This would not be possible with joined-up clips of video. Another advantage over regular video is that the avatar is truly 3-dimensional. The user can watch her from the angle he prefers.

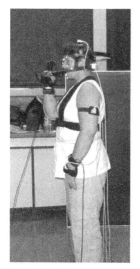

Fig.1. Signer with motion tracking devices   **Fig.2.** Avatar in a web page

4   From Motion Data to a Web Page

Technically, a large set of motion-files and the avatar software are enough to see the avatar Visia sign the sequence. However, this is not user-friendly and it is not in the right form to send it over the Internet.

The problem of sending information over the Internet is solved in two steps. On the one hand, the avatar software and the motion-files are stored at the user's PC. On the other hand, the playlist is encoded in a special language (SiGML, Signing Gesture Markup Language, developed in ViSiCAST) and enclosed in a regular HTML-page.

In order to view and control the avatar in a user-friendly way, we created a clear and simple interface. The software for visualisation of the signing has been developed to run as an Active X control. This has permitted the creation of versions compatible with Microsoft/Windows applications – particularly Internet Explorer. Based on the experiences that we have with deaf people and deaf people with limited literacy skills, we designed an Internet-page with the weather forecast as shown in Figure 2. On the right hand side, you see the avatar Visia. There are 3 buttons, similarly labelled as buttons on a video-recorder and most movie-players on the PC. The functions are: to start Visia signing, to pause at a certain point, and to end the signing altogether. A sliding-bar indicates how far the forecast has proceeded. That same slider can also be used to jump to a particular point in order to skip or repeat a part of the forecast. Next to Visia is the same forecast in text. This is meant for the significant group of people for whom sign language and text are complementary sources of information.

5   Conclusion

The weather forecast application is the first prototype application that combines techniques and knowledge from virtual human animation, motion capturing, and language processing involving both written and sign languages. This prototype shows that a virtual human can convey information in sign language, and that a limited set of separately captured motions can be used flexibly to express varying information within a certain domain. With the development of more complex language processing and techniques for synthetic signing (both also in ViSiCAST), applications can be built for larger domains and more complex sentences.

Acknowledgements

We are grateful for funding from the EU, under the Framework V IST Programme (Grant IST-1999-10500). We are also grateful to colleagues, both here at IvD, and at partner institutions, for their part in the work described above.

References

1.   Elliott, R., Glauert, J.R.W., Kennaway, J.R., Marshall, I.: The Development of Language Processing Support for the ViSiCAST Project. Presented at 4th International ACM SIGCAPH Conference on Assistive Technologies (ASSETS 2000) Washington, November 2000.
2.   Kennaway, R.: Synthetic Animation of Deaf Signing Gestures. Presented at International Gesture Workshop 2001, City University, London, April 2001.
3.   Bangham, J.A., Cox, S.J., Lincoln, M., Marshall, I., Tutt, M., Wells, M.: Signing for the Deaf using Virtual Humans. Presented at IEE Seminar on "Speech and language processing for disabled and elderly people", London, April 2000.

Iconicity in Sign Language:
A Theoretical and Methodological Point of View

Marie-Anne Sallandre and Christian Cuxac

Department of Linguistics, Paris VIII University
2 rue de la Liberté, F-93526 saint-Denis Cedex 02 - France
sallandre@yahoo.com, ccuxac@univ-paris8.fr

Abstract. This research was carried out within the framework of the linguistic theory of iconicity and cognitive grammar for French Sign Language (FSL). In this paper we briefly explain some crucial elements used to analyse any Sign Language (SL), especially transfer operations, which appear to make up the core of a spatial grammar. Then we present examples taken from our video database of deaf native speakers engaged in narrative activities. Finally we discuss the difficulty as well as the importance of studying highly iconic occurrences in uninterrupted spontaneous FSL discourse.

1   Introduction

French Sign Language (FSL) is based on a standard lexicon, which is a group of discrete and stabilised signs which can be found in FSL dictionaries [1]. This standard lexicon is widely studied by linguistic and computer researchers, in different Sign Languages (SL). But the originality of any Sign Language, as compared with Spoken Languages, is the possibility of having recourse to other structures which are endowed with highly iconic value and which function more or less independently of the standard lexicon. These structures are quite similar from one sign language to another, as Fusellier-Souza [2] has shown for Brazilian Sign Language for example.

In order to analyse the level and the nature of iconicity in FSL discourse, a video database was elaborated using deaf native speakers as informants. In this paper we will comment on a few video examples of this data, and will try to suggest some ideas for the further analysis of primitive elements which could help to characterise highly iconic structures in FSL discourse.

2   A Linguistic Model Based on Iconicity

We propose following the hypothesis by Cuxac [3, 4] concerning the distinction between these two different approaches. In this sense, the primary iconisation is split into two sub-branches, according to whether or not the process of iconisation serves the express aim of representing experience iconically. We shall term this "iconic

I. Wachsmuth and T. Sowa (Eds.): GW 2001, LNAI 2298, pp. 173-180, 2002.
© Springer-Verlag Berlin Heidelberg 2002

intent". This process of iconisation represents the perceptual world, thanks to the strong iconic resemblance of the forms to what they represent.

On the one hand, the formation of standard signs "without iconic intent" permits the meaning which is attributed a general value. Furthermore, the iconicity established in the discrete signs is preserved. On the other hand, the iconic intent, characterised by the meaning given a specific value, allows a range of meaningful choices in the larger iconic structure activated by the transfer operations. These structures are called "Highly Iconic Structures" (HIS).

Figure one illustrates these two sub-branches resulting of the process of iconisation and sketches a model for an iconic grammar of Sign Language :

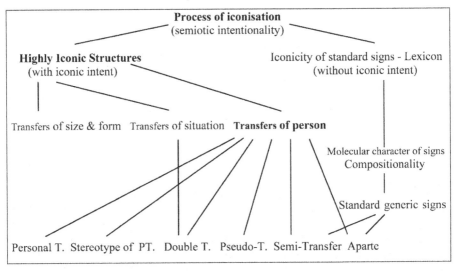

Fig. 1. Model of an iconic grammar for Sign Language

2.1   Imagistic Iconicity: A Relevant Framework for Sign Language

The most common form of iconicity in Sign Language is imagistic[1], because of visual perception ; that is, when there is a formal resemblance between the sign and what is being referred in the extra-linguistic world of experience. Ex: [HOUSE] in FSL is signed by describing a roof (a part of the concrete object). Brennan [6] and Taub [7] showed many examples which prove that iconicity and metaphor are very useful methods for signs formation and neologisms.

2.2   Highly Iconic Structures

These are not discrete signs and their signifying form can hardly be transcribed into a written protocol. Their function serves rather to represent the monstrative mode of

---

[1] Imagistic iconicity can be completed by diagrammatic iconicity, according to Haiman's conception [5] : a type of syntactical iconicity (word order for example) whose relationship to Sign Language has yet to be investigated.

"like this" combined with "as if", showing and acting out while telling (personal transfers and double transfers, as shown below).

In HIS, the signer gives an imagistic reconstitution of experience. The monstrative dimension "like this" can be activated at any moment through showing and imitating (as if I were the person of whom I speak, whatever his actions might be).

About twenty years ago, even if Sign Language researchers affirmed that HIS belonged more specifically to pantomime, they began to consider HIS as linguistic structures [8].

2.3   Transfer Operations

Highly Iconic Structures are clearly reflected in transfer operations : *transfers of form & size* (showing and describing the form or size of an object, without any process involved); *transfers of situation* (showing a situation, as if we see the scene in the distance)[2] and *transfers of person (*with process and roles).

These transfers are the visible traces of cognitive operations, which consist in transferring the real world into the four-dimensionality (the three dimensions of space plus the dimension of time) of signed discourse.

2.4   Transfer of Person

In transfers of person, the signer "disappears" and "becomes" a protagonist in the narrative (human, animal or thing). His gestures correspond to the gestures effected by the character he refers to, and whose place he has taken [3]. Thus, the signer-narrator can embody a little boy, a dog, a tree, and so on. These types of extremely iconic structures can be divided into different transfers of person arranged along a continuum, starting from a high degree (complete transfer) to a low degree (partial transfer) : *personal transfer* (complete role playing); *stereotype of personal transfer* (borrowing gestures from the culture of the hearing); *double transfer* (combining a personal transfer for acting and a situational transfer for locative information or for a second character, simultaneously); *pseudo-transfer* (describing a character with his acts, almost like a personal transfer but with less involvement of corporal energy); *aparte* (that is, an aside - as in the theatrical convention, admitting standard signs); *semi-transfer* (partial role playing, accompanied by brief standard signs).

Transfers of person are the most complex form of Highly Iconic Structures, compared with transfers of form & size, which express no process, and transfers of situation, which can only express motion and locative processes. This is why we decided to describe and analyse them [10].

In international literature for SL, transfers of person are often considered as *role playing* ; they permit to express different *points of view* [11]. Engberg-Pedersen [12] analyses some phenomena in Danish Sign Language that can be used to express a specific point of view. She speaks about *shifts* and *perspectives* : the term *role shifting* has been used to describe how signers take on a referent's identity in certain types of signing. But in this case the term *shifts* refers only to reported speech (direct

---

[2] For other SL researchers, transfers of form & size and transfers of situation (a part of them) can be considered as *classifiers* [9].

and indirect discourse) and focuses on the pronominal system. The terms *blend* and *surrogate* used by Liddell [13] should be more appropriate in dealing with cognitive operations like transfers of person.

3   Linguistic Primitives to Distinguish Transfers

We would like to discuss the methodological and technical problems we encountered in establishing the level of iconicity and how to distinguish one transfer from another.

3.1   The Data: Methodology and Examples

Our study is based on a series of alternations involving transfers observed in a videotaped corpus by three deaf native speakers, and supplemented by elicitations from consultants[3]. We ourselves filmed signers engaged in different narrative activities : three narratives used in second language learning [15], based on simple pictures of a complete story, and personal narratives with a free choice of theme and way of being told [10,16]. A new video corpus will help us to complete our present results.

Fig. 2. "To climb on a rock " by two different signers of FSL

In the narrative, the sentence "to climb on a rock " is signed with a transfer of situation by both signers. The dominant hand expresses the process of climbing while the immobile hand expresses a locative (the rock). The gaze is towards the locative, the facial expression reflects the effort of climbing.

This sentence "to climb the tree" is signed with a personal transfer, a complete role taking. The whole body is involved in the process of climbing and the gaze is towards a nest (on the branch), not towards the addressee. The two hands have a claw handshape, to imitate the cat's paw.

---

[3] We would like to insist on professional ethics when one is involved in SL research, which are echoed in Braffort's remarks in this volume [14]. First, they must be born deaf and belong to the deaf community. Especially if they are aware of the cultural context, their Sign Language could be a very imagistic one, far from the dominant written and spoken language structures. Second, signers must have confidence in the researcher to be able to produce signs naturally. Finally, given these conditions, a "natural" SL can emerge (versus an artefact of language, composed only of lexical signs) with many occurrences of Highly Iconic Structures.

Fig. 3. Drawing n° 4 [15] "The cat climbs the tree" by two different signers

Fig. 4. Drawing n° 5 [15]"The cow puts a bandage around horse's foreleg" by one signer

This double transfer (two transfers, most complex operation) illustrate "the horse falls and hurts himself" (body, mimicry and left hand : personal transfer) and "the cow puts a bandage around its foreleg" (right hand : situational transfer).

3.2   Table Presenting a Synthesis of Transfers: Linguistic Primitives

In the following table we have combined all our observations concerning the manifestation of each type of transfer. It was compiled based on the analysis of a dozen examples of FSL, some of which are illustrated above. The aim here was to distinguish both the directly measurable formal criteria (such as the direction of the gaze), as well as more semantically-oriented criteria (such as the type of iconicity involved).

Which primitives are useful for linguists ? Actually, the crucial question is to determine whether semantic or articulatory criteria are the most relevant, or if we have to combine both. Moreover, we can observe that these criteria are problematic, because different types of iconic structures appear alternatively and can be superimposed [17]. These narrative practices, which are difficult to duplicate in oral discourse, are frequently found in Sign Language and figure even as a truly cultural behaviour [3].

Table 1. Linguistic Primitives for transfers

Primitives	Transfers							
	Transfer of form & size	Transfer of situation	Transfers of person					
			Personal transfer	Double transfer	Transfer stereotype	Pseudo-transfer	Aparte	Semi-transfer
Gaze	Towards hands (form)	Towards hands	Everywhere except towards interlocutor			Towards interlocutor		Everywhere
Facial expression	To express form and size	To express form, size and process	Of the character being embodied		Of the character being embodied (with reference to hearing culture)	Of the character being embodied (less dramatic)	The signer's	
Iconicity of signs	HIS	HIS	HIS[4]	HIS	HIS	HIS		HIS + standard signs
Process	No	Yes	Yes					
Process embodied	No	No	Yes					

4   Conclusion

In this paper, we have tried to show that transfers appear to make up the core of a spatial grammar, due to their highly economical structure.

But there are still many unsolved problems with respect to articulating a unified theory including both types of signing (standard or with Highly Iconic Structures). In particular, we try to determine when it is preferable for the signer to use HIS, rather than a neutral description (with standard signs). Furthermore, the mode of communication chosen when signing depends on the capabilities of each person.

In this research we also wanted to show that Highly Iconic Structures are genuine syntactic structures, like structures which deal with a standard lexicon. We cannot properly understand SL discourse without the production of signs manifesting HIS, yet this type of extreme iconicity is particularly difficult to handle because its signs are produced simultaneously and not consecutively.

The main contribution of this work, we feel, is to develop a hierarchy between linguistic structures and to articulate an iconic grammar for French Sign Language.

---

[4] Standard signs are possible in dialogs (when a signer embodies two characters who are talking to each other, for example).

Acknowledgements

The authors would like to express their gratitude to the deaf signers Frederic Girardin, Jean-Yves Augros and Simon Attia. Many thanks to Deirdre Madden and Steven Schaefer for corrections and comments on previous versions of this manuscript.

References

1. Moody, B.: La Langue des Signes. Dictionnaire bilingue LSF/Français. Volume 2 & 3, Editions IVT, Paris (1997)
2. Fusellier-Souza, I. : Analyse descriptive de trois récits en Langue des Signes Brésilienne (LIBRAS). Master Degree report, Linguistics department, Paris VIII, Saint Denis. (1998)
3. Cuxac, C.: La LSF, les voies de l'iconicité. Faits de Langues, Ophrys, Paris (2000)
4. Cuxac, C.: French Sign Language : Proposition of a Structural Explanation by Iconicity. In Braffort, A. et al. (eds.) : Gesture-Based Communication in Human-Computer Interaction. International Gesture Workshop, LNAI 1739, Springer-Verlag, Berlin Heidelberg New York (1999) 165-184.
5. Haiman, J. (ed.): Iconicity in Syntax. J. Benjamins, Amsterdam (1985)
6. Brennan, M.: Word Formation in British Sign Language. University of Stockholm. (1990)
7. Taub S.: Language from the Body: Iconicity and Metaphor in American Sign Language. Cambridge University Press. (2001)
8. Cuxac, C.: Esquisse d'une typologie des Langues des Signes. Journée d'études n°10, 4 juin 1983 : Autour de la Langue des Signes. Université René Descartes, Paris (1985), 35-60.
9. Emmorey, K.: Language, cognition, and the brain : Insights from sign language research. . Lawrence Erlbaum Associates (2001).
10. Sallandre, M.A.: La dynamique des transferts de personne en Langue des Signes Française. DEA report (post graduate), Linguistics Department, Paris VIII, Saint-Denis (1999).
11. Poulin, C., Miller, C. : On Narrative Discourse and Point of View in Quebec Sign Language. In Emmorey, K., Reilly, J. (eds.) : Language, Gesture, and Space. (1995) 117-132.
12. Engberg-Pedersen, E. : Point of View Expressed Through Shifters. In Emmorey, K., Reilly, J. (eds.) : Language, Gesture, and Space. (1995) 133-154.
13. Liddell, S.K. : Grounded blends, gestures, and conceptual shifts. Cognitive Linguistics 9-3. Walter de Gruyter, New York (1998) 283-314
14. Braffort, A. : Research on Computer Science and Sign Language : Ethical Aspects. In Wachsmuth I.& Sowa T. (eds.). LNAI 2298 (Gesture Workshop 2001, London), Springer-Verlag (2002). To be published in this volume.
15. Berman, R., Slobin D. : Relating events in narrative : a cross-linguistic developmental study. Lawrence Erlbaum Associates, Hillsdale, NJ. (1994)

16. Sallandre, M.-A.: Va-et-vient de l'iconicité en Langue des Signes Française. A.I.L.E. 15 (Acquisition et Interaction en Langue Etrangère) (2001) 37-59.
17. Cuxac, C., Fusellier-Souza, I., Sallandre, M.-A. : Iconicité et catégorisations dans les langues des signes. Sémiotiques 16. Didier Erudition, Paris (1999) 143-166

Notation System and Statistical Analysis of NMS in JSL

Kazuyuki Kanda[1], Akira Ichikawa[2], Yuji Nagashima[3], Yushi Kato[4],
Mina Terauchi[5], Daisuke Hara[6], and Masanobu Sato[7]

[1] Chukyo University, 102-2 Yagoto-honcho, Showa-ku,
Nagoya-shi,Aichi, 466-8666, Japan
kanda@lets.chukyo-u.ac.jp
[2] Chiba University, 1-33 Yayoi-cho, Inage-ku, Chiba-shi,Chiba, 263-8522, Japan
ichikawa@ics.tj.chiba-u.ac.jp
[3] Kogakuin University, 2665-1 Nakano-machi, Hachiouji-shi, Tokyo, 192-0015, Japan
nagasima@cc.kogakuin.ac.jp
[4] Testing Organization for Proficiency of Sign Language(TOPSL)
12-9, Nihonbashi-kobuna-cho, Chuo-ku, Tokyo, 103-0024, Japan
ykato@shuwaken.org
[5] Polytechnic University,4-1-1, Hashimoto-dai, Sagamihara-shi,
Kanagawa, 229-1196, Japan
terauchi@uitec.ac.jp
[6] Aichi Medical University Nagakute-cho, Aichi-gun, Aichi, 488-1195, Japan
daisuke@aichi-med-u.ac.jp
[7] Tokyo University of Technology 1401-1, Katakura-machi
Hachiouji-shi,Tokyo, 192-0982, Japan
msato@cc.teu.ac.jp

Abstract. To describe non-manual signals (NMS's) of Japanese Sign Language (JSL), we have developed the notational system sIGNDEX. The notation describes both JSL words and NMS's. We specify characteristics of sIGNDEX in detail. We have also made a linguistic corpus that contains 100 JSL utterances. We show how sIGNDEX successfully describes not only manual signs but also NMS's that appear in the corpus. Using the results of the descriptions, we conducted statistical analyses of NMS's, which provide us with intriguing facts about frequencies and correlations of NMS's.

1 Introduction

We can record strings of oral languages with letters such as alphabetical letters and Chinese characters. They facilitate linguistic research. Some systems/ methods are now available for sign language research. HamNoSys [1] developed at Hamburg University and Stokoe method [2] in the U.S. are two of them. They represent signs as combinations of handshapes, movements, locations and others with some special symbols. But just as alphabets and Chinese characters do not represent shapes of the mouth or positions of the tongue, it is not necessary to describe constituents of signs to represent them. This fact led us to develop

I. Wachsmuth and T. Sowa (Eds.): GW 2001, LNAI 2298, pp. 181–192, 2002.

sIGNDEX. We coined the name "sIGNDEX", which is a compound of "sign" and "index".

Our study started by assigning labels of sIGNDEX to the JSL signs. We briefly discuss the label assignments in 2.1. (See [3] and [4] for the details.) In this paper, we show how sIGNDEX describes non-manual signs (hereafter, NMS's). We described 100 utterances by using sIGNDEX to confirm its effectiveness and validity(see Section 3). We also discuss statistical analyses of NMS's with sIGNDEX.

2   Fundamentals of sIGNDEX

2.1   Basic Concepts of sIGNDEX

The following are basic concepts of sIGNDEX:

(1) sIGNDEX is a sign-level notational system.
   A label of sIGNDEX is assigned to a sign as a whole, and not to each constituent of the sign or morpheme. Note that sIGNDEX is a set of labels, and not Japanese translations. To emphasize this notion, we write the sIGNDEX labels with the first letter lowercased and the rest capitalized. For instance, a sign representing "book" has the sIGNDEX label "hON". We do not have to use a space to demarcate sIGNDEX labels because lowercased letters always indicate a boundary of labels.
(2) sIGNDEX does not need special fonts or devices.
   sIGNDEX uses letters and symbols available for all types of computers. This facilitates access to sIGNDEX. sIGNDEX also observes computer environments based on SGML (one of ISO standards).
(3) sIGNDEX can describe NMS's.
   NMS's are of great importance for sign languages. sIGNDEX are strong at describing NMS. We discuss this matter in 2.2.
(4) sIGNDEX is good for sign languages in the world.
   sIGNDEX can describe signs in a particular sign language if a dictionary for the language is given. This is because sIGNDEX uses letters and symbols available on common keyboards.

The following is an example utterance represented by sIGNDEX. Note that the representation does not contain NMS's.

   (a) dOCHIRA kOOCHA kOOHIIdOCHIRA//

This represents a JSL utterance meaning "Which do you like, tea or coffee?". aNATA, dOCHIRA, kOOCHA, and kOOHII are labels for "you", "which", "tea", and "coffee" respectively.

2.2   NMS Representations in sIGNDEX

NMS's are parts of a sign language, and it is important to examine them. Labels for NMS's are represented by two letters for categories followed by one letter for subcategorization. Currently, we have fourteen categories for NMS's as follows. See more detailed descriptions about symbols in Appendix.

(1) Pointing: pT   10 symbols (Pointing to the pronoun space)
(2) Eye sight: eS   8 symbols (The direction of eye sight)
(3) Jaw: jW   7 symbols (The direction of jaw, or indication jaw stress)
(4) Eyebrow: eB   4 symbols (Movement of eyebrow)
(5) Eye: eY   6 symbols (Opening and movement of eyes)
(6) Corner of lips: cL   2 symbols (Movement of lip corners)
(7) Cheek: cK   2 symbols (Swelling of cheek)
(8) Tongue: tN   5 symbols (Showing of tongue)
(9) Head: hD   11 symbols (Nodding, tilting, turning of head)
(10) Mouthing: mO   18 symbols (Opening of mouth or speech)
(11) Lips: iP   5 symbols (Puckering, pouting, sucking, etc)
(12) Teeth: tH   2 symbols (Showing, biting)
(13) Shoulder: sH   3 symbols (Direction of shoulder to the pronoun space)
(14) Body posture: bP   8 symbols (Lean of body)

sIGNDEX symbols and rules used for describing JSL utterances are shown in Table 1. Using the rules in Table 1, we will have new NMS labels or delete some of them if necessary. The utterance (a) can be rewritten as in (b), incorporating NMS labels.

(b) aNATA-PT2 dOCHIRA+@eBU+@eYO+hDS+mOS-DOCCHI
kOOCHA+hDN+mOS-KOOCHA
kOOHII+hDN+mOS-KOOHII
dOCHIRA+eS2+hDS+hDTF+mOS-DOCCHI
+ &[ ] +@@eBU+@@eYO+eYB//

The extent of the NMS is represented by two symbols "@" and "@@". The symbols "@" and "@@" mark the onset and the end point of the NMS respectively. Thus, we can get the following information from the utterance.

(1) Eyebrow raising (eBU) and eye opening (eYO) co-occur with dOCHIRA ("which") and spread to the time when another dOCHIRA is executed.
(2) Three head noddings (hDN) co-occur with the first dOCHIRA, kOOCHA ("coffee"), and kOOHII ("tea") respectively.
(3) kOOCHA and kOOHII are accompanied by mouthing (mOS) respectively. Eyesight (eS2), head shaking (hDS), and forward head tilt (hDF) occur with the second dOCHIRA.

3   Working Process of Analyses

We made a corpus to examine NMS's in JSL. First, we drew up a scenario and had a native signer of JSL express JSL utterances. This did not work well,

Table 1. List of Symbols Used in sIGNDEX V.2 Notational System

Item of Contents	Symbol	Explanation
sIGNDEX	<sIGNDEX >	Shift code for HTML
sIGNDEX Ending	</sIGNDEX >	Shift code for HTML
Sequential Combination	No symbol	
Simultaneous Combination	+	
Optional Combination	( )	
Comments	:	Comments for the symbol
	&	Assimilation (forward,backward,mixed)
Preservation	&[ ]	Residual form or assimilated word(hold [ ])
	&[ ]-W	Hold weak hand as [ ]
Top of the Sign Word	♯	Turning point for words by the intuition of observer
Pause	/	
End of Utterance	//	
Beginning of NMS	@Symbol of NMS	Used in case of words crossing
End of NMS	@@Symbol of NMS	
Cross over Combination		@Symbol1+@Symbol2... + @@Symbol1+@@Symbol2 ...

however, because NMS's that should have occurred in natural JSL utterances did not occur. We, then, chose to have the native speaker spontaneously express JSL utterances. Namely, we showed her a certain sign as a keyword and had her express JSL utterance including it.

We repeated this procedure until we gained 100 utterances. Besides, we developed a supportive tool that helps extract and examine NMS's from the 100 utterances mentioned above. This tool is called Sign language Analysis and Indication Tool (sAIT) [5]. There are some tools to help analyze visual data of sign languages such as $SignStream^{TM}$ [6],Anvil [7] but they are not good to satisfy our needs. For example, we sometimes need Japanese to write Annotations. Also, we need tools that can work on Windows, because our computer environments are exclusively Windows-based. These requests made us develop a supportive tool specialized for our purposes. The lower part of Figure 1 shows an example of the two-dimentional notation of sAIT.

The two-dimensional notation of sIGNDEX results from the fact that the tool that we developed uses the two-dimensional space. sIGNDEX was originally a one-dimensional notational system, which is preferable when we develop a database by using a computer. But the two-dimensional notation of sIGNDEX is more useful when we look at relations between word labels and NMS's symbols,

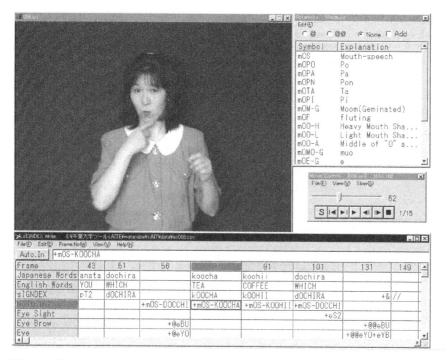

Fig. 1. The upper leftlarge window shows video recorded in avi files, which is controlled with buttons in the middle right rectangle control box. We can insert frame numbers of video into the lower chart along with other pieces of information such as sIGNDEX labels and NMS sybols

and the duration of such labels and symbols. Examples of both notations are shown in Figure 2.

4   Results of NMS Analyses

We described 100 utterances mentioned above by using sIGNDEX to confirm its effectivenss and validity.Using these results, we conducted statistical analyses of NMS's. We have finished analyzing 90 utterances so far. We discuss results of the data analyses. In the analyses, the 90 utterances have 610 signs and 1503 NMS's, which are classified into 73 categories. Table 2 shows relative frequencies of fifteen of the most frequent signs. Note that functional signs are ranked high in frequency in Table 2. This result coincides with the results of a previous study made by some of the aouthers [8].

Table 3 shows the numbers of occurrence and relative frequencies of both NMS categories and NMS symbols. Figure 3 shows 10 of the most frequent NMS's, and Figure 4 shows frequencies of the NMS categories. It follows from the tables and figures that the number of the NMS's in the "eY" category (i.e.,

Japanese sentence = Koocha to koohii docchi ga ii.
English equivalent = Which do you like, Tea or Coffee?
sIGNDEX   = aNATA-PT2 dOCHIRA+@eBU+@eYO+hDS+mOS-DOCCHI
                  kOOCHA+hDN+mOS-KOOCHA
                  kOOHII+hDN+mOS-KOOHII
                  dOCHIRA+eS2+hDS+hDTF+mOS-DOCCHI
                  + &[ ] +@@eBU+@@eYO+eYB//

	aNATA	dOCHIRA	kOOCHA	kOOHII	dOCHIRA
	PT2	@eBU+@eYO+hDS+mOS-DOCCHI	hDN+mOS-KOOCHA	hDN+mOS-KOOHII	eS2+hDS+hDTF+mOS-DOCCHI+&[]+@@eBU+@@eYO+eYB
Pointing: pT					
Head shaking: hDS					
Head nodding: hDN					
Head tilting to forward: hDTF					
Eye sight to personal pronoun: eS2					
Eye brow up: eBU					
Eye open: eYO					
Eye blink: eYB					
Mouthing: mOS		DOCCHI	KOOCHA	KOOHII	DOCCHI
Preservation: &					

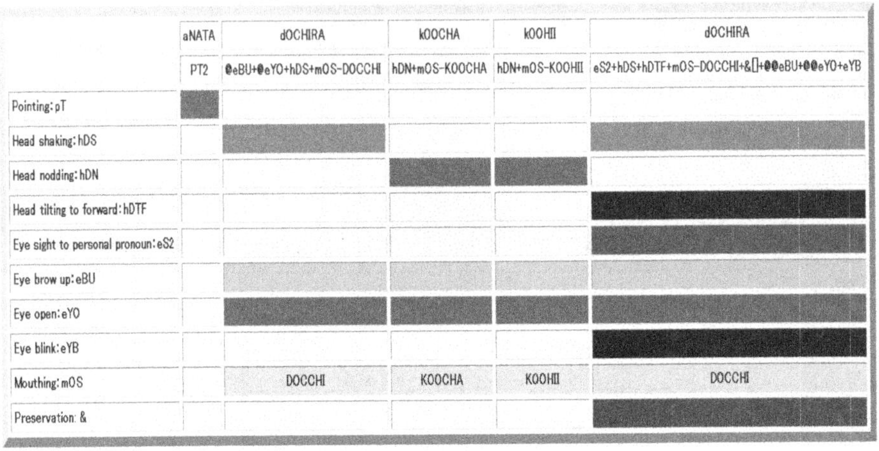

Fig. 2. The upper part shows an example of sIGNDEX's one-dimentional expressions while the lower part the one of the two-dimensional expressions. In the lower chart, the first line has word-level sIGNDEX symbols, and the second line strings of NMS symbols. The word-level sIGNDEX symbol and the string of NMS symbols in the same column synchronize. Colored squares represent NMS's that synchronize with manual sings in the first line

eye movement) is the largest or 26.0 % of all NMS's. The number of the NMS's in the "hD" category (i.e., head movement) is the second largest, or amounts to 21.6 %.

According to the NMS classification, the "eYB(eyebrow)" element is the most frequent and amounts to 18.1 %, and "mOS(mouth-speech)" is the second most frequent NMS and amounts to 13.5 %. The third most frequent NMS is "hDN". Note that "mOS" accompanies more than 30 % of manually articulated signs.

As for eYB(eye blinking) and hDN(nodding), the former occurs 273 times and the latter 174 times in our corpus, 34 of them synchronized with each other; 34 is equivalent to 19.5% of all the hDN's. It is predicted that there should be more cases where two or more NMS's partially overlap. This result coincides with the result of the analysis regarding the relation between blinking and nodding by Takata et al. [9]

Table 2. Relative Frequencies of Frequently Used Signs

Rank	sIGNDEX Word	English Equivalent	Relative Frequency %
1	wATASHI	me	7.2
2	yOBIKAKE	hey	4.4
3	aNATA	you	3.0
4	iKU	go	2.0
5	nAI	nothing	1.8
6	yOI	good	1.6
7	nANI	what	1.3
8	mIRU	look	1.1
8	pT1		1.1
10	gENZAI	now	1.0
10	oOI	many	1.0
10	sURU	do	1.0
13	dAIJOUBU	okay	0.8
13	pT1B		0.8
13	pT1N		0.8

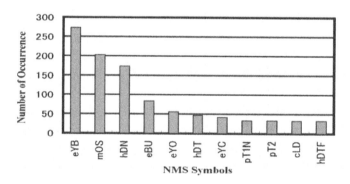

Fig. 3. Frequencies of NMS Symbols

5   Conclusion and Future Work

We collected JSL utterances and described them by sIGNDEX. While we described manual sings, we devised labels and symbols for, and described MNS's as well as manual signs. We made statistical analyses of MNS's that appear in the utterances. As a result we found that eye movements, head movements and mouthings occurred frequently. We have also found that eye blinking and head nodding correlate with each other in their occurrence. The results suggest that we should examine the correlates of one NMS with another, and also those of NMS's with manual signs such as words and morphemes.

Table 3. Frequencies and Relative Frequencies of NMS Categories and Symbols

Category	Number of Occurrence	Relative Frequency %	NMS Symbol	Number of Appearance	Relative Requency in Category %
pT	145	10.0	pT1	8	5.5
			pT1N	34	23.4
			pT1B	27	18.6
			pT1C	3	2.1
			pT1S	1	0.7
			pT2	34	23.4
			pT3	23	15.9
			pTS	0	0.0
			pTW	0	0.0
			pTR	15	10.3
eS	85	5.7	eS2	10	11.8
			eS3	4	4.7
			eSU	18	21.2
			eSD	19	22.4
			eSR	28	32.9
			eSA	0	0.0
			eST	5	5.9
			eSTS	1	1.2
jW	39	2.6	jWT	17	43.6
			jWP	13	33.3
			jWD	4	10.3
			jWS	3	7.7
			jW2	1	2.6
			jW3	0	0.0
			jWR	1	2.6
eB	124	8.3	eBU	84	67.7
			eBD	7	5.6
			eBC	32	25.8
			eBS	1	0.8
eY	391	26.0	eYO	56	14.3
			eYC	42	10.7
			eYS	13	3.3
			eYG	3	0.8
			eYT	4	1.0
			eYB	273	69.8
cL	40	2.7	cLD	33	82.5
			cLP	7	17.5
cK	3	0.2	cKP	3	100.0
			cKB	0	0.0
tN	21	1.4	tNS	11	52.4
			tNSS	5	23.8
			tNSM	3	14.3
			tNSF	2	9.5
			tNIN	0	0.0

Category	Number of Occurrence	Relative Frequency %	NMS Symbol	Number of Appearance	Relative Requency in Category %
hD	325	21.6	hDN	174	53.5
			hDS	27	8.3
			hDT	47	14.5
			hDTF	33	10.2
			hDTB	6	1.8
			hDL	21	6.5
			hDI	1	0.3
			hDC	10	3.1
			hDF	0	0.0
			hDR	3	0.9
			hAD	3	0.9
mO	230	15.3	mOS	203	88.3
			mOPO	1	0.4
			mOPA	9	3.9
			mOPN	0	0.0
			mOTA	0	0.0
			mOPI	1	0.4
			mOM-G	9	3.9
			mOF	0	0.0
			mOO-H	1	0.4
			mOO-L	0	0.0
			mOO-A	2	0.9
			mOMO-G	0	0.0
			mOE-G	0	0.0
			mOA-L	0	0.0
			mOA-G	1	0.4
			mOPO-G	1	0.4
			mOP-R	1	0.4
			mOSW	1	0.4
iP	53	3.5	iPP-P	9	17.0
			iPP-H	15	28.3
			iPP-L	1	1.9
			iPD	19	35.8
			iPS	9	17.0
tH	3	0.2	tHS	3	100.0
			tHB	0	0.0
sH	7	0.5	sH2	0	0.0
			sH3	0	0.0
			sHI	7	100.0
bP	37	2.5	bPF	29	78.4
			bPB	2	5.4
			bPS	1	2.7
			bPW	1	2.7
			bPFW	2	5.4
			bPFS	1	2.7
			bPBW	1	2.7
			bPBS	0	0.0

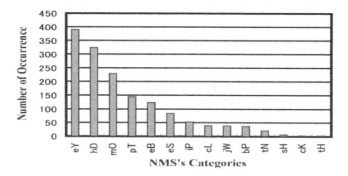

Fig. 4. Frequencies of NMS Categories

We hope that sIGNDEX will develop so as to separately describe the phenomena of and the functions and meanings of NMS's. We also hope that sIGNDEX will incorporate devises that can record the time of occurrence of the manual and non-manual sign.

Acknowledgements

The research environment is getting more convenient. Owing to new inventions, we were able to collect and analyze sign language data in a short period. We were supported by the excellent consultants of JSL native signers. We express our greatest gratitude to them. This project was partly supported by Grant-in-Aid for Scientific Research (A)(1) Project Number:11308007 and (C) Project Number:11680399.

References

1. Prillwitz, S., Leven, R., Zienert, H., Hamke, T., Henning, J.: HamNoSys Version 2.0; Hamburg Notation System for Sign Languages; An Introductory Guide, volume 5 of International Studies on Sign Language and Communication of the Deaf, Signum Press, Hamburg(1989) 181
2. Friedman, L. A.:On the Other Hand; Academic Press (1977) 181
3. Kanda, K., Ichikawa, A., Nagashima, Y., Kato, Y., Terauchi, M.: Signdex V.1 & V.2 The Computerized Dictionary of Japanese Sign Language; Proceedings of TISLR(7 th International Conference on Theoretical Issues in Sign Language Research),p.195(2000) 182
4. Kanda, K., Ichikawa, A., Nagashima, Y., Kato, Y., Terauchi, M., Hara, D., Sato, M.: Notation System and Statistical Analysis of NMS in JSL; Proceedings of HCI, Vol.1, pp.25-27(2001) 182
5. Watanabe, I., Horiuchi, Y., Ichikawa, A.: An notation Support Tool of JSL Utterances; Technical Report of IEICE, WIT99-46, pp.131-136(2000) 184
6. MacLaughlin, D., Neidle, C., Greenfield, D.:$SignStream^{TM}$ User's Guide Version 2.0; Report No.9 American Sign Language Linguistic Research Project(2000) 184

7. Kipp,M.:Anvil - A Generic Annotation Tool for Multimodal Dialogue; Proceedings of Eurospeech 2001, Aalborg, pp. 1367-1370(2001) 184
8. Nagashima, Y., Kanda, K., Sato, M., Terauchi, M.: JSL Electronical Dictionary and its Application to Morphological Analysis; Proceedings of The Human Interface Symposium 2000, pp.189-192(2000) 185
9. Takata, Y., Nagashima, Y., Seki, Y., Muto, D., Lu, S. Igi, S., Matsuo,H.: Analysis of Segmented Factors for Japanese Sign Language Recognition; Trans. of Human Interface Society, Vol.2, No.3, pp.29-36(2000) 186

Appendix

Table 4. List of Symbols Used in sIGNDEX V.2 Notational System(No.1)

Item of Contents	Symbols	Explanation
Pointing :pT	pT1	Pronoun space for 1st person
	pT1N	Point to one's nose (I do ...)
	pT1B	Point to one's bust (I am ...)
	pT1C	Point to one's chin
	pT1S	Point to one's shoulder
	pT2	Pronoun space for 2nd person
	pT3	Pronoun space for 3rd person
	pTS	Point to the strong hand by one's weak hand
	pTW	Point to the weak hand by one's strong hand
	pTR	Point to the reference space
Eye sight:eS	eS2	Pronoun space for 2nd person
	eS3	Pronoun space for 3rd person
	eSU	Eye sight up
	eSD	Eye sight down
	eSR	Eye sight to reference space
	eSA	Concius avoiding of Eye sight
	eST	Eye sight tracing
	eSTS	Eye sight tracing to strong hand
Jaw:jW	jWT	Jaw thrusted
	jWP	Jaw depressed
	jWD	Non stressed Jaw down (Contrast to jWS)
	jWS	Stressed : Pout jaw(Contrast to jWD) (Mouthing accompanied: +mO$)
	jW2	Point to 2nd pronoun space by one's Jaw
	jW3	Point to 3rd pronoun space by one's Jaw
	jWR	Reference
Eyebrow:eB	eBU	Eyebrow up
	eBD	Eyebrow down
	eBC	Eyebrow closing
	eBS	Eyebrow moving separated
Eye:eY	eYO	Eye open
	eYC	Eye close
	eYS	Eye shut
	eYG	Eye glisten
	eYT	Eye triangled
	eYB	Eye blinking
Corner of Lips:cL	cLD	Lip corners depressed
	cLP	Lip corners pulled
Cheek:cK	cKP	Cheeks puffed
	cKB	Cheek bulged
Tongue:tN	tNS	Tongue showing
	tNSS	Tongue showed slightly in the mouth
	tNSM	Tongue showed medially
	tNSF	Tongue showed fully (Stick out one's tongue)
	tNIN	Stick in tongue in front of one's teeth

Table 5. List of Symbols Used in sIGNDEX V.2 Notational System(No.2)

Item of Contents	Symbols	Explanation
Head:hD	hDN	Head nodding
	hDS	Head shaking
	hDT	Head tilting
	hDTF	Head tilting forward
	hDTB	Head tilting backward
	hDL	Head turned laterally
	hDI	Head turned ipsi-laterally
	hDC	Head chasing
	hDF	Far
	hDR	Reference
	hAD	Anti-reference
Mouthing:mO	mOS	Mouth-speech
	mOPO	"Po"
	mOPA	"Pa"
	mOPN	"Pon"
	mOTA	"Ta"
	mOPI	"Pi"
	mOM-G	"Moom"(Geminated)
	mOF	fluting
	mOO-H	Heavy mouth shape "O"
	mOO-L	Light mouth shape "O"
	mOO-A	Middle of "O" and "A"
	mOMO-G	"muo"
	mOE-G	"e"
	mOA-L	Light "A"
	mOA-G	"AA"
	mOPO-G	"Poo"
	mOP-R	"papapapa"(repeat of "Pa")
	mOSW	Schwa
Lips:iP	iPP-P	Pucker
	iPP-H	Pout heavily
	iPP-L	Pout lightly
	iPD	Lips drawn back
	iPS	Suck
Teeth:tH	tHS	Showing teeth
	tHB	Bite
Shoulder:sH	sH2	Turn one's shoulder to the 2nd pronoun space
	sH3	Turn one's shoulder to the 3rd pronoun space
	sHI	Sholder in
Body Posture:bP	bPF	Turn the body posture forward
	bPB	Turn the body posture backward
	bPS	Turn the body posture ipsilaterally
	bPW	Turn the body posture contralaterally
	bPFW	Turn the body posture forward contralaterally
	bPFS	Turn the body posture forward ipsilaterally
	bPBW	Turn the body posture backward contralaterally
	bPBS	Turn the body posture backward ipsilaterally

Head Movements and Negation in Greek Sign Language

Klimis Antzakas and Bencie Woll

Dept. of Language and Communication Science, City University
London, EC1V 0HB
Tel: 020 70408816.
K.Antzakas@city.ac.uk

Abstract This paper is part of a study examining how negation is marked in Greek Sign Language (GSL). Head movements which are reported to mark negation in other sign languages have been examined to see if they are also used in GSL along with negation sings and signs with incorporated negation. Of particular interest is the analysis of the backward tilt of the head which is distinct for marking negation in GSL.

1   Introduction

Backwards tilt of the head seems to operate in GSL as an analogue of the headshake in other sign languages. Headshake is one of the most frequently reported head movements used in sign languages. A headshake can be used to negate a sentence or a single sign. It is often accompanied by negation facial expression which means a) wrinkling the nose, b) raising the upper lip, c) depressing the lip corners and d) raising the chin and so on. The use of headshake as a negation marker has been reported in American Sign Language, British Sign Language, Swedish Sign Language, Sign Language of the Netherlands, Argentinean Sign Language, Brazilian Sign Language, Chilean Sign Language and International Sign. Additionally, Sutton-Spence and Woll (1999) note that in BSL, a 'negation turn' of the head is used by signers. In this head movement the head makes a half turn which accompanies a negation sign but is not used to negate a whole sentence.

A pilot study was carried out in order to confirm these observations. Three Deaf informants were videotaped in free conversation, a structured 'interview' and signing stories. The informants were Deaf male adults with Deaf parents. The pilot study confirmed the use of head movements, negation signs and signs with incorporated negation. The head movements were coded as H1, H2 and H3 as follows:

H1. The head tilts backwards

H2. The head shakes from side to side. This is a repeated movement of the head. As in backward tilt of the head, headshake can vary in size, speed and duration of the movement. There are both individual differences and differences due to a stronger or weaker expression of negation.

I. Wachsmuth and T. Sowa (Eds.): GW 2001, LNAI 2298, pp. 193-196, 2002.
© Springer-Verlag Berlin Heidelberg 2002

H3. The head makes a half movement to one side only and then moves back to the initial position. As in backward tilt of the head and headshake, half turn of the head can vary in size and duration of the movement.

The study also confirmed the use of negation signs and signs with incorporated negation in GSL. Negation signs can be translated as 'no', 'not', 'not yet', 'nothing or nobody', 'never', 'won't', 'not fair' etc (Sutton-Spence & Woll, 1999; Baker & Cokely, 1980). Negative incorporation is described by Woodward (1974) as "several verbs that may be negated by a bound outward twisting movement of the moving hand(s) from the place where the sign is made" According to Sutton-Spence and Woll (1999) these verbs are often verbs of experience or sensation. Examples include: 'have-not', 'like-not', 'want-not', 'know-not', 'disagree', 'believe-not', 'should-not' etc (Sutton-Spence & Woll, 1999; Deuchar, 1984; Baker & Cokely, 1980; Woodward, 1974).

Additionally two videotapes of free conversations (in Deaf clubs, social events in the National Institute for the Protection of the Deaf, in public places etc.) and stories in GSL were coded. The duration of these videotapes was approximately 4h.46min. These videotapes contained 720 tokens of negation in GSL. The informants who signed stories in GSL were seven Deaf adults, six male and one female. The parents of five of the informants were hearing and the parents of the other two were Deaf. The codification and definition of the tokens permits the division of the tokens of negation into two main groups:

a)   Tokens of manual negation. These are the cases where a signer produces a negation sign or a sign of incorporated negation. In these tokens non-manual features may also appear. In this category the main characteristic is the use of manual negation.

b)   Tokens of Non-Manual negation. These are the cases where a signer expresses negation by using only non-manual features.

2   Data Analysis

The following analysis has been based on data collected by the videotapes with free conversations and stories.

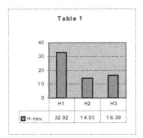

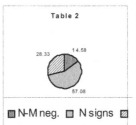

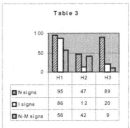

Table 1 shows that head movements are used to mark negation in more than the half of the overall data (59.58% of 720 tokens). The rest of the tokens (40.42%) are not accompanied by any head movement. Table 2 represents the percentages of negation signs (N signs), signs with incorporated negation (I signs) and Non-Manual negation (N-M neg.) of the overall data.

In table 3 the number of occurrences of the different head movements in relation to different tokens of negation are represented. A chi square test was conducted in order to examine if the number differences of occurrences of head movements are not significant (null hypothesis), or if these differences are significant (alternative hypothesis). The value of chi square was 67.81. The tabulated value for $p<0.001$ at 4 degrees of freedom was 18.46. Therefore there was a statistically significant difference (at $p<0.001$) between the numbers of occurrences of head movements in different types of negation. This indicates that head movements are defined by a linguistic pattern related to the different types of negation and that are not distributed randomly among the different types of negation.

This view is reinforced by the examination of the association of specific signs with specific head movements. Head movements tend to correspond to handshape movement. Thus, when the movement of the handshape is upwards the corresponding head movement is a backward tilt (H1).

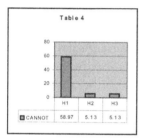
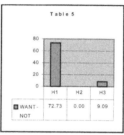
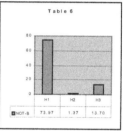

Signs CANNOT (Table 4), WANT-NOT (Table 5) and NOT-B (Table 5) were associated more with backward tilt of the head than with headshake or half turn of the head. On the other hand when the movement of the sign is side to side then the corresponding head movement is a headshake or a half movement of the head to one side.

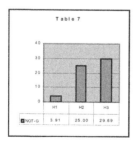
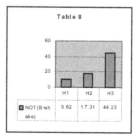

In signs NOT-G(Table 7) and NOT(B+shaking) (Table 8) the corresponding head movement is a headshake or a half movement of the head to one side (H2 or H3).

3   Conclusion

The study confirmed that GSL uses the following head movements as negation markers: a) backwards tilt of the head (H1), b) headshake from side to side (H2) and

c) half shake-movement of the head to one side only (H3). The existence of a linguistic pattern which determines the occurrences of the head movements according to the type of negation was supported by this evidence. Examination of head movements in relation to particular signs indicates that a head movement tends to agree with the movement of the sign. This movement agreement is determined by GSL phonology.

There are indications that the Greek Deaf community has adopted the backward tilt of the head from the Greek hearing community. Eibl-Eibesfeldt (1970), Morris (1977, 1979), and De Jorio (2000) have reported that this gesture is used in Greece and Naples by hearing people in order to express negation. This gesture has been transformed by Greek Deaf people to a linguistic feature and to one of the major non-manual negation markers in GSL. To the best of our knowledge backward tilt of the head has not been reported in other sign languages until now. Personal communication confirmed the use of the gesture by hearing communities in the eastern Mediterranean (south Italy and Israel) but not by the corresponding Deaf communities.

These are only preliminary findings. These observations need further research in order to be confirmed. The study of phonology, grammar, syntax and scope of negation in GSL will clarify head movement and sign movement agreement and will give an insight to the analysis of negation in GSL and in sign languages generally.

References

1. Baker, C., Cokely, D.: American Sign Language: A Teacher's resource text on grammar and culture. Silver Spring, Md.: T. J. Publishers, Inc (1980).
2. Deuchar, M.: British Sign Language. London, Routledge &Kegan Paul (1984).
3. Eibl-Eibesfeldt, I.: Ethology: The Biology of Behavior. New York (1970).
4. De Jorio, A.: Gesture In Naples And Gesture In Classical Antiquity. In: Kendon A. (eds): A translation of La Mimica Degli Antichi Investigata Nel Gestire Napoletano. Indiana University Press (2000).
5. Morris, D.: Manwatcing A Field Guide to Human Behaviour. London, Cape (1977).
6. Morris, D., et al.: Gestures: Their Origin and Distribution. London, Cape (1979).
7. Sutton-Spence, R. & Woll, B.: The Linguistics of British Sign Language: An Introduction. Cambridge University Press (1999).
8. Woodward, J.: Implication variation in American Sign Language: Negative incorporation. Sign Language Studies, Vol 5, (1974) 20-30.

Study on Semantic Representations
of French Sign Language Sentences

Fanch Lejeune[1,2], Annelies Braffort[1], and Jean-Pierre Desclés[2]

[1] LIMSI/CNRS, Bat 508
BP 133, F-91 403 Orsay cedex, France
{Fanch.Lejeune,Annelies.Braffort}@limsi.fr
[2] LALIC
96 bd Raspail, F-75 005 Paris, France
Jean-pierre.Descles@paris4.sorbonne.fr

Abstract. This study addresses the problem of semantic representation in French Sign Language (FSL) sentences. We studied in particular static situations (spatial localisations) and situations denoted by motion verbs. The aim is to propose some models which could be implemented and integrated in computing systems dedicated to Sign Language (SL). According to the FSL functioning, we suggest a framework using representations based on cognitive grammars.

1 Introduction

Linguistic research dedicated to SL is quite recent (since the early 1960s in USA) and the first researchers' aim was to prove that SL were fully-fledged language, like spoken languages. Thus, SL were studied within theoretical frameworks related to spoken language. Now, new theoretical frameworks has appeared and propose some models taking into account the *iconicity* of SL [1,2,3].

Most of the computing systems dedicated to SL don't take into account this iconicity except few studies limited to directional verbs [4,5] and pointing gestures [4]. The modelling of semantic and syntactic rules of SL remains quite unexplored.

In this context, the aim of our study is to understand the functioning of some *semantic relations* in FSL and to suggest some models which could be implemented and integrated in computing systems.

2 French Sign Language Functioning

Due to the gestural-visual channel, the spatial component of the language is essential in SL. An interesting point relates to the relevant and structured use of space to express semantic relations. Thus, spatial relations of localisation between entities are

I. Wachsmuth and T. Sowa (Eds.): GW 2001, LNAI 2298, pp. 197-201, 2002.
© Springer-Verlag Berlin Heidelberg 2002

generally established without using specifics signs. This is the sign order and the representation in space which allows interpretation.

Moreover, one can quote the *classifier* functions. Classifiers are gestures that give information about salient characteristics of the entities by using iconicity. For example, in the sentence "*I give an apple to the boy*", the handshape of the classifier represents the apple shape, while its movement gives the verb conjugation (see [4] for a detailed description of this sentence).

Others significant properties must be taken into account: property of conveying simultaneously several information, use of gaze, iconicity [1,6].

3   Theoretical Framework

In this context, we consider that *cognitive grammars* form an ideal theoretical framework for the analysis of SL. Under the hypothesis that the language activity is not autonomous, they suggest representations to describe lexicon which have a certain relevance from the point of view of cognition, and are built on the basis of visual perception [7], [8].

Our approach relies on the *Cognitive and Applicative Grammar* theory which proposes formal structures, named *Semantico-Cognitive Schemes* (SCS), to represent meaning of verbs. They allow verbs to be described as a composition of primitives based on visual perception and action [7]. These structures seem to be very close to the spatial organization of the sentences in FSL [6]. The SCS can be either *static* (denoting properties, positions in space and time), *cinematic* (denoting motion or change of state), or *dynamic* (expressing intentional actions). As example, the generic SCS used for the verbs which express motion is given in Fig. 1.

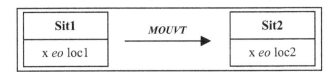

Fig. 1. Generic SCS for verbs of motion

This SCS encodes the transition (given by the primitive *MOUVT*) from a first situation (*Sit1*) to another one (*Sit2*). The spatial primitive *eo*, in combination with topological specification, locates the objects (*x*) with regards to the interior, exterior, frontier or closure of locations *(loc1, loc2)*.

Considering that each language has its own cognitive organisation with instantiation and combination of such schemes, we suggest to use this model for the representation of sentences in FSL, adapting it to the specificities of this language.

4   Analysis Method and Examples

Our first video corpus of FSL sentences describes different kind of situations: spatial localisations (e.g. water *on* a table, a man *in* the water, a ball *under* a car), cinematic and dynamic situations expressed by several motion actions (e.g. the cat *jump* on the table, the wind *opens* the door, the man *put* the cat on the table).

For each sentence, we extract the information which is significant to describe the meaning and the interaction between the signs. In particular, for motion actions, our attention focuses on the movement parameter. We retain the handshapes used during the movement, the locations of the beginning and the end of the movement, and the quality of motion. For example in a possible realisation of the sentence *"the cat jump on the table"*, the left hand is an horizontal *"flat hand"* classifier, which represents the location of the top of the table, and the right hand is a *"X"* classifier, which represents the cat's legs. The movement begins from a location indicated by a pointing gesture and finishes *on* the left hand (representing the top of the table).

In a second step, for each sentence, we build a representation of the meaning from a global point of view. From this representation, we design a *SCS* representation using pertinent primitives. We give as example the simple sentence *"the cat is inside the car"* (see [9] for a detailed description of this sentence), which only represents a spatial localisation (only one situation in the SCS). The "C" classifier, persistent until the end of the sentence, represents the car as a container and its location. In the localisation relation sign, the cat, represented by a "X' classifier, is localised with regards to the car position. Table 1 shows the order of the different linguistic operations giving the meaning of the sentence.

Table 1. Linguistic operations for "the cat is inside the car"

N°	Linguistic operations	Corresponding signs
1	Landmark	Standard sign (car)
2	Relevant property of the landmark	Classifier (C)
3	Localised entity	Standard sign (cat)
4	Relevant property of the localised entity	Classifier (X)
5	Localisation relation	Relation between the 2 classifiers (X and C)

Fig. 2 gives a global representation of this sentence: The first part provides information on the context of the relation. The car is a landmark, the cat is located with regards to this landmark. The classifier used for the car represents a container, because we are interested by the inside of the car. The second part of Fig. 2 is an image showing the relation given by the two classifiers "C" and "X": the cat is *inside* the car.

This representation can be viewed as an instantiation of the relation <x *eo IN* loc> expressing that the entity (x) is located with regards to the interior (*eo IN*) of the location (loc).

Landmark : car	**Classifier** : C	**Property** : container
Localised : cat	**Classifier** : X	

Fig. 2. Global representation of "the cat is inside the car"

5   Conclusion and Perspectives

According to this method, we have proceeded for a hundred of sentences. During the next months, the suggested representations will be integrated in a computing system, in order to facilitate the transposition of a spoken language representation to a gestural language representation. For that we have to explicit how to construct new representations in the cognitive system of SL and to precise the language encoding by using grammatical rules.

References

1.   Cuxac C.: *French Sign language: Proposition of a structural Explanation by Iconicity.* In Braffort A. et al. (eds.): Gesture-Based Communication in Human-Computer Interaction, LNAI 1739, Springer, pp.165-184 (1999).
2.   Taub, S. F.: Language from the body : iconicity and metaphor in American sign language. Cambridge University Press (2001).
3.   Brennan, M.: *In Sight - In Mind: Visual Encoding in Sign Language.* (Keynote Address) in Loades, C. (ed) Proceedings of the Australian and New Zealand Conference for Educators of the Deaf, Adelaide (1997).
4.   Braffort A.: *ARGo: An Architecture for Sign Language Recognition and Interpretation.* In P. Harling et al. (eds.): Progress in Gestural Interaction, Springer, pp.17-30 (1996).
5.   Sagawa H. and Takeuchi M.: *A Method for Analysing Spatial Relationships Between Words in Sign Language Recognition.* In Braffort A. et al. (eds.): Gesture-Based Communication in Human-Computer Interaction, LNAI 1739, Springer, p.197-209 (1999).
6.   Risler A.: Ancrage perceptivo-pratique des catégories du langage et localisme cognitif à travers l'étude de la motivation des signes et de la spatialisation des relations sémantiques. PhD in linguistics, Toulouse-le-Mirail University (2000). In French.

7. Desclés J.-P. et al.: *Sémantique cognitive de l'action : 1. Contexte Théorique*. In Langages n°132 "Cognition, catégorisation, langage" (1998). In French.
8. Langacker R.-W.: Foundations of Cognitive Grammar. Vol I: Theoretical Prerequisites. Stanford University Press (1987).
9. Braffort A.: *Research on computing science and sign language: ethical aspects.* In Wachsmuth I. and Sowa T. (eds): gesture and Sign language in Human-Computer Interaction, LNAI 2298, Springer (2002).

SignWriting-Based Sign Language Processing

Antônio Carlos da Rocha Costa and Graçaliz Pereira Dimuro

Escola de Informática, Universidade Católica de Pelotas
96.010-000 Pelotas, RS, Brazil
{rocha,liz}@atlas.ucpel.tche.br

Abstract. This paper[1] proposes an approach to the computer processing of deaf sign languages that uses `SignWriting` as the writing system
for deaf sign languages, and `SWML` (`SignWriting Markup Language`) as
its computer encoding. Every kind of language and document processing (storage and retrieval, analysis and generation, translation, spell-
checking, search, animation, dictionary automation, etc.) can be applied
to sign language texts and phrases when they are written in `SignWriting`
and encoded in `SWML`. This opens the whole area of deaf sign languages
to the methods and techniques of *text-oriented* computational linguistics.

Keywords: Sign language processing, SignWriting, SWML.

1   Introduction

The `SignWriting` system[2] is a practical writing system for deaf sign languages.
It is composed of a set of intuitive grahical-schematic symbols and of simple rules
for combining such symbols to represent signs. It was conceived to be used by
deaf people in their daily lives, for the same purposes hearing people commonly
use written oral languages (taking notes, reading books and newspapers, learning
at school, making contracts, etc.)

The `SignWriting` system is easy to learn and use. Thus, the development
of software supporting such system makes concrete the expectation of the appearance of a large community of deaf people interested in communicating with
computers in written sign languages, as well as it makes concrete the possibility
of the creation of large electronic corpora of *sign language texts*.

That opens the opportunity to immediately adapt and apply to sign languages the whole spectrum of text-oriented methods and techniques of computational linguistics, that exist today. That is an approach to sign language
processing which is to be contrasted to more usual approach of trying to process
sign languages in an *image-oriented* way, requiring image processing techniques
that are mostly still in development.

An appropriate encoding of the `SignWriting` symbols is necessary, in order
to allow the computer storage and processing of sign language documents, as
well as the use of written sign languages for direct human-computer interaction.

---

[1] Work with financial support from CNPq and FAPERGS.
[2] http://www.signwriting.org

I. Wachsmuth and T. Sowa (Eds.): GW 2001, LNAI 2298, pp. 202–205, 2002.
© Springer-Verlag Berlin Heidelberg 2002

That is the purpose of SWML (`SignWriting Markup Language`), an XML-based language that is being developed[3] to allow the computer-platform independent representation of sign language texts written in `SignWriting`, and to allow the interoperability of `SignWriting`-based sign language processing systems.

This short paper is as follows: section 2 hints on the `SignWriting` system. Section 3 explains the current version of SWML. Section 4 pictures the overall scenario of the `SignWriting`-based sign language processing approach.

2 The `SignWriting` System

Valerie Sutton, the inventor of the `SignWriting` system [1], took the stance that sign language notation should be visually driven and graphically displayed. Such stance came from her previous experience with the development of a *visual language* for choreographic movements[4], the `DanceWriting` system [2].

There are various groups of graphical symbols in the `SignWriting` system, corresponding to the various linguistic features of signs.

For example, figure 1 shows the sign for BRAZIL in LIBRAS, the Brazilian Sign Language. It's a right hand, in B-shape, pointing upwards, palm facing left, moving downwards so as to suggest the contour of the eastern coast of the country, over the Atlantic.

Fig. 1. The sign for BRAZIL in LIBRAS, written in `SignWriting`

3 SignWriting Markup Language

The `SignWriting Markup Language` (SWML) is an XML-based language created to allow the computer encoding of `SignWriting` and the interoperability of `SignWriting`-based sign language processing systems. The current version is `version1.0-draft2`, whose DTD is available at `http://swml.ucpel.tche.br`. Its main features are the following:

⬦ SWML can represent both `SignWriting` texts and dictionaries, as they are generated by the `SignWriter` program [3].
⬦ For every sign in the text or dictionary, there is a `<sign_box>` comprising the set of `<symbol>`s that together represent the sign.
⬦ For every `<symbol>` in a `<sign_box>`, a `"number"` attribute identifies the basic `<shape>` of the symbol, and attributes `"x"` and `"y"` identify the coordinates of the symbol within the `<sign_box>`.

---

[3] `http://swml.ucpel.tche.br`
[4] `http://www.dancewriting.org`

⋄ For every symbol, a set of attributes ("variation", "fill", "rotation" and "flop") identify the <transformation>s to which the basic symbol shape was subjected before it was included in sign box.

⋄ The final result is that the various features of any sign can be extracted from its representation in an SWML-encoded file, and its image reconstructed, if necessary.

Below is the essential part of the SWML encoding of the sign BRAZIL:

```
<?xml version="1.0" ?>
<swml version="1.0-d2" symbolset="SSS-1995">
    <generator>
        <name>SignWriter</name>
        <version>4.3</version>
    </generator>
    <sw_text>
        <sign_box>
            <!-- the B hand -->
            <symbol x="8" y="13">
                <shape number="21" fill="1" variation="1" />
                <transform flop="0" rotation="0" />
            </symbol>
            <!-- the movement -->
            <symbol x="7" y="25">
                <shape number="108" fill="0" variation="1" />
                <transform flop="1" rotation="4" />
            </symbol>
        </sign_box>
    </sw_text>
</swml>
```

4 SignWriting and Sign Language Processing

We use the term *sign language processing* to denote the application of the text-oriented methods and techniques of *computational linguistics* to deaf sign languages expressed in written form.

Such methods were originally developed to process oral languages presented in *written form*. That was quite a natural start, given both the easiness with which oral (western) languages could be represented in computer systems, and the socially determined dominance of oral languages. The extension of that work to non-western languages posed important technical problems, but has not changed the very conceptual foundation of the area, because it still targeted only languages of the oral modality.

Dealing with written forms of languages of the visual modality is interesting both because it may challenge the conceptual framework of current text-oriented methods and techniques of computational linguistics, and because it puts new technological problems for the human-computer interface area (ways to easily write signs, etc.).

The Gesture Workshop series is one of the forums where the problem of processing gestures and sign languages is tackled and discussed. Some works presented in those workshops dealt with notations for sign languages, but the notations were either linguistically oriented (e.g., based on the *Stokoe* system or on HamNoSys[5]) or computationally oriented (i.e., modelled after some programming language).

As far as we know, no other work proposed to process sign languages using signs written in a form which is at the same time practical for the common (deaf) computer user and adequate for computer processing. To attempt that is that we have engaged in the area of sign language processing.

Some initial software developments and experimentation with algorithms that we are carrying out are the following [4]:

⋄ New tools do generate and exhibit sign language texts, working either as stand-alone applications (new SignWriting editors) or in web-based environments (for webpages written in sign languages, sign language electronic mail and chats).
⋄ On-line dictionary systems for sign languages, with queries using the written forms of the signs.
⋄ Sign language document repositories, with document search and retrieval based on pieces of sign language texts.
⋄ Automated animation of written sign language sentences and dialogs.

5   Conclusion

A SignWriting-based approach to sign language processing is possible. From the point of view of the common (deaf) computer users, that approach may be highly practical and useful. The SWML format is a means for the computer encoding of sign languages texts and for the interoperability of sign language processing systems. It opens the area of deaf sign languages to the current text-oriented methods and techniques of computational linguistics.

References

1. Sutton, V. *Lessons in SignWriting - Textbook and Workbook*. Deaf Action Committee for SignWriting, La Jolla, Ca., 1999. (2nd ed.)   203
2. Sutton, V. *Sutton Movement Shorthand: a Quick Visual Easy-to-Learn Method of Recording Dance Movement - Book One: The Classical Ballet Key*. The Movement Shorthand Society, Irvine, Ca., 1973.   203
3. Sutton, V. and Gleaves, R. *SignWriter - The world's first sign language processor*. Deaf Action Committee for SignWriting, La Jolla, Ca., 1995.   203
4. Costa, A. C. R. and Dimuro, G. P. *Supporting Deaf Sign Languages in Written Form on the Web*. The SignWriting Journal, number 0, July 2001. Available on-line at http://sw-journal.ucpel.tche.br.   205

---

[5] http://www.sign-lang.uni-hamburg.de/Projects/HamNoSys.html

Visual Attention towards Gestures in Face-to-Face Interaction vs. on Screen[*]

Marianne Gullberg[1] and Kenneth Holmqvist[2]

[1] Max Planck Institute for Psycholinguistics
PO Box 310, NL-6500 AH Nijmegen, The Netherlands
marianne.gullberg@mpi.nl
[2] Lund University Cognitive Science
Kungshuset Lundagård, S-22222 Lund, Sweden
kenneth@lucs.lu.se

Abstract. Previous eye-tracking studies of whether recipients look at speakers' gestures have yielded conflicting results but have also differed in method. This study aims to isolate the effect of the medium of presentation on recipients' fixation behaviour towards speakers' gestures by comparing fixations of the same gestures either performed live in a face-to-face condition or presented on video ceteris paribus. The results show that although fewer gestures are fixated on video, fixation behaviour towards gestures is largely similar across conditions. In discussing the effect of the absence of a live interlocutor vs. the projection size as a source for this reduction, we touch on some underlying mechanisms governing gesture fixations. The results are pertinent to man-machine interface issues as well as to the ecological validity of video-based paradigms needed to study the relationship between visual and cognitive attention to gestures and Sign Language.

1 Introduction

Previous studies of visual attention to gestures based on eye-tracking have yielded very different results concerning if and when recipients look at gestures. In a study of gesture fixations in live face-to-face interaction, Gullberg & Holmqvist [1] found that recipients fixated only a minority of speakers' gestures (8.8%), chiefly perceiving gestures through peripheral vision and instead maintaining eye or face contact. Recipients only fixated gestures if they were performed in peripheral gesture space or were fixated by speakers themselves (autofixation). The dominance of the face corroborated findings from earlier studies of gaze in interaction conducted without eye-trackers [2, 3, 4]. We hypothesised that gestural performance features compete with (culture-specific) social norms for maintained eye contact in determining

---

[*] We gratefully acknowledge the financial support of Birgit and Gad Rausing's Foundation for Research in the Humanities, and of the Max Planck Society.

I. Wachsmuth and T. Sowa (Eds.): GW 2001, LNAI 2298, pp. 206-214, 2002.

whether gestures are fixated in live interaction. In Western-European societies maintained eye contact is considered as a mark of sustained interest, attention, and involvement [3, 5]. In the absence of any social pressure for eye contact, as in a video setting, fixation behaviour towards gestures might therefore change. And indeed, Nobe et al. [6, 7] found that recipients who were shown gestures on video fixated the majority of gestures (70-75%).

However, the differences found in these studies are difficult to assess given that the studies differ on a range of methodological points, including the medium of presentation (live face-to-face vs. on screen), the naturalness of gestures (natural vs. synthesised or otherwise manipulated gestures), the embeddedness of gestures in discourse (sustained discourse vs. isolated gestures listed), the type of gesture (spontaneous co-speech gestures vs. emblematic gestures), and agent (human vs. anthropomorphic agents). This study therefore aims to isolate the effect of the medium of presentation by comparing fixations of naturally occurring, spontaneous co-speech gestures [8] in story retellings under two conditions keeping all other features constant. We expect the comparison between a live condition and a video condition to yield two possible types of effects. The presence/absence of the interlocutor will reveal the interactional, social effects on fixation behaviour towards gestures. In contrast, the reduced projection size will reveal more 'mechanical' effects related to the capacities of peripheral vision. We asked the following specific questions:

- do recipients fixate the speaker's face less often on video than in a live condition?
- do recipients fixate more gestures overall on video than live?
- do recipients fixate different gestures on video than live? Or put differently, is the impact of the gestural performance features that determine fixations different on video than in a live condition?

The answers to these questions are pertinent to the study of the relationship between visual and cognitive attention to gestures and Sign Language, or the relationship between fixations and information processing. Although there is growing evidence that recipients attend to and retain information expressed in gestures (e.g. [9, 10, 11]), the relationship between visual attention to gestures and cognitive uptake of gestural information is poorly understood. This issue needs to be studied in video-based experimental designs that allow the repeated display of the same gestures to different recipients. However, video-based designs risk compromising ecological validity as long as we do not know what effect the medium of presentation has on recipients' fixation patterns. Moreover, the answers to these questions should be of interest to the field of man-machine interaction. The construction of interfaces with agents gesticulating on screen should benefit from insights into how humans attend to other humans gesticulating on screen as a first step.

2   Procedure

In the live condition, 20 native speakers of Swedish, unacquainted prior to the experiment, were randomly assigned the role of speaker or recipient. The speakers memorised a printed cartoon and were then told to convey the story as well as they could to the recipients who would have to answer questions about it later. Similarly, the recipients were asked to make sure they understood the story, since they would have to answer questions about it later. The recipients were seated 180 cm away from the speakers (measured back to back) facing them, and were wearing a head-mounted SMI iView© eye-tracker (see Figure 1). The output data from the eye-tracker consist of a video recording of the recipient's field of vision (i.e. the speaker), and a superimposed video recording of the recipient's fixations (see Figure 2).

   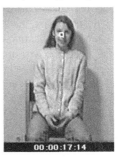

Fig.1. The SMI iView© headmounted eye-tracker

Fig. 2. Example of output from the eye-tracker: the recipient's field of vision (=the speaker) and the recipient's fixation marker (=the white circle) superimposed

While retelling the stories to the recipients, the speakers were simultaneously video recorded with a separate video camera placed behind the recipients. These video recordings of the speakers served as stimuli in the video condition. 20 new Swedish recipients were shown these video recordings on a video screen, 2 new recipients per original speaker. Note that this design allowed us to collect fixation data for exactly the same gestures presented live and on video. In addition, the design ensured that the gestures shown on video were 'natural' since they were performed by speakers facing a live recipient; the gestures were thus not performed 'for the camera'. The new recipients were seated 110 cm away from a 28" video screen. The projection size and the distance between recipient and screen decreased all angels by 52.3% of the original size. The SMI iView© remote set was placed between the recipient and the video screen. The instructions to the recipients were the same as in the live condition.

A post-test-questionnaire was distributed to all subjects to ensure that gesture was not identified as the target of study. The eye-tracker appears not to have disrupted interaction (cf. [1]). All subjects, speakers and recipients alike, declared that the equipment did not disturb them. The speakers' speech and gestural behaviour did not differ quantitatively or qualitatively from data collected in an identical situation without eye-trackers [12]. Recipients' fixation data include fixations of body parts that the subjects might have avoided to fixate had they been concerned about the equipment. This in turn suggests that the recipients tended to forget about the apparatus and behaved naturally.

3   Data Treatment

Speech was transcribed and checked for demonstrative expressions referring directly or indirectly to the gestures, e.g. 'he held it like this'. Such demonstrative expressions function as an interactional deictic device by which speakers can direct recipients' attentions towards the gestures [6, 13]. No such deictic expressions were present in the data.

Both a temporal and a spatial criterion were used to identify fixations. The fixation marker had to remain in an area the size of the marker itself for at least 120 ms (=3 video frames) in order to count as a fixation. The fixation data were also coded for the target object fixated, e.g. the right hand gesturing, the shoulder, an object in the room, etc. (see Figure 3).

The gesture data were coded for the three performance features that have been found to attract fixations in previous studies:

a) place of articulation in gesture space using McNeill's [8] schema of space (e.g. centre-centre, peripheral right). For the purpose of calculations, all cases of centre-centre and centre were collapsed, as were all peripherals, leaving two broad categories Central and Peripheral.

b) autofixation, or whether or not speakers fixate their own gestures (see Figure 4). All cases of autofixation were enacted gestures (character viewpoint in McNeill's terms, [8]), where the speakers acted as a character in the story.

c) presence vs. absence of hold, i.e. a momentary cessation of movement in a gesture [14, 15]. Post-stroke holds, i.e. cessations of movement after the hand has reached the endpoint of a trajectory, were found to attract fixations in the video-based studies by Nobe et al. [6]. We wanted to test their relevance in a live setting. Note that non-hold means all other phases of the gesture phrase, i.e. preparation, stroke, or retraction.

The data is considered to have a binomial distribution. Proportions of fixated gestures are calculated relative to produced gestures in the various categories, whereby the quantitative variations in gesture production are neutralised. We employ a test of significance of differences between proportions that is mathematically equivalent to the chi-square test under one degree of freedom.

Fig. 3. Example of a fixation coded for target Body part (low abdomen)

Fig. 4. Example of autofixated gesture. The gesture is also fixated by the recipient (=white circle)

4   Results

Do Recipients Fixate the Speaker's Face less Often on Video than Live?

Table 1 shows that recipients spend less time fixating the face in the video condition than in the live condition, although not significantly so (from on average 95.6% of the time in the live condition to 88.7% in the video condition). Despite this decrease, the face nevertheless clearly dominates as a fixation target in both conditions.

Table 1. Average time in percent spent fixating face vs. outside face across conditions

	Live	Video	Comparison
Average time on face %	95.6	88.7	n.s.

Do Recipients Fixate More Gestures Overall on Video than Live?

Table 2 indicates that only a minority of gestures are fixated in either condition, on average 7.4% vs. 3% of all gestures. The amount of fixated gestures in the live condition corresponds well to the findings in Gullberg & Holmqvist [1] (8.8%). Contrary to expectations, however, recipients fixate significantly *fewer* gestures in the video condition than in the live condition.

Table 2. Proportion of fixated gestures across conditions

	Live	Video	Comparison
# gestures	364	734[*]	
# fixated gestures	27	22	
Fixated gestures %	7.4	3	p=.0025

The combination of more time spent outside the face but less time on gestures in the video condition means that other targets receive more fixations. Table 3 lists the significant changes in extra-facial fixation targets between the conditions. In the video condition, the amount of extra-facial fixations landing on gestures decreases significantly (from 13.8% to 3.6%), as well as fixations of objects in the room (from 45.9% to 21.2%). In contrast, fixations of body parts other than gesticulating hands or arms, i.e. immobile body parts such as the abdomen, the chest area, etc., increase significantly (from 35.2% to 57.4%).

---

[*] Note that the total number of gestures in the video condition exceeds the number of gestures in the live condition times two ((364*2)=728 vs. 734). There was a technical problem with the recording of one of the live speakers, such that it could not be projected in the video condition. Instead, we projected the recording of another live speaker, who produced roughly the same amount of gestures in the live condition as the original speaker.

Table 3. Proportions of extra-facial fixation on three targets across conditions

Extra-facial fixation targets %	Live	Video	Comparison
Gestures	13.8	3.6	p≤.025
Objects in the room	45.9	21.2	p≤.01
Immobile body parts	35.2	57.4	p≤.05

Do Recipients Fixate Different Gestures on Video than Live?

Table 4 shows the extent to which the three gestural performance features attract fixations across the conditions. Despite the overall decrease in gesture fixations in the video condition, the proportional attraction strength of the features is maintained such that the gestural performance features can be said to operate across conditions. As can be seen in Table 4, there is no effect for Place of articulation in either condition, meaning that Peripheral gestures are *not* fixated more often than central gestures. In contrast, both Holds and Autofixations attract significantly more fixations than other gestures in both conditions.

Table 4. Fixated gestures in % with respect to performance features across conditions

Performance features	Live		Video		Live vs. video
	fix %		fix %		
Peripheral/Central	8/7	n.s.	4/2	n.s.	n.s.
Hold/no Hold	33/4	p≤.01	15/2	p≤.025	n.s.
Autofix/no Autofix	23/5	p≤.05	8/2	p≤.05	n.s.

5   General Discussion and Conclusion

The results show that fixation behaviour is both similar and different across the conditions. The similarities include the face dominance and the tendency to fixate only a minority of gestures. Similarly, the two articulatory features that reliably attract fixations, Holds and Autofixations, operate in both conditions. The differences consist of a significant reduction of face fixations and a significant increase in fixations of body parts in the video condition. In addition, there is an unexpected overall significant reduction in gesture fixation rate. This decrease is the only significant difference in fixation behaviour towards gestures across the conditions. The question then is what causes it.

We originally hypothesised that the absence of a live interlocutor on video would lead to relaxed social pressure for maintained eye contact with a reduction of face fixations and an increase in gesture fixations as a direct result. The first prediction was borne out, since the results indicate that the norm for face fixations is indeed relaxed. However, recipients at liberty to fixate whatever they want do not fixate more gestures. Instead they fixate more immobile body parts, i.e. they allow themselves to fixate areas that are probably not 'socially acceptable' fixation targets in a live condition (cf. [5]). The general change in face and body fixations thus seems clearly socially motivated. A tentative social explanation for the decrease in gesture fixations, related to the tendency to fixate only 'socially acceptable' targets in a live condition, could be that fixations of gestures are the only 'legitimate' reason to fixate something other than the speaker's face in a live condition. In a video situation without any social constraints on fixation behaviour, recipients can ignore gestures to a greater extent and only fixate their 'real' targets of interest.

However, the decrease in fixation rate may also be related to the reduced projection size of the video condition. The arguments related to this 'mechanical' factor are associated with the capacities of peripheral vision by which fixations occur when the target is too far removed from the fixation point. Under this reading, the video condition would lead to fewer gesture fixations simply because the reduced visual field means that recipients can rely on their peripheral vision to a greater extent than live. In the video condition at hand, the projection size decreased all angels by 52.3% of the original size. If the reduction in gesture fixations were determined by size projection alone, the reduction in fixation rate between conditions should match this number [16]. However, the general reduction in gesture fixation between the conditions is 67.9% (from 7.4% of all gestures fixated live to 3% of all gestures fixated on video), which leaves 15.3% of the reduction unexplained. When the individual gestural performance features are considered, we find that the reduction in fixation rate for Peripherals (50.2%) and Holds (55.7%) is close to the value of the size reduction. In contrast, for Autofixation the reduction in fixation rate is somewhat greater (64%), leaving 11.7% of the reduction unexplained. These figures suggest that projection size accounts for the brunt of the reduction, but that different gestures are affected differently by the move from a live to a video condition such that some gesture types may be affected by social factors as well. Ultimately, this differential effect may reflect the fact that different gestures are fixated for different reasons.

As seen above, the reduction of fixations of Holds is well accounted for by the size reduction. This suggests that the tendency to fixate Holds is associated with the capacities of peripheral vision. In a live condition, Holds in fact represent a challenge to peripheral vision, which is good at motion detection but bad at fine-grained detail. As long as gestures are moving, peripheral vision is sufficient for detecting (and processing) the broader gestural information (location, direction, or size) even when gestures are performed in the periphery. This is, incidentally, the likeliest explanation for why peripherally performed gestures are in fact not fixated more often than centrally performed ones in this study. However, when a gesture ceases to move as in a hold, peripheral vision is challenged. In a live condition where recipients mainly fixate the speaker's face, their peripheral vision may tell them that the speaker's hand is not in a resting position. Peripheral vision is insufficient to retrieve information from a gesture in hold, as it can pick up neither motion nor configurational detail such as hand shape from it. Speakers may therefore need to fixate holds in order to retrieve any gestural information at all. In the video condition, on the other hand, the distance between the gesture and the fixation of the face on the screen is presumably short enough for peripheral vision to operate efficiently despite the lack of movement. If holds are fixated for reasons of limitations to peripheral vision, then they may be affected by size changes in the video condition.

In contrast, the size reduction did not account for the entire decrease in fixation rate for Autofixations. Autofixation is not a purely articulatory feature, and is probably fixated for different reasons than Holds. It is not the gesture itself that attracts the fixation, but the fact that the speaker's gaze is on it. Speakers' gaze serves as a powerful social cue to joint attention [17,18,19]. As a consequence, Autofixation may be perceived as a means for the speaker to direct recipients' attention towards gestures [20,21]. In this sense, Autofixation is a social phenomenon. Not to co-fixate an autofixated gesture in a live condition would be socially inept. It is in fact common for speakers who autofixate their gestures to look back up on the recipient to ensure that joint attention has indeed been established. In a video condition, there is no such social pressure to follow an Autofixation. The decrease in fixations of Autofixations on video may thus be partly due to the increased capacities of peripheral vision following from the reduced projection size, and partly to the absence of a live interlocutor.

What are the implications of these findings for the ecological validity of a video-based paradigm used to study gesture perception and gesture processing? The social change between conditions, i.e. the absence of a live interlocutor, mainly affects face and body fixations, and not (or only partly) gesture fixations. In contrast, the mechanical change, i.e. the reduction in projection size, does affect gesture fixations, leading to fewer gestures being fixated overall. A life-size projection should eliminate most of this reduction in gesture fixation rate. We therefore suggest that, as a first evaluation, a video-based paradigm need *not* compromise ecological validity. However, in order to isolate the true domains of the social and the mechanical size effect for all gesture types, and especially for autofixated gestures, a direct comparison should be made between a live condition and a video condition where stimuli are projected life-size. Such a study should also reveal more about the general underlying mechanisms governing gesture fixations. In particular, it should inform us on the relationship between fixations prompted by or leading to information retrieval and processing, and fixations that occur merely as a reaction to changes in the field of vision.

References

1. Gullberg, M., Holmqvist, K.: Keeping an Eye on Gestures: Visual Perception of Gestures in Face-to-Face Communication. Pragmatics & Cognition 7 (1999) 35-63

2. Argyle, M.: The Psychology of Interpersonal Behaviour. Penguin, Harmondsworth (1976)

3. Kendon, A.: Conducting Interaction. Cambridge University Press, Cambridge (1990)

4. Nielsen, G.: Studies in Self Confrontation: Viewing a Sound Motion Picture of Self and Another Person in a Stressful Dyadic Interaction. Munksgaard, Copenhagen (1962)

5. Fehr, B., Exline, R.: Social Visual Interaction: A Conceptual and Literature Review. In: Siegman, A., Feldstein, S. (eds.): Nonverbal behavior and communication. 2nd ed. Erlbaum, Hillsdale, NJ (1987) 225-326

6. Nobe, S., Hayamizu, S., Hasegawa, O., Takahashi, H.: Are Listeners Paying Attention to the Hand Gestures of an Anthropomorphic Agent? An Evaluation Using a Gaze Tracking Method. In: Wachsmuth, I., Fröhlich, M. (eds.): Gesture and Sign Language in Human-Computer Interaction. Springer Verlag, Berlin (1998) 49-59

7. Nobe, S., Hayamizu, S., Hasegawa, O., Takahashi, H.: Hand Gestures of an Anthropomorphic Agent: Listeners' Eye Fixation and Comprehension. Cognitive Studies. Bulletin of the Japanese Cognitive Science Society 7 (2000) 86-92
8. McNeill, D.: Hand and Mind. Chicago University Press, Chicago (1992)
9. Cassell, J., McNeill, D., McCullough, K. E.: Speech-Gesture Mismatches: Evidence for one Underlying Representation of Linguistic and Nonlinguistic Information. Pragmatics & Cognition 7 (1999) 1-33
10. Beattie, G., Shovelton, H.: Do Iconic Hand Gestures Really Contribute Anything to the Semantic Information Conveyed by Speech? Semiotica 123 (1999) 1-30
11. Beattie, G., Shovelton, H.: Mapping the Range of Information Contained in the Iconic Hand Gestures that Accompany Spontaneous Speech. J. of Lang. and Social Psy. 18 (1999) 438-462
12. Gullberg, M.: Gesture as a Communication Strategy in Second Language Discourse. Lund University Press, Lund (1998)
13. Streeck, J., Knapp, M. L.: The Interaction of Visual and Verbal Features in Human Communication. In: Poyatos, F. (ed.): Advances in Nonverbal Communication. Benjamins, Amsterdam (1992) 3-23
14. Kendon, A.: Some Relationships between Body Motion and Speech: An Analysis of an Example. In: Siegman, A.W., Pope, B. (eds.): Studies in Dyadic Communication. Pergamon, New York (1972) 177-210
15. Kendon, A.: Gesticulation and Speech: Two Aspects of the Process of Utterance. In: Key, M. (ed.): The Relationship of Verbal and Nonverbal Communication. Mouton, The Hague (1980) 207-227
16. Latham, K., Whitaker, D.: A Comparison of Word Recognition and Reading Performance in Foveal and Peripheral Vision. Vision Research 37 (1996) 2665-2674
17. Baron-Cohen, S.: The Eye Direction Detector (EDD) and the Shared Attention Mechanism (SAM): Two cases for evolutionary Psychology. In: Moore, C., Dunham, P.J. (eds.): Joint attention. Erlbaum, Hillsdale (1995) 41-59
18. Langton, S. R. H., Watt, R. J., Bruce, V.: Do the eyes have it? Cues to the direction of social attention. Trends in Cognitive Sciences 4 (2000) 50-59
19. Tomasello, M.: Joint attention as social cognition. In: Moore, C., Dunham, P. J. (eds.): Joint attention. Erlbaum, Hillsdale (1995) 103-130
20. Streeck, J.: Gesture as Communication I: Its Coordination with Gaze and Speech. Communication Monographs 60 (1993) 275-299
21. Goodwin, C.: Gestures as a resource for the organization of mutual orientation. Semiotica 62 (1986) 29-49

Labeling of Gestures in SmartKom –
The Coding System

Silke Steininger, Bernd Lindemann, and Thorsten Paetzold

Institute of Phonetics and Speech Communication
Schellingstr. 3, 80799 Munich, Germany
kstein@phonetik-uni-muenchen.de

Abstract. The SmartKom project is concerned with the development of
an intelligent computer-user interface that allows almost natural
communication and gesture input. For the training of the gesture
analyzer, data is collected in so called Wizard-of-Oz experiments.
Recordings of subjects are made and labeled off-line with respect to the
gestures that were used. This article is concerned with the coding of
these gestures. The presented concept is the first step in the
development of a practical gesture coding system specifically designed
for the description of communicative and non-communicative gestures
that typically show up in human-machine dialogues. After a short
overview over the development process and the special requirements
for the project, the labels will be described in detail. We will conclude
with a short outline of open points and differences to the well-known
taxonomy for gestures of Ekman [1]. Keywords: Human-Machine
interaction, annotation of corpora, multimodal dialogue systems,
gesture coding system.

1 The SmartKom Project

The goal of the SmartKom[1] project is the development of an intelligent computer-user
interface that allows almost natural communication between human and machine. The
system does not only allow input in the form of natural speech but also in the form of
gestures. Additionally the emotional state of the user is analyzed via his/her facial
expression and prosody of speech. The output of the system comprises a graphic user
interface and synthesized language. The graphic output is realized as a computer
screen that is projected onto a graph tablet.

To explore how users interact with a machine, data is collected in so-called
Wizard-of-Oz experiments: The subjects have to solve certain tasks with the help of
the system (e.g. planning a trip to the cinema). They are made believe that the system
they interact with is already fully functional. Actually, many functions are only
simulated by two "wizards", who control the system from a separate room. The

---

[1] http://smartkom.dfki.de/index.html

I. Wachsmuth and T. Sowa (Eds.): GW 2001, LNAI 2298, pp. 215-227, 2002.

different functionalities of the system are developed by different partners of the project. The Institute of Phonetics and Speech Communication in Munich is responsible for the collection and annotation of the multimodal data and the evaluation of the system.

1.1   Collection of Multimodal Data

In each Wizard-of-Oz session spontaneous speech, facial expression and gestures of the subjects are recorded with different microphones, two digital cameras (face and sideview hip to head) and an infrared sensitive camera (from a gesture recognizer: SIVIT/Siemens) which captures the hand gestures (2-dimensional) in the plane of the graphical output. Additionally, the output to the display is logged into a slow frame video stream.

Fig. 1. The 4-view-videostream for the coding of the gestures

Each subject is recorded in two sessions of about 4.5 minutes length each. To facilitate the coding, the display recording is overlaid with the stream of the infrared sensitive camera (see figure 1) and combined with front view, side view and display recording alone.[2] For the gestural coding the program Interact[3] is used.

2   Starting Points of the Development Process

2.1   Practical Requirements

The coded gestures in the project are needed for the development of a gesture recognizer, as well as for a model which is able to predict typical human-machine interactions. Therefore, it is of interest which gestures are used in this special context and in which way. The coding system thus has to provide material for the training of

---

[2] The front view is added for the labeling of the facial expression. It does only play a marginal role for the gesture coding.

[3] http://www.mangold.de/

the gesture analyzer, but it also has to mark gestural episodes that are perhaps too complex to be useful yet for automatic detection but may be interesting with respect to the nonverbal communicative behavior in general. The following requirements are the result of these considerations:

1. The labels should refer to the *functional level*[4], not the morphological level. For theoretical reasons we want to use a functional coding system (see below). However, the decision is also made for practical reasons since the coding of the morphological form of gestures ("phenomenal" systems as well as "structural" classification systems) and body movements is exceedingly time consuming.
2. The labels should be *selective*. Functional codes (as indirect measurements) are not as exact as direct methods, therefore exceptional care has to be taken to find labels that are well-defined, easy to observe and unproblematic to discriminate by means of objective (communicable) criteria.
3. The coding system should be *fast and easy to use*.
4. The resulting label file *should facilitate automatic processing* (a consistent file structure, consistent coding, non-ambiguous symbols, ASCII, parsability). Preferably it should be easy to read.[5]
5. The main labels and most modifiers should be realized as *codes*, not as annotations, in order to heighten consistency. Annotations (free comments and descriptions that don't follow a strict rule) are more flexible, but codes (predefined labels from a fixed set) increase the conformity between labelers.

2.2   Theoretical Considerations

We assume that existing taxonomies are not very well suited for our task of labeling human-machine interaction because they were developed with respect to human-human interaction [4,5,6,7,8,9]. Others are too specific for our need of a practical system (for example methods for transcribing American Sign Language - see SignStream[6] or methods for transcribing dance [10]). So how should a system look like that is exactly tailored to our context?

In order to be able to study the nonverbal aspects of human-machine interaction one needs labels that correspond to units that have meaning in a (human-machine) dialogue. It can be expected that gestures are used differently when the dialogue partner is a machine that communicates with synthesized speech and graphic output - similar to the changes that take place in speech in such a context [11]. For example, a lot more deictic gestures can be expected and it is not clear if other illustrative gestures will show up at all.

Because the function or meaning of the gesture within the communication process is the interesting point in our context, we decided to define labels on a functional

---

[4] "Functional code" or "functional unit" is sometimes defined differently by different authors. We use the term in accordance with Faßnacht [2] for a unit that is defined with regard to its effect or its context.

[5] Many of the practical criteria were adopted from the transliteration conventions for speech in SmartKom, see [3].

[6] http://web.bu.edu/asllrp/SignStream/

level.[7] Functional systems emphasize the connection with speech [8] which is of special interest to us. Additionally, a morphological classification system (with a great amount of categories) is ill suited to allow statistical analyses and inter-individual comparisons [7]. Most importantly, with morphological labels it is not possible to signify the difference between communicative and non-communicative gestures.

In order to still be able to catch some of the morphological information we decided to include morphological modifiers. They roughly describe the form, duration and stroke of the gesture. By including some aspects of structural or "micro" coding systems we hopefully can work around some of the disadvantages of a purely functional system. But since the morphological descriptions are not the main criterion in defining the labels, we benefit from the fact that the labels highlight the point which is most interesting for us: The intent with which the gesture is performed.

3   Definition of the Coding System

To match our practical needs and theoretical considerations with the context that we have to deal with (a multimodal human-machine dialogue), we did not start with explicit definitions or an existing taxonomy for gestures, but with a heuristic analysis of a number of recordings of human-machine dialogues. Such a free, unsystematic observation can lead to valuable insights for the definition of a system that allows systematic observations [2]. The recordings that served as the basis for the analysis are described below.

No definite taxonomy was used as a basis for the analysis, but the theoretical frame outlined below. Two observers[8] separately listed every episode that they could identify as a "functional unit". Then they tried to assign a label to the unit with respect to the obvious meaning of the gesture ("meaning" with respect to the communication process). For obvious reasons, it was not attempted to ascertain the true intent with which the gesture was performed but the intent that could be identified by an observer. Identifiable units with no obvious meaning for the communication process were listed as well. The main questions for the list of units were: Which functional/non-functional units (with respect to the dialogue) can be identified (e.g. "this", "no", "back", "non functional" etc.)? What is the best way to categorize the observed units? What is the best way to describe the units? Is there a reference on the display - if so which one? Which broad morphological form does the unit have? How can we define beginning and end of a given unit? Is there a reference word in the audio channel and if so which one?

The observers made their judgments independently. After completing the list, the identified gestures were discussed. Every label that could not be operationalized satisfactorily was removed. Similar labels were combined. Three broad categories emerged, each with several sub-categories, called labels. After this, some additional

---

[7] Well-known functional classifications are from Ekman and Friesen [7] (the newest version being described in [1]) and from McNeill [9]. Older systems are from Efron [4], Krout [12], Rosenfeld [13], Mahl [14], Freedman et al. [15] and Sainsbury & Wood [16].

[8] One of the observers was trained in observing and judging movements (a dancing master) the other one had no previous experience in the field (a student of German and English).

labels were added that had not been observed but were thought probable to appear or had to be added in order to complete the categories. Before giving a detailed description of how the labels have been defined so far, we will describe the data that served as the basis for the analysis.

3.1   The Empirical Basis

3.1.1   The Subjects

The data we used consisted of 70 sessions of about 4.5 min length each. 35 voluntary, naive subjects were recorded and paid a small recompense. 18 subjects were female, 17 male. They were between the age of 19 and 60. Occupations: 24 students (including 2 Ph.D. students, 1 pupil) and 11 employed persons (including 1 retired person).

3.1.2   The Task

A full account of the Wizard-of-Oz procedure cannot be given here, it will be published separately.[9] Instead we give a short overview of the task and the performance of the subjects. The subjects were carefully instructed. It was not shown to them what sort of gestures could be analyzed, it was only pointed out that the system understood movements of the hand which were performed on or above the display. The subjects had to imagine themselves being in Heidelberg (a German town) and using a new information booth for the first time. They were told to "Plan a trip to the cinema this evening". After the instruction two sessions were recorded (with a short break in between). Afterwards the subjects were led into a different room, where they were interviewed. They were, for example, asked about problems, what they had liked and disliked about the system, if it seemed more like a computer or a machine to them etc. Most subjects thought that the use of the system was easy. About two third thought the system was more like a machine, the rest judged it to be either something in between or rather more human. Even of the latter subjects none reported suspicions that it actually was human.

4   The Coding System

In the following the system that emerged from the qualitative analysis of the data and the first refining step will be described. It serves as the basis for the second step of the development process: The actual labeling of a set of dialogues and the calculation of quality measures, which will be undertaken if enough additional data is available.

4.1   Overview

A label, that belongs to one of three superordinate categories, is assigned to each identified gesture. The label is complemented by several modifiers. Three modifiers

---

[9] For information on the graphic user interface see "Evoking Gestures in SmartKom - Design of the Graphic User Interface" (Beringer) in this volume.

refer to aspects of time (beginning, end, stroke), while four refer to aspects of content (morphology, reference word, reference zone, object). If necessary, an identified gesture is specified by a comment. Beginning and end are marked as points in time, the stroke is marked as a period. Explanatory notes can be added (see below). The reference word is an annotation, i.e. the word in the audio channel that corresponds to the gesture is quoted. Additionally, it is marked whether the reference word showed up before, simultaneously or after the gesture. For "morphology" there exist codes, as well as for "reference zone". "Object" specifies if the targeted object was hit or not.

4.2   Definition of the Unit

The length of the segment that someone identifies as a single gesture and the chosen category are arbitrary in the sense that they don't result from observation, but conform to the theoretical foundation (such a theoretical foundation can also be the common understanding of what a "gesture" is). For a more complete discussion of the problem of the definition of a unit see Faßnacht [2]. Because the coding conventions for SmartKom are especially designed for the labeling of gestures during the human-machine dialogue, we decided to define our units from a functional definition with three broad categories: Is the gesture an interaction with the system (an "interactional gesture"), a preparation of an interaction (a "supporting gesture") or something else (a "residual gesture")? The only (functional and supporting) gestures that are labeled are the ones that enter the so called "cubus", the field of the display and the room above the display where the SIVIT gesture recognizer records data (it corresponds roughly to the border of the display).[10] Additionally, gestures that are interesting (but take place outside the »cubus«) are coded in the category of residual gestures.

4.3   Definitions

Each gesture is assigned to one of the three following categories: Interactional gesture (I-gesture), SUpporting gesture (U-gesture)[11] and Residual gesture (R-gesture). The criterion for this assignment is the intention of the subject. However, the goal of the labeling process is not to label the "true" intention of the subject. We think it is quite sufficient to retrieve the intention that gets through to the communication partner (in this case the system).

Interactional Gesture: The interactional gesture is (possibly together with the verbal output) the means of the interaction with the computer. When a subject uses a hand/arm-movement to request something from the computer this gesture is called interactional. A second type of interactional gesture is the confirmation of a question from the system.

Supporting Gesture: The supporting gesture occurs in the phase when a request is prepared. It signifies the gestural support of a "solo-action" of the user (like reading or

---

[10] In reality the "cubus" has the form of a pyramid. For practical purposes of the coding this can be ignored however.

[11] We called the supporting gesture U-gesture for reasons of consistency with the originally German name "Unterstützende Geste".

searching), e.g. an action that is not an interaction with the system (interaction in the sense of communication). Before a gesture is assigned to the category "supporting gesture", and to distinguish it from an "interactive gesture", at least one (or both) of the following prerequisites has to be confirmed as true:

a)   The gesture is accompanied by verbal comments from the user that make the preparational character of the underlying cognitive process obvious (e.g. "hm ... where is... here?... well...")

b)   The gesture is accompanied by a facial expression that makes the preparational character of the underlying cognitive process obvious (like lip movements, searching eye movements, frowning).

In order to be able to differentiate a supporting gesture from a residual gesture at least one (or both) of the following two prerequisites have to be given:

a)   A linkage between gesture and speech (both streams refer to the same topic).

b)   An identifiable focus, i.e. gaze and gesture fall on the same spot.

Residual Gesture: This category subsumes all gestures that take place within the cubus, but do not belong to one of the above categories. The few of the labeled gestures that take place outside of the cubus belong to this category, too. A residual gesture does not prepare a request (at least not obviously) and is not a request or confirmation. A residual gesture is either an emotional gesture or an unidentifiable gesture.

When an unidentifiable gesture is given, there exists

a)   no linkage between gesture and speech

b)   no identifiable focus, i.e. gaze and gesture don't fall on the same spot.

Label: The label signifies the function of the gesture within the communication process. To facilitate the assignment, the name contains one of the prefixes I-, U- and R-, which stand for the three categories.

*   The following I-labels exist: I-circle (+), I-circle (-), I-point (long +), I-point (long -), I-point (short +), I-point (short -), I-free.

*   The U-labels: U-continual (read), U-continual (search), U-continual (count), U-continual (ponder), U-punctual (read), U-punctual (ponder).

*   The R-labels: R-emotional (+ cubus), R-emotional (- cubus), R-unidentifiable (+ cubus).

Morphology: The term "morphology" here signifies which hand and which finger was used to perform the gesture. Some peculiarities (like double pointing) are noted as comment (see below).

Reference Word: This modifier signifies the word that is spoken in correspondence to the gesture. It is known that "hand gestures co-occur with their semantically parallel linguistic units" [9]. "Correspondence" therefore means that the word occurs shortly before, simultaneously or shortly after the gesture and gesture and word are linked with regard to the content.

Reference Zone: This modifier signifies the region of the display to which the gesture refers - the middle or one of the four corners (see figure 1). It can also be "none" or "whole display". "Whole display" is labeled if the hand/finger moves through at least two reference zones.

Stroke: This modifier signifies the "culmination point" of the gesture, i.e. the "most energetic" or "most important" part of a gesture that is often aligned with the intonationally most prominent syllable of the accompanying speech segment [17], [9]. Its beginning and end are labeled.

Beginning: The beginning of a gesture is determined with the help of the following criteria: 1. Hand enters cubus, 2. End of previous gesture or 3. Hand moves after pausing within the cubus for a certain time.

End: The end of a gesture is determined with the help of the following criteria: 1. Hand leaves cubus, 2. Beginning of following gesture, 3. Hand stops moving within the cubus.

Comment: Comments about the stroke, beginning, end or peculiarities that cannot be noted anywhere else.

4.4   Descriptions of the Labels

4.4.1   I-Label

I-circle (+), I-circle (-): A *continuous* movement with one hand that specifies one or more objects on the display. The display is touched (+)/not touched (-). The whole hand or parts of the hand circumscribe one or more objects/a certain region on the display fully or almost fully or stroke over it/them. It doesn't have to be a circular movement. It is always a well directed movement. At least during a part of the gesture the gaze of the subject is aimed at the chosen region. Possibly, the intent to select something becomes apparent from the words that are spoken.

I-point (long +), I-point (long -): A *selective* movement of one hand that specifies an object on the display, with the display being touched (+)/not touched (-). The stroke of the movement (i.e. the pointing of the hand at the object) is of *extended* duration (20 frames or more)[12]. The whole hand or parts of the hand are pointed at a certain spot on the display. The movement is well-directed. At least during a part of the gesture the gaze of the subject is aimed at the chosen region. Possibly, the intent to select something becomes apparent from the words that are spoken.

I-point (short +), I-point (short -): As I-point (long +/-), but the stroke of the movement (i.e. the period, during which the hand is pointed at the object) is not extended/is *very short* (up to 19 frames).

I-free: A movement of the hand/the hands which takes place above the display and signifies a wish or command of the user (for example to go to another page). This gesture can vary considerably with respect to its morphology. For example a "pageturn" can look like waving or the mimicking of turning a page. A "stop" can be

---

[12] The timespan of 20 frames was chosen because it separated the qualitatively different long and short pointing gestures very well.

a decisive waving with both hands. The display is not touched. One or both hands make a movement above the display which can be identified as a unit. The movement obviously has a function in the communication process (the context makes the intent of the user obvious). The subject's gaze is directed towards the display.

Fig. 2. A pointing gesture during which the display is not touched - I-point (short -)

4.4.2   U-Label

U-continual (read): A *continuous* movement of the hand above or on the display. No object is specified. At least during a part of the movement the subject is obviously *reading*. The whole hand or parts of the hand move continuously above or on the display, possibly sometimes pausing shortly in-between. At least during a part of the gesture the gaze of the subject follows a textline. There is no circling of a region. The movement can be well-directed, but also tentative. The gaze of the subject is at least for a part of the gesture directed towards the text. At least during a part of the movement the subject obviously reads, i.e. the text is read aloud or the lips move.

U-continual (count): A *continuous* movement of the hand above or on the display. No object is specified. At least during a part of the movement the subject is obviously *counting*. The whole hand or parts of the hand move continuously above or on the display, possibly sometimes pausing shortly in-between. The movement does not stop on the counted objects (a stop is counted as an interactional gesture - a selection). A full circling of the region does not take place. The movement can be well-directed, but also tentative. At least during a part of the gesture the gaze of the subject is aimed at the spot where the hand is placed. At least during a part of the movement the subject obviously counts, i.e. numbers are spoken or formed with the lips.

U-continual (search): A *continuous* movement of the hand above or on the display that moves over *a large part of the display*. No object is specified. The whole hand or parts of the hand move continuously above or on the display, possibly sometimes pausing shortly in-between. The hand moves through at least two reference zones. The movement can be straight or curved. An interactional gesture can follow but this is not mandatory. If a full circling takes place it is obvious from the context that the circled object is not selected (e.g. from a verbal comment, "should I choose this?"). The movement can be well-directed, but also tentative. At least during a part of the gesture the gaze of the subject is aimed at the display (towards the spot where the hand is placed). From the context it is obvious that the subject is looking for something or preparing a request, i.e. through the spoken words and/or a following interactional gesture (selection).

U-continual (ponder): A *continuous* movement of the hand above or on the display that is focused on a certain part of the display. No object is specified. Like U-continual (search) with the difference that the hand/finger moves only within one reference zone. From the context it is obvious that the subject is preparing a request, i.e. through the spoken words and/or a following interactional gesture (selection).

U-point (read): A *selective* movement of the hand, which is aimed at a text. The duration is short. It is obvious that the subject does not want to select the text. The subject obviously *reads* the text. The whole hand or parts of the hand are aimed at a specific spot on the display (text). The movement is well-directed. The gaze of the subject rests on the text. It is obvious that the subject reads, i.e. the text is read aloud or the lips move. From the context it is obvious that the movement is not a selection, as it is verified by the spoken words of the subject.

U-point (ponder): A *selective* movement of the hand to a spot on the display. The duration is short. The whole hand or parts of the hand are aimed at a specific part of the display. The movement is well-directed. The gaze of the subject rests on the object/region. The subject does not read. From the context it is obvious that the movement is not a selection, as it is verified by the spoken words of the subject.

4.4.3   R-Label

R-emotional (cubus +): A movement of one or both hands within the cubus with an obviously emotional content. The movement is neither a functional nor a supporting gesture. The movement of the hand/the hands form a unit. The gesture conveys the impression of a certain emotion (anger/irritation, joy/amusement, surprise, puzzlement).

Fig. 3. An emotional gesture (puzzlement) - R-emotional (cubus +)

R-emotional (cubus -): Like R-emotional (cubus +), but outside of the cubus.

R-unidentifiable (cubus +): A movement of one or both hands within the cubus that cannot be assigned to any of the other labels.

5   Discussion

We have mentioned before that we did not want to adapt existing taxonomies of body movements in the literature. Some systems were too detailed to fulfill our need for a fast and easy to use system or were too general for our specific task of coding gestures in the context of a human-machine dialogue. Additionally, these systems are used for human-human interaction and we felt that the interaction with a machine was a very different form of dialogue than a dialogue with a human being [18]. Therefore, we decided to develop a system from scratch, although it of course has a lot of similarities with existing taxonomies. To discuss the peculiarities of our system we point out some differences and similarities to one of the best known taxonomies, the one from Ekman [1].

Ekman distinguishes emblems (movements with precise meaning), illustrators (movements that illustrate speech), manipulators (manipulations of the face/body that are on the edge of consciousness like scratching, stroking etc.), regulators (movements that have the sole purpose of regulating the speech-flow) and emotional expressions. We don't code manipulators and regulators, because they are not of interest in our context. Furthermore, our system is not well suited to code such gestures. This is true for all relatively complex gestures (see I-free). The inclusion of spatial movements (depict spatial relationship) for example could be problematic. They would be of great interest in our context ("no, no - this big"). If and how such gestures can be included in a framework like ours has to remain open as long as there is so little data on gestures during a human-machine dialogue.

Emblems could be included, but we decided against it. Although emblems and illustrators can be distinguished reliably [19], there are severe difficulties with the less obvious cases [8]. Therefore we will code emblems (if they show up) with the system described above and include the information about the "emblematic meaning" as a comment. We think that it is more useful to specify the meaning of a gesture within a certain context than to try and find an "absolute" meaning. Emotional Expressions are also difficult to detect reliably. We included a label because of the interest for the project, but we assume that the inter-labeler reliability will be relatively low.

The focus of our system are illustrators. However, we classify illustrators somewhat differently than Ekman: For example deictic movements (pointing gestures) seem to correspond (widely) to our "interactional gestures". However, we included deictic gestures in the category of supporting gestures as well. Why was this done? As has already been mentioned, one of our main goals is to define meaningful units within a human-machine dialogue. In this situation a deictic gesture can have the meaning of a selection ("ok, I want this"), but in some cases people point at something while they are still making up their minds if they really want the thing indicated ("hm, should I go to this cinema") or point to figure something out ("ah - this restaurant is near my place - or isn't it?"). It is important to know how often these different kinds of deictic gestures show up and if and how they can be distinguished from real selections. Ideographs (sketch a direction of thought) correspond (widely) to our "supporting gestures". Ideographs normally take place in front of the thorax. We assume that the supporting gestures we describe are such "thought sketching" gestures. Since the subjects think about the things presented on the display, they move their hands in this area (and not in front of their body as in a human-human interaction). We suppose that a display encourages the subjects "to think with the hands" (i.e. trace a thought with their hands). This can be compared to the "thinking aloud" one sometimes uses to help make up one's mind. A fatal situation for a gesture analyzer that has to disambiguate these gestures from real selections!

6 Conclusion

We have presented a system for the coding of gestures in the context of human-machine interaction. The system is the first step in the development process of a fast and easy to use system for the practical needs of applied research. We did not start from an existing taxonomy because we think that gestures that are used during the interaction with a multimodal machine need a description system that is specifically designed for the context. Therefore we analyzed a set of human-machine dialogues that were recorded with Wizard-of-Oz experiments. The emerging categories are good candidates for selective labels and will be tested for their quality criteria in the second step of the development process. The interaction with the system emerged as an interesting discrimination criterion: Does the user want something from the system or does she react to an output from the system (interaction)? Does she prepare such an interaction (supporting gesture)? For a gesture analyzer it is vital to be able to distinguish a deictic gesture that is a command from a deictic gesture that is not a command. The coding system we presented shows a way to examine which differences there are between "conversational" and "non conversational" gestures. Another interesting question for future research is, how a human-machine dialogue and a human-human dialogue differ with regard to "interactional" and "non-interactional" gestures. Human-human interaction has so far been the focus for analyzing gestures in the context of speech - we hope that the new context of human-machine interaction will bring new insights for the topic.

Acknowledgments

This research is being supported by the German Federal Ministry of Education and Research, grant no. 01 IL 905. We give our thanks to the SmartKom group of the Institute of Phonetics in Munich that provided the Wizard-of-Oz data. Many thanks to Florian Schiel and Urban Hofstetter for helpful comments on the document.

References

1. Ekman, P. (1999): Emotional and conversational nonverbal signals. In Messing, L. S., & Campbell, R. (Eds.), Gesture, speech and sign (p. 45-55). New York: Oxford University Press.
2. Faßnacht, G. (1979): Systematische Verhaltensbeobachtung. München: Reinhardt.

3. Beringer, N., Oppermann, D., & Burger, S. (2000): Transliteration spontan-sprachlicher Daten – Lexikon der Transliterationskonventionen. SmartKom Technisches Dokument Nr. 2.
4. Efron, D. (1941): Gesture and environment. Morningside Heights, NY: King's Crown Press.
5. Bales, R. F. (1970):Personality and interpersonal behavior. New York: Holt.
6. McGrew, W. C. (1972):An ethological study of children's behavior. New York: Academic Press.
7. Ekman, P., & Friesen, W. V. (1972): Hand movements. Journal of communication, 22, p.353-374.
8. Wallbott, H. G. (1982): Bewegungsstil und Bewegungsqualität. Weinheim, Basel: Beltz.
9. McNeill, D. (1992): Hand and Mind: What Gestures Reveal about Thought. Chicago: University of Chicago Press.
10. Laban, R. (1956): Principles of Dance and Movement Notation. London.
11. Jönsson, A., & Dahlbäck, N. (1988): Talking to a computer is not like talking to your best friend. Proc. of the First Scandinavian Conference on Artificial Intelligence, Tromso, Norway, p. 297- 307.
12. Krout, M. H. (1935): Autistic gestures: An experimental study in symbolic movement. Psychological Monographs, 46, p.119-120.
13. Rosenfeld, H. M. (1966): Instrumental affiliative functions of facial and gestural expressions. Journal of Personality and Social Psychology, 4, p. 65-72.
14. Mahl, G. F. (1977): Body movement, ideation, and verbalization during psycho-analysis. In: N. Freedman & S. Grand (Eds.), Communicative structures and psychic structures. New York: Plenum Press, p. 291-310.
15. Freedman, N., Blass, T., Rifkin, A. & Quitkin, F. (1973): Body movement and the verbal encoding of aggressive affect. Journal of Personality and Social Psychology, p. 72-83.
16. Sainsbury, P., & Wood, E. (1977): Measuring gesture: Its cultural and clinical correlates. Psychological Medicine, 7, p. 63-72.
17. Kendon, A. (1980): Gesticulation and speech: two aspects of the process. In M. R. Key (ed.), The Relation Between Verbal and Nonverbal Communication. The Hague: Mouton.
18. Dahlbäck, N., Jönsson A., & Ahrenberg, L. (1993): Wizard of Oz Studies - Why and How. Knowledge-Based Systems, 6 (4), p. 258-266.
19. Friesen, W. V., Ekman, P., & Wallbott, H.G. (1979): Measuring hand movements. Journal of Nonverbal Behavior, 1, P. 97-112.

Evoking Gestures in SmartKom –
Design of the Graphical User Interface

Nicole Beringer

Institute of Phonetics and Speech Communication
Schellingstr. 3, 80799 Munich, Germany
beringer@phonetik.uni-muenchen.de

Abstract. The aim of the SmartKom project[1] is to develop an intelligent, multimodal computer-user interface which can deal with various kinds of input and allows a quasi-natural communication between user and machine. This contribution is concerned with the design of the Graphical User Interface (GUI) of the Wizard-of-Oz recordings. Our special interest was to create a display different from known internet applications which could motivate the users to communicate with the machine also via gestures. The following sections give a short overview about the system itself followed by a detailed description of the used methods and the results in the first phase. To conclude, a short overwiew of the methods we implemented in the second phase is given. Keywords: GUI-design, human-machine interaction, multimodal dialogue systems, gestural input

1   Introduction

Within the SmartKom project - a project on dialogue systems - particular attention has been paid to multimodality, i.e. to provide a quasi-natural communication of user and machine via speech or gestures. To develop a gesture analyzer, data is required, preferably realistic data. Our institute is responsible for the recording and annotation of such data.

Since the planned dialogue system does not already exist the data collection has been realized using the Wizard-of-Oz techique: A simulation of the system by humans (the so-called wizards) while the subjects are influenced by instructions suggesting that they are interacting with an existing machine. Our subjects are recorded while solving some small tasks in the SmartKom domains, the recordings are annotated afterwards with regard to speech, facial expression, emotional state and gestures.

To get usable data for the development of the several modules which build up the system itself, we had to provide material for the simulation. In this context the design of the GUI that we use in the Wizard-of-Oz recordings particularly with regard to evoking "interactional[2]" and/or non-interactional gestures is presented.

---

[1]   [1] http://smartkom.dfki.de/index.html
[2]   For further details of the coding of gestures please refer to [3]

I. Wachsmuth and T. Sowa (Eds.): GW 2001, LNAI 2298, pp. 228-240, 2002.
© Springer-Verlag Berlin Heidelberg 2002

We start with a short outline of the SmartKom project including the data collection, the recruitment of the subjects and the task. Then we describe in detail the design of the GUI which is used in the recording setup. The main focus lies on finding a layout which elicits gestures and on designing the GUI appropriately.

Finally, results and future work are discussed in the last section.

2   The SmartKom Project

Within the SmartKom project a consortium of 12 industrial and university partners develops a multimodal dialogue system which allows various kinds of input, e.g. speech, gestures or facial expression. The output of the system is presented with a graphical user interface - a computer screen projected onto a graphic tablet - and with synthesized speech.

SmartKom is being developed for three application scenarios, each with different requirements:

1.   SmartKom Home/Office to operate machines at home (e.g. TV, workstation, radio),
2.   SmartKom Public to have a public access to the internet or other services,
3.   SmartKom Mobile as a mobile assistant.

The system understands input in the form of natural speech as well as in the form of gestures. In order to "react" properly to the intentions of the user the emotional status is analyzed via the facial expression and the prosody of speech.

One of the requirements of the project is to develop new modalities and new techniques. For this reason gesture recognition and recognition of facial expression is implemented in order to get more naturalness in the human-machine-dialogues instead of using for example a touch screen where only pointing gestures could be made.

It is clear, of course, that multimodal interaction between humans is very rich and we know little yet about many quantitative and qualitative aspects so that any technical simulation of a multimodal inter-human interaction has to be restricted to some of the more or less known aspects. As a first step to multimodal dialogues SmartKom combines three different input facilities (gesture, facial expression, speech) in the man-machine-dialogue, which allow to proof some already known aspects of e.g. interactional gestures (for definitions please refer to [3]), and to empirically find new ways of interaction (e.g. how can we track the emotional state of the user via facial expression?) within the restricted modalities gesture, speech and facial expression.

Although the automatic recognition of gestures and facial expression can be done in all three scenarios, our investigation was restricted to the SmartKom Public scenario when we started our Wizard-of-Oz Data Collection.

To get the right idea of our work presented here, please bear in mind, that it is **not** our aim here to develop some kind of automatic gesture recognition[3] as in [9] but to

---

[3] This work is done by another group of the consortium. For principles and development of the SIVIT gesture recognizer used in SmartKom please refer to [8], [1]

find some new ways of using interactional gestures via graphical design in the Wizard-of-Oz Data Collection for later use in the "real" SmartKom system.

2.1   Multimodal Data Collection

In the first phase of the project we are collecting data for the following three different main purposes:

1. The training of speech, gesture and emotion recognizers.
2. The development of user-, language-, dialogue-models etc. and of a speech synthesis module.
3. The general evaluation of the behavior of the subjects in the interaction with the machine.

In each session of a Wizard-of-Oz experiment the spontaneous speech, the facial expression and the gestures of the subjects are recorded.

The technical equipment is as follows:
Audio:
*   a microphone array of 4 Sennheiser ME 104
*   a directional microphone (Sennheiser ME66/KG) and
*   (alternating) a headset or a clip-on-microphone (Sennheiser ME 104).

Video:
*   a fixed positioned digital camera captures the face of the subjects (facial expression).
*   a second digital camera captures the gestures in a side view of the subject (upper body)
*   and an infrared sensitive camera (from a gesture recognizer: SIVIT/Siemens) captures the hand gestures (2-dimensional) in the plane of the output projection.

Other:
*   The coordinates of pointing gestures on the work space are recorded (SIVIT)
*   as well as the inputs of a pen on the graph tablet.
*   Graphical output as video stream

2.1.1   Preparations/Technical Details

Each subject is recorded twice in sessions of about 4.5 minutes length. The recordings are done directly onto harddisc. They are synchronized manually and all layers including the audio and the gesture annotations are aligned in a QuickTime frame[4] which is distributed via DVDs to the project partners for further investigation.

2.1.2   The Task (SmartKom Public)

For this investigation we analyzed the already recorded dialogues in the SmartKom Public scenario.

---

[4]   [2] http://www.apple.com/quicktime/

The subjects were carefully instructed how to test a new prototype of a dialogue system. It was suggested that the system could understand spoken language and gestures without giving a detailed explanation either on the functioning or on the gestures the subjects could use. The aim was to encourage the subjects to experiment with their gestures as well as with the system itself.

While working with the system for the first - and in the second session for the second-time they had to imagine being in Heidelberg (a German town). Their job was to solve a task like planning a trip to the cinema / restaurant in the evening, finding a hotel, programming their VCR at home or arranging a sightseeing tour. They also were encouraged to try different ways of communicating with the system. After the instruction two sessions were recorded (with a short break in between). Afterwards, in a different room, the subjects were questioned if problems had appeared, which aspects of the system they approved of and which they resented, and if the system all in all gave more the impression of being a human or a machine.

2.1.3   The Corpus

At the time this investigation was made we had recorded 78 sessions of about 4.5 min multimodal material. Not all possible tasks were offered yet. The first phase comprises the cinema /restaurant information. For the evaluation of the achieved goals not all sessions could be used because of the lack of the gesture coordinates in some recordings. Therefore, we analyzed 41 labelled dialogues including the SIVIT-coordinates.

2.1.4   The Subjects

35 voluntary, naive subjects have been recorded and paid for their participation. The 18 females and 17 males were between the age of 16 and 60. Among them were 24 students (including 2 PhD students, 1 pupil) and 11 employed persons (including 1 retired person).

3   Design of the Wizard-of-Oz Graphical User Interface

3.1   Idea

As was mentioned before, one of the SmartKom principles is to provide an intelligent man-machine dialogue system which allows maximal naturalness in conversation. One main approach is to give the user the opportunity to use interactional gestures while using the SmartKom system. For this purpose, gestures other than pointing gestures which are nowadays taken for granted in the use of computers (touch screens, mouse click[5]) should be provided in the system. To get some idea for the further development of the gesture recognition within SmartKom it is necessary to know which gestures are indeed possible in the use of such a kind of multimodal system.

---

[5] Interpreting the mouse as the simulated hand on your screen

Therefore, the main purpose of creating the Wizard-of-Oz GUI is to simulate a working system that differs from the usual computer interfaces, especially from internet applications for two reasons:

1. Users should not be limited to known computer interaction. They should get a possibility to try something new.
2. To simulate a "new" way of man-machine-interaction we had to give the system a new look, which should not be suspected as a WOZ system by the user.

Moreover the graphical design is mainly based on series of webpages that correspond to the most likely way subjects are expected to solve the problem given to them. Care was taken to give them a different look and feel to standard webpages. We wanted the subjects not only to use simple "click" gestures but above all new kinds of gestural interaction. Apart from that the design had to be easy to handle for the wizards. Together with the demand to present a completely new working system, the GUI had to provide the possibility to affect gestures while the subjects used the Wizard-of-Oz system. The goals have been achieved by

1. avoiding visible weblinks
2. creating "unusual" buttons
3. showing self-moving elements

We hoped to encourage the users to interact "multimodally" with the system instead of simply using spoken requests.

By implementing our ideas we hoped to find gestures which are actively used by the subjects in order to interact with the system instead of or in combination with speech, e.g. pointing to graphical elements.

But the subjects were free to use any gesture they wanted. Besides interactional gestures, which are actively used to communicate with the system, speech accompanying gestures or new reactions towards the system, e.g. waving of the hands were also possible.

Although, of course, we already had some gestures in mind while creating the GUI, we were open to any kind of gestures. Apart from that not all design elements had a intended gestural function when they were implemented but - as we found out - elicited gestures unexpectedly from the subjects. Sometimes those elements were combined with a function in the second phase. We hoped people would interact with the system via these new elements more often, as soon as they found out how to get the information they wanted while using them.

3.2   Realisation – First Phase

The following paragraphs give an overview of the graphical elements used in the first phase, namely the design of the cinema and restaurant information. Of course, interactional gestures, in particular, request or confirmation can be evoked in many different ways. Non-interactional gestures are not that easy to evoke by graphical design. They are more or less side effects.

Although it is not easy to imagine the dynamic User Interface that subjects were confronted with while testing the system we try to give an idea of how interactional gestures can be evoked by describing the following (static) figures.

Figure 1 gives a selection of the GUI we provided for the several tasks (also including self-moving elements which are currently recorded and annotated).

3.2.1   Invisible Weblinks

As you can see in figure 1 the GUI provides general information about cinemas or restaurants. All display pages are presented without visible links, i.e. links are not marked by a special colour and are not underlined.

The visible user interface is divided into four frames:

SmartKom Assistant	Bubbles for Orientation
Help	Task Information

The SmartKom Assistant in the left upper corner gives information about the current state of the system: inactive (sleeping), active (eyes open), working (thinking), error (k.o.). The Help-Frame provides advice for the use of the system in general or in particular situations. The information presented in the Task-Information-Frame appears in form of headlines in the frame above and these again in form of balloons.

The subject gets information to read which can be illustrated by photos similar to movie ads, TV guides, touring guides or menu cards. The subject's task is to choose the preferred information from the ones presented.

Without weblinks the subjects have two possibilities to get further information:

- they use speech
- they indicate text or images

If the latter is used, subjects are neither influenced by the design nor by the instruction they got. They are free to use any kind of gesture in their requests.

Since we are mainly interested in evoking gestures we experimented with the design of

Unusual Buttons

By "unusual" buttons we understand any form of icons which lead to further information.

These can be   - mapsmenu cards
                - units of time
                - images
                - Flash applications

All those graphical elements are created in such a way that different kinds of gestures can be used to manipulate them to prevent subjects from simply "clicking" on them like they do while working with touch screens.

Examples of unusual buttons can be seen in figures 1 (part 3) and 2. Maps invite to zoom in or to circle the interesting region (left lower corner of part 3 in Fig. 1). To zoom in, people could touch or surround areas. To get more information about the places in question they could either point at the names on the left handside or the corresponding spot on the map.

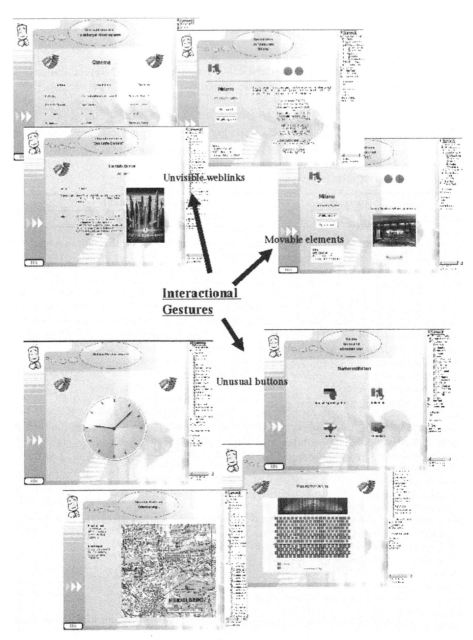

Fig. 1. Overview over the possibilities to evoke gestures by graphical output in phase 1: part 1 invisible weblinks (left upper corner), part 2: movable elements (right upper corner) partly realized in the first phase, part 3: unusual buttons (lower part)

To get more information about the timetable of movies or tv the subjects had to touch one of the time units marked on the page (fig. 1 left upper corner of part 3). Alternatively of course, they could circle a time slot to indicate the chosen period.

In order to make a reservation for a movie in one of the cinemas provided the subjects had to choose seats in the Flash application in the right lower corner of part 3 in figure 1. (They had the possibilities to press each seat separately or simultaneously or to circle the chosen seats).

Figure 2 shows a possibility of an "unusual" button which we had not foreseen. The triangles on the left encouraged some subjects to press (which of course didn't lead to any success!). Because of this observation we decided to intersperse graphical elements without any functions to see if the subjects reacted to them. If so, the relevant elements could be included in future tasks.

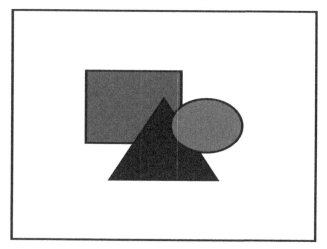

Fig. 2. System Front Page

3.2.2   Movable Elements

The application for moving around in a restaurant via webcam could be activated by touching or strolling over the picture in the respective place to indicate that one wished to look at another part of the room. This is realized by a QuickTime application (figure 1: part 2)

4   Results

4.1   Interactional Gestures

To evaluate our design we analyzed 41 of 78 sessions which were annotated using the coding system developed by Steininger et al. [3]. As can be seen in Table 1, there were indeed gestures evoked by our GUI. Of course, it is not possible to evaluate which of the gestures would also have been evoked with another kind of display in an explorative study like this one. Nevertheless, the results are an interesting first step towards answering the question how the use of gestures looks with regard to a multimodal GUI we use in SmartKom. Most of the gestures we observed belonged to the broad category of pointing (I-point). We observed the following gestures with which the infra-red gesture recognizer of the final system will be trained:

- pointing with one or more fingers
- pointing with the whole hand
- using both hands

For our purpose the gestures are not divided into the category long and short, or into the display-touching and non-display-touching category that are described in [3].

Apart from pointing we observed the following gestures

- circling objects or regions (I-circle) [6]
- new forms of interactional gestures like "no", "go back", mostly realized as a kind of waving of the hands. (I-free)[7]

As can be seen in table 1 most of the gestures were some kind of pointing, although there can be found some free and some circling gestures. While analysing the design, we found that those gestures occured with "self-moving" elements (i.e. the assistant) which belong to unusual buttons in phase 1, changable elements (zoom in with maps) and movable elements like the QuickTime-Application. Unfortunately, the latter was provided on one page which only occured twice in the recordings. Maps seem to elicit particularly large variation in the use of gestures.

The 36 gestures of the "missing reference display" in Table 1 occured in sessions where only the SIVIT-output was annotated. The basical GUI could not be reproduced, so no information about the gestural context can be given.

Table 1. Gestures found while interacting with the different types of graphical design. The coloumns describe the interactional (I-) gestures "free", "circle", and "point". For detailed description of the gesture coding, please refer to [3]

interface		I-free	I-circle	I-point
invisible Weblinks		2		51
unusual buttons	maps	5	4	48
	Reservation (movie)		2	9
	clock		1	9
	flag			5
	assistant	1	1	27
Movable Elements				1
Missing reference	display	2		34
Number of	occurences	10	8	184

---

[6] I-circle gestures therefore include realizations with (+) or without (-) display contact

[7] I-free gestures vary widely in the realizations and describe any interactional gesture which cannot be classified within the other categories

4.2   Non-interactional Gestures

Although our Wizard-of-Oz user interface is primarily designed to evoke interactional gestures, it is also possible to evoke non-interactional gestures. Unfortunately, these cannot be planned and are thus a side product of the User Interface and the wizards' reactions. We often observed delayed pointing gestures as well as residual gestures (outside the interface area) in situations where subjects tended to review their plans.

The **non-interactional** gestures were mostly a reaction to the wizards' output. They were labelled as "supporting" (the preparation of an interaction like searching) or "rubbish" gestures depending on the context and obvious intention of the user.

We did not analyze their occurence with regard to the GUI because they vary too much and are too dependent on other factors apart from the graphical design so that no conclusion can be drawn from their co-occurence with the special features of the GUI.

4.3   Questionnaire/Interview

Finally, while evaluating the interviews and observing the recordings we found out that most of the subjects tended to interact with the system via speech. They reported that it seemed to be unusual to gesture without a human dialog partner. They judged it as not necessary if the system understands them via voice input. Without a clear demonstration of the recognizable gestures during the instruction, some of the subjects were inhibited from using gestures. Finally, some wished to have a clear emphasis on the elements which could be activated by gestures, which of course would have contradicted our intention to find out what gestures are produced spontaneously and in which context.

5   Ongoing Work

The Wizard-of-Oz User Interface is still under construction for further tasks. We have already shown some examples of what kind of gestures showed up in such a situation and we are currently searching for new ways of evoking gestures by the design. The more recordings we have, the easier it will become to find new objects for the display to encourage the subjects to produce interactional gestures. We plan to examine this question systematically in order to improve and enlarge our knowledge of how to evoke different gestures by the design of the GUI. fIn view of the results of phase one we now are occupied with the realisation of

1.   movable elements
2.   providing elements which are likely to be moved in some way by the subjects
3.   unusual buttons

and as a specialization within the unusual buttons

4.   unusual maps

Figure 3 shows the GUI as it is realized at the moment.

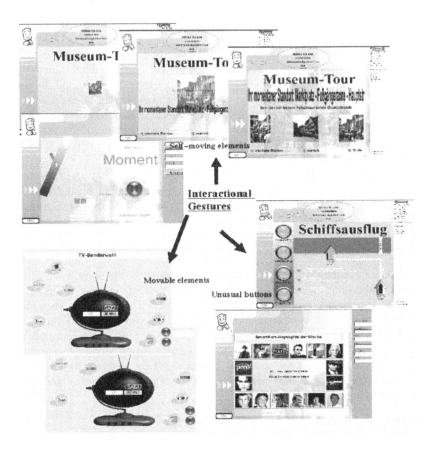

Fig. 3. part 1: Self-moving elements pop up when activated, some of the several stages can be seen (upper part of the figure); part 2: Movable elements: channels can be chosen e.g. via "drag and drop" (left lower corner); part 3: unusual buttons: choose your favourite movie or actors (right lower corner)

5.1.1   Unusual Buttons

Apart from invisible weblinks unusual buttons were frequently used in the first phase, therefore we enlarged this method while realizing the future tasks TV and Tourism.

Examples are the actor selection (recordings in the tv scenario starting soon)the incremental maps (in the tourism scenario currently being recorded). These allow only the selection in question to be seen in a more or less detailed fashion.- and actor selection.

Pressing, circling etc. one of the actors or programmes in the latter (right upper corner of part 3 (fig. 3)) leads to more information about the films the chosen actor played in for the week in question or about the contents of the chosen programme including the starting times.

Using incremental maps (Fig3. Part 1, right lower part) instead of the standard ones we used in the cinema task, seemed not to inhibit subjects from using gestures as far as could be seen yet (six pointing gestures in 6 sessions so far), although the conditions are somewhat different.

5.1.2   Movable Elements

Objects movable on the display from one point to another (figure 3 part 2) should evoke "drag (and drop)" gestures.

To indicate the selection of the channel one or more of the transparent balls representing the TV channels has to be dragged and dropped into the monitor. The selection can be done by circling as well.

5.1.3   Self-Moving Elements

It is known from compatibility experiments, e.g. [4, 5] that people not only imitate the movements of their dialog partner [6] but also tend to adjust their movements to the location and speed of objects [7].

This fact can be of use for the design of the Wizard-of-Oz Graphical User Interface (GUI).

By this method the subject is encouraged to "imitate" the movement in general. From psychological experiments it is known that subjects tend to adjust the direction and speed of their own movements to observed movements. Therefore, we provided moving images, as can be seen in part 1 of figure 3. The idea is to motivate subjects to use more movements themselves (either an interactional or a non-interactional gesture) while being confronted with a moving display. Part 1 of figure 1 gives an idea of how the dynamic website looks like in the several stages of the call before it ends in the static version.

Although we actually found more gestures during the communication with the system in the first few recordings if self-moving elements were present, people differed in their opinions towards the design (ratings varied from "good" to "too turbulent").

6   Conclusion

We presented a method for the gathering of material about gesture use in a new multimodal dialogue situation, namely a dialogue between a human and a machine that could interpret gestures. It was shown how it is possible to get information on multimodal interaction while simulating a not yet existing system via Wizard-of-Oz experiments which are recorded and annotated.

Our main purpose was to desccribe a method for the collection of data about gestural input. Some observations we made during the collection of the data seem to be interesting for further investigation:

An unusual GUI design can motivate users to experiment with gestures that are not normally used while interacting with a computer - apart from the standard ways to use touch pads (pointing which therefore seems to be a natural way of interacting with a GUI like the one described), the subjects could not know if other input modes (e.g. circling) led to any success, nevertheless some tried it.

Even complex gestures like waving for going back were sometimes used to give a comment to the system. As far as we can tell from the recently annotated material people generally seemed to use even more gestures with the new display mode (3 interactional free gestures, 12 pointing gestures in 6 sessions so far).

Apart from interactional gestures many non-interactional gestures were used that seem to be less influenced by the GUI at first sight.

As mentioned, people felt somewhat inhibited in interacting with the SmartKom (Wizard-of-Oz) system due to the lack of introductory information about the possible gestures. Nevertheless, many interacted with gestures and even used new forms of gestures. Therefore, it can be assumed that the multimodal communication with a machine can highly be influenced by the design of a non-standard GUI as all our experimental data has shown so far. Evoking gestures via graphical design seems to be a method which could lead to even more interesting results!

Acknowledgements

This research is being supported by the German Federal Ministry of Educcation and Research, grant no. 01IL 905. We give our thanks to the SmartKom group of the Institute of Phonetics in Munich, especially the display group. Many thanks to Silke Steininger and Florian Schiel for helpful comments on the document.

References

1. http://smartkom.dfki.de/index.html
2. http://www.apple.com/quicktime/
3. S. Steininger, B. Lindemann, T. Paetzold: Labeling of Gestures in SmartKom - The Coding System, same volume
4. Stins, John-F.; Michaels, Claire-F.: Strategy differences in oscillatory tracking: Stimulus-hand versus stimulus-manipulandum coupling. Journal-of-Experimental-Psychology:-Human-Perception-and-Performance. 1999 Dec; Vol 25(6): 1793-1812
5. Stins, John-F.; Michaels, Claire-F.: Stimulus-response compatibility is information-action compatibility. Ecological-Psychology.1997; Vol 9(1): 25-45
6. Bandura: Social learning theory of identificatory processes, in Goslin, Handbook of socialization theory and research. Chicago, 1969
7. Worringham, Charles-J.; Beringer, Dennis-B.: Directional stimulus-response compatibility: A test of three alternative principles. Ergonomics.1998 Jun; Vol 41(6): 864-880
8. http://www.siemens.com/page/1,3771,263380-1-19_0_0-162,00.html
9. R. Gherbi, A. Braffort: Interpretation of pointing gestures: The PoG system. In : Gesture-Based Communication in Human-Computer Interaction, LNAI 1739, A. Braffort, R. Gherbi, S. Gibet, J. Richardson et D. Teil Eds., Springer Pub., 1999

Quantitative Analysis
of Non-obvious Performer Gestures

Marcelo M. Wanderley

Faculty of Music, McGill University
555, Sherbrooke Street West, H3Q 1E3, Montreal, Canada
mwanderley@acm.org

Abstract. This article presents preliminary quantitative results from movement analysis of several clarinet performers with respect to non-obvious or ancillary gestures produced while playing a piece. The comparison of various performances of a piece by the same clarinetist shows a high consistency of movement patterns. Different clarinetists show different overall patterns, although clear similarities may be found, suggesting the existence of various levels of information in the resulting movement. The relationship of these non-obvious gestures to material/physiological, structural and interpretative parameters is highlighted.

1 Introduction

Musical performance is an interesting field for the analysis of human-human communication due to its expressive content. Not only traditional musical parameters are conveyed – melody, rhythm, articulation – but also information on different levels, such as a high content of emotion [5], contribute to the overall listening experience.

The analysis of musical performance is therefore a broad research subject based on knowledge and methods from several domains, as shown in large reviews of the different studies in this area [4] [7]. These studies have analyzed the various aspects of musical production by performers of all levels of skill and have presented general findings that help us understand the highly developed motor and cognitive resources employed by musicians.

In musical performance, the term *gesture* may have very different meanings [1]. For instance, *musical gestures* usually indicate abstract musical features that do not present clear links to physical movements. On the other hand, *performer gestures* usually designate the ensemble of an instrumentalist's performance technique. Performer gestures therefore encompass both physical movements intended to produce sounds on an instrument (called *effective* or *instrumental gestures*), and other movements that do not have a straight link to the generation of sound, but are nevertheless an integral part of musical performance [8]. The latter have been termed *ancillary*, *accompanist* or *non-obvious gestures*.

Among the various works on musical performance, only very few studies have analyzed the gestural behavior of instrumentalists from the point of view

I. Wachsmuth and T. Sowa (Eds.): GW 2001, LNAI 2298, pp. 241–253, 2002.

of ancillary gestures. These works have mostly focused on piano playing [3], piano and violin [2], and more recently on the clarinet [8].

Delalande [3] studied the gestural behavior of late pianist Glenn Gould from video recordings of his performances. He was able to show a strong correlation between expressive or, as he termed, *accompanist gestures*, to the score being played. Davidson [2] presented a study on the perception of visual cues related to expressive gestures of four violin players as well as of a pianist. She showed that the visual information available from different performances (standard, exaggerated and deadpan) could even surpass the auditory information when related to the perception of expressive musical features.

In [8], it has been shown that, unlike the case of piano playing, expressive gestures performed by wind instrumentalists do influence the resulting sound. This influence is the result of the displacement of the sound sources (the open holes and bell of a clarinet) with respect to a fixed close microphone in a reverberant environment [9]. Therefore, when performing computer simulations of an acoustic wind instrument, the modeling of these effects may contribute to a naturalness found in certain real situations, expressed by temporal fluctuations of sinusoidal sound partials' amplitudes[1].

But in order to model these effects, there is a need to first understand the underlying mechanism of these movements. To my knowledge, no work has been presented on the measurement of performer ancillary gestures in general, and specifically in the case of clarinet performances.

In this article, preliminary quantitative results are presented and discussed, with the focus on three main points related to the analysis of clarinetists ancillary gestures:

1. *The production of ancillary gestures*, i.e., whether it is common to most players to move the instrument while playing, what is the magnitude order of these movements and whether it is possible to identify basic movement patterns.
2. *The repeatability of ancillary gestures*, i.e., whether one clarinetist repeats the same movements while playing one piece at different times, and what can be the correlation between movements during different performances of the same piece.
3. *The possible production of similar movement patterns by different performers*, i.e., whether there exist similarities in the movements of different performers.

1.1 Quantitative Data Analysis – Methods and Materials

Movement data acquisition was performed in collaboration with the Faculty of Movement Studies at the Free University of Amsterdam and the Music, Mind and Machine Group at the NICI, Nijmegen, the Netherlands.

---

[1] May also be true for the case of the violin and other instruments where performer movements cause displacements of the sound sources, i.e. the instrument itself.

 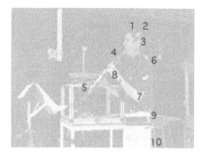

Fig. 1. Player 1 at the movement data acquisition session at the Free University Amsterdam. The positions of each marker are displayed in the right picture

An Optotrak system was used to measure players' movements during performance. It consists of a three-dimensional highly accurate acquisition system based on infrared (IR) markers tracked by three IR cameras. In this study, between eight and ten markers were used, the performers standing at a distance of around 3 to 4 meters away from the cameras. The sampling rate used for each marker was 100 Hz, each marker giving a three dimensional coordinate at each sample time (horizontal, vertical and sagittal coordinates).

Four clarinet players have been recorded: one French player (Player 1) and three Dutch players (Players 2, 3 and 4). All of them are professional musicians with many years experience in both solo, chamber and/or orchestral performances.

Figure 1 shows the first player and the placement of the markers during the recording session at the Free University Amsterdam.

The pieces selected consisted of basic standard clarinet repertoire: solo pieces and sonatas for clarinet and piano, both classical and contemporary. The list of recorded pieces is the following:

1. Domaines, by Pierre Boulez[2]. Original. Cahiers A to E and F (incomplete).
2. First Clarinet Sonata, by Johannes Brahms. First movement (incomplete).
3. Clarinet Sonata, by Francis Poulenc. First and second movements (both incomplete).
4. Three Clarinet Pieces, by Igor Stravinsky. First and Second Pieces.

In this paper two of the recorded pieces are analyzed in detail: *Domaines*, by Pierre Boulez, cahier A, and an excerpt of the first movement of the clarinet sonata, by J. Brahms[3].

The players were asked to play these pieces three times: a *standard* performance, an *expressive* performance, and a *performance trying not to move the*

---

[2] Player 1 only.

[3] All pieces were recorded without piano accompaniment. Similarly, no metronome was used in the recordings.

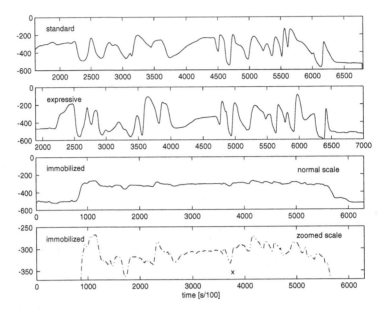

Fig. 2. Different performances of Domaines, cahier A by the first player. The fourth plot presents the third performance with a zoomed scale. Vertical axes show the bell vertical position in millimeters, the horizontal axis shows time in cents of seconds

instrument at all[4]. These different performances were repeated to get supplementary data[5]. Finally, the players also performed all pieces seated, in order to verify their gestural behavior in different contexts.

2   Production of Ancillary Gestures

Figure 2 shows the standard performance of *Domaines*, cahier A. It displays the vertical movement of the marker placed on the clarinet bell (number 5). From top to bottom: the standard performance, expressive performance and immobilized performance, the latter presented twice: the first time in the same (vertical) scale as the previous ones and then again with a zoomed vertical scale.

Figure 2 shows that there exists a strong correlation between vertical movements of the bell during the standard and expressive performances. Also, it is apparent that the first player *can play without significant movements*, as shown in the third plot. Furthermore, analyzing the fourth plot, one can notice that even for a very reduced movement amplitude, certain movement patterns tend

---

[4] This performance is different from Davidson's *deadpan*, where *deadpan* meant *inexpressive*. I am here interested in knowing whether it is possible for a performer to play a piece without producing expressive gestures.

[5] Pieces 2, 3, and 4.

to be reproduced. For instance, notice the downward movement around 38 s (marked with an 'x') that corresponds to the beginning of the fourth section of the piece.

Figure 3 depicts the vertical, horizontal and sagittal coordinates of the instrument's bell for the standard performance of Domaines by the first player. Ancillary gestures were produced in all three directions, presenting fairly large amplitudes.

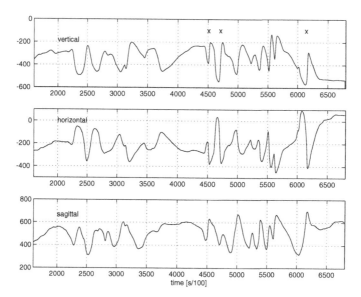

Fig. 3. *Domaines*, cahier A, first player. Standard performance, vertical, horizontal and sagittal movements of the instrument's bell (in millimeters). Notice the three fast upward gestures in the first plot (marked with an 'x')

2.1   Movement Patterns

One point to be studied refers to the possible existence of individual movement patterns and whether these could be produced by different players. In [8] three movement patterns were proposed:

1. Changes in posture (usually at the beginning of phrases).
2. Slow continuous gestures.
3. Fast sweeping movements of the bell.

Postural Adjustments Figure 4 shows the fourth player performing the first movement of the clarinet sonata (expressive performance) where extreme postural adjustments are performed. Postural adjustments were frequent in the performances of some players, but not all [1]. This suggests a predominantly idiosyncratic characteristic of these movements.

Fig. 4. Three shots showing different postures during the (expressive) performance of the clarinet sonata by the fourth player

Slow Continuous Movements The production of slow gestures, both slow upward movements during sustained notes and circular horizontal patterns, can be found in several performances. Specifically considering circular patterns, these may vary in amplitude from one performer to another. For instance, figure 5 shows two performances of an excerpt of the clarinet sonata by the first and fourth players. One can notice circular patterns with different amplitudes throughout the performances.

Fast Upward Movements Examples of fast upward gestures are present in the last musical unit of cahier A, corresponding to a note played crescendo (*mf* to *ff*). This can be seen in figure 3 around 62 seconds. Other fast upward gestures performed by the first player are also marked with an 'x' in the first plot of figure 3. Comparing the different performances of the two pieces, these gestures were not found for the clarinet sonata, indicating a dependence on the structure of the piece being performed.

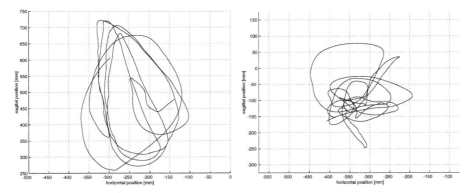

Fig. 5. Circular movement patterns (horizontal plane) during the first several seconds of the clarinet sonata. *Left:* Player 1, *Right:* Player 4

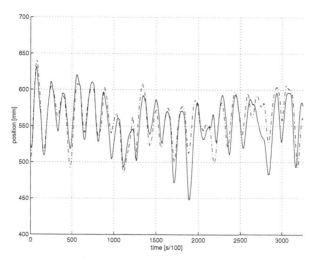

Fig. 6. Two standard performances of the clarinet sonata by the second player

3   Repetition of Equivalent Movements

This section presents detailed analysis of data from multiple performances of the same piece. For instance, figure 6 shows the superposed vertical movement of the marker placed on the clarinet bell for two performances of the clarinet sonata by the second player. Note the consistency of timing during more than 30 seconds of playing and the similarity of the movements throughout the piece.

This can also be seen in the performances of other players. For instance, figure 7 shows two standard performances of the clarinet sonata by the third player. Generally, one may again notice the temporal consistency between these two performances, although there are more significant differences in timing is this case.

Considering now the spatial relationship between the two performances depicted in figure 7, only twice were there substantial differences in the movement pattern. It is interesting to notice that these differences in spatial movement patterns are performed at a certain time and that the player soon afterwards returns to a common movement pattern. Therefore, these different *strategies* seem to correspond to interpretative choices during the performance, since its temporal evolution was unchanged.

Correlation Coefficients In order to provide a quantitative idea of the similarity between the different performances depicted in the previous section, we have calculated correlation coefficients for the two performances depicted in figure 6[6] for the three coordinates (cf. figure 8).

---

[6] This is not performed for the other players because of more significant temporal variations between performances. This could be expected, since as pointed out, there

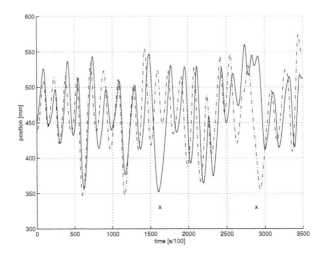

Fig. 7. Two standard performances of the clarinet sonata by the third player. Notice the different movement strategies in the two performances (marked with and 'x')

Table 1. Correlation coefficients for the two performances depicted in figure 8

Correlation coefficient	Player 2
Vertical movement	0.7885
Horizontal movement	0.7456
Sagittal movement	0.4598

Correlation coefficients were calculated for a total of 32.5 s of movement data (3250 samples), from the beginning of the first note to just after the end of the last note of the excerpt. Analyzing the results, it can be seen that both the vertical and horizontal movements are highly correlated, with the sagittal movement presenting a smaller coefficient. But if one looks carefully to the movement patterns in the sagittal direction, it can be seen that there exist several movements that are similar but inversed in direction, i.e. upward in one performance and downward in the other. This means that, although different in absolute values, these movements are similar if compared in relative terms and seem to represent different interpretative strategies in the two performances.

is no accompaniment, and because of the total duration of the excerpts. A solution is presented in section 4.

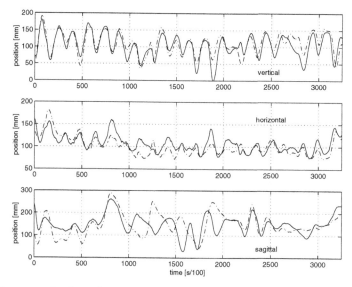

Fig. 8. Two standard performances of the clarinet sonata by the second player. From top to bottom: vertical, horizontal and sagittal coordinates

4   Comparison of Different Performances: Time Warping

Since the data collection was performed without restrictions related to timing (no reference from a metronome), the temporal evolution of different performances by the same player can vary. These variations, although almost negligible if compared to the total duration of the excerpts, can prevent an exact examination of the spatial patterns of movement actually performed.

In order to overcome this situation, a technique widely used in speech recognition has been used, time warping. Figure 9 shows three standard performances of the clarinet sonata by the second player, where the files have been time-warped to a *reference score* to reduce time variations across performances and allow an exact comparison of each moment in all performances.

In this figure, the score is plotted as vertical lines: solid lines correspond to the first note in each bar, while dashed lines correspond to the other notes in the score placed *exactly at the moment they occur for the case of a mechanical performance*, i.e. with no deviation from absolute tempo.

One can see that at the beginning of each bar, the second performer is *at the same point* in an upward part of the movement. This is an interesting finding as it suggests a kind of beat-keeping function of the gesture. It represents a dependence of the movement on the piece's structural characteristics, in this case its tempo.

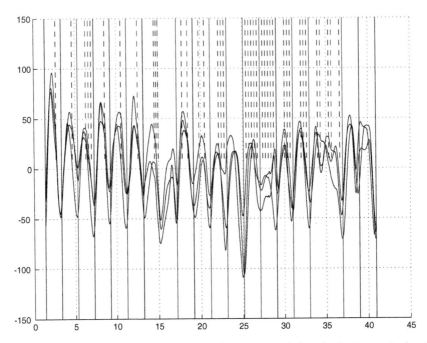

Fig. 9. Three time-warped standard performances of the clarinet sonata by the second player. The X axis presents time in seconds. All curves have been referenced to zero millimeters

5   Performances by Different Players

Finally, let us analyze the third question raised in the introduction of this article, i.e., whether there are equivalent movements across multiple performers.

Figure 10 shows that there is no one-to-one similarity between the movements of the four players, as could be expected. Important differences concerning the amplitude of movements as well as the temporal characteristics of each performance can be seen. Nevertheless, performances of some players can be closer than others considering, for instance, the temporal evolution of the movement. As an example, the last two performances (Players 2 and 3) show a similar number of dips, while the first performance presents half that number, i.e., the first player swings the instrument's bell at roughly half the frequency of the other two.

This point is interesting as it indicates a rhythmic component of the total movement, related to the performance of the piece. In fact, this is a common feature of all performances of the sonata, but absent from the performance of Domaines, since the tempo is not as well defined.

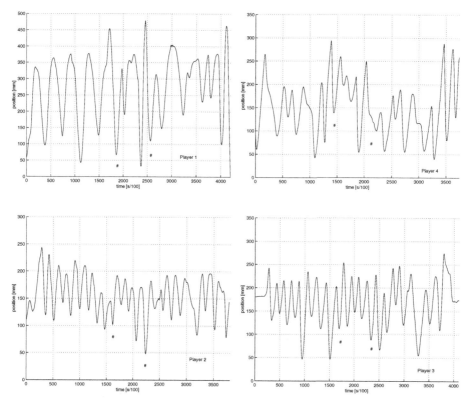

Fig. 10. Four performances of the clarinet sonata. Points marked with an '#' indicate moments where the performers breathe. Notice that the performers have the tendency to be in similar positions when breathing

6   Discussion: Implication of Empirical Findings

As indicated above, there is evidence that ancillary gestures performed by wind instrumentalists are related to musical features in different levels. In fact, three main levels can be suggested:

1. *Material/Physiological,* e.g. the influence of respiration, fingering, ergonomics of the instrument, etc. It can be seen, for instance, that each time performers breathe there is a tendency to bring the instrument down to a vertical position and soon afterwards start an upward movement again.
2. *Structural,* i.e., dependent on the characteristics of the piece being performed. We have seen that all performers presented a rhythmic pattern associated to the piece for the case of the clarinet sonata. Although there are various differences in these movements, there exist similarities that cannot be explained by randomness.

3. *Interpretative*, i.e., related to the mental model of the piece developed by the performer and/or to specific interpretation conditions. For instance, in figure 7 the performer changes the movement pattern twice, at around 17 s and at 28 s, although the rhythmic component remains the same in both performances. The interpretative level will likely present a strong idiosyncratic component.

7   Conclusions

From what has been presented, general conclusions regarding clarinetists ancillary gestures can be drawn:

1. Although ancillary gestures were common in all performances, clarinet players are *eventually* able to play a piece almost without expressive movements, even though it has been stressed that this causes difficulties in respiration[7]. Therefore it is possible to disregard physiological influences or aspects of the playing technique as the only reason for the production of ancillary gestures.
2. For the case of the same *expert* clarinet player performing one piece at different times, one can reasonably consider that there will likely be a strong correlation between the player's movements at the same points in the score in the different performances. This fact suggests a strong relation between *what* is being played and *how* it is played. Therefore it is evident that ancillary gestures by clarinet players are not randomly produced or just a visual effect. On the contrary, the quantitative results presented in this article tend to indicate that these gestures are an integral part of the performance process.
3. Quantitative data from performances of different players showed that most ancillary gestures present a high proportion of idiosyncratic movements, although movement features related to structural characteristics of the piece (e.g. tempo) and to material/physiological aspects tend to be largely invariant across performances of different musicians. A three level typology of ancillary gestures according to their possible origin was proposed in order to take into account similarities and differences among the different gestural patterns.

Finally, apart from a better understanding of musical performance itself, another interest of this type of research is to extrapolate findings presented in the musical context to contexts related to human-computer interaction. Understanding the behavior of people manipulating instruments in expressive communicational situations such as music can lead to better designs of computer interfaces, not only for musical use, but also in general. This may eventually help offset the lack of expressive or affective content [6] in current human-computer interfaces.

---

[7] Player 1 considered that breathing was unnatural since it had to be done with the upper part of the chest.

Acknowledgments

My warmest thanks to all performers for their collaboration and to P. Desain, P. Beek and E. Schoonderwaldt for the possibility to perform the measurements. X. Rodet and the Analysis-Synthesis Team at IRCAM provided the necessary means to accomplish this research. N. Orio provided the time-warped files and T. Hélie the Matlab code for figure 9. Also thanks to P. Dutrieu, P. Depalle, R. Kirk, D. Matzkin, H. Mossel, O. Ramspek, L. Wouters, and R. McKenzie for several comments. Part of this research was funded by a doctoral scholarship from CNPq, Brazil.

References

1. C. Cadoz and M. M. Wanderley. *Trends in Gestural Control of Music*, chapter Gesture-Music. Ircam - Centre Pompidou, 2000. 241, 245
2. J.-W. Davidson. Visual Perception of Performance Manner in the Movements of Solo Musicians. In *Psychology of Music*, volume 21, pages 103–113, 1993. 242
3. F. Delalande. La gestique de Gould. In *Glenn Gould Pluriel*, pages 85–111. Louise Courteau, éditrice, inc., 1988. 242
4. A. Gabrielsson. Music Performance. In *The Psychology of Music*, pages 501–602, 1999. Diana Deutsch, editor. 2nd edition. 241
5. S. Hashimoto. Kansei as the Third Target of Information Processing and Related Topics in Japan. In *Proc. KANSEI - The Technology of Emotion Workshop*, pages 101–104, 1997. 241
6. R. Picard. *Affective Computing*. MIT Press, 1997. 252
7. J. Sloboda. Music Performance. In *The Psychology of Music*. Academic Press Inc., 1982. Diana Deutsch, editor. 241
8. M. M. Wanderley. Non-Obvious Performer Gestures in Instrumental Music. In *Gesture Based Communication in Human-Computer Interaction*, pages 37–48. Springer-Verlag, 1999. A. Braffort, R. Gherbi, S. Gibet, J. Richardson, and D. Teil. editors. 241, 242, 245
9. M. M. Wanderley, P. Depalle, and O. Warusfel. Improving Instrumental Sound Synthesis by Modeling the Effects of Performer Gesture. In *Proc. of the 1999 International Computer Music Conference. San Francisco, Calif.: International Computer Music Association*, pages 418–421, 1999. 242

Interactional Structure Applied to the Identification and Generation of Visual Interactive Behavior: Robots that (Usually) Follow the Rules

Bernard Ogden[1], Kerstin Dautenhahn[1], and Penny Stribling[2]

[1] Adaptive Systems Research Group, Department of Computer Science
University of Hertfordshire, Hatfield, Herts, UK
{bernard|kerstin}@aurora-project.com,
Project home page: http://www.aurora-project.com
Department home page: http://homepages.feis.herts.ac.uk/nehaniv/ASRG.html
[2] School of Psychology and Counselling, Whitelands College
University of Surrey Roehampton, West Hill, London, UK
P.Stribling@btinternet.com

Abstract. This chapter outlines the application of interactional structures observed by various researchers to the development of artificial interactive agents. The original work from which these structures are drawn has been carried out by researchers in a range of fields including anthropology, sociology and social psychology: the 'local approach' described in this paper draws particularly on conversation analysis. We briefly discuss the application of heuristics derived from this work to the development of an interaction tracking system and, in more detail, discuss the use of this work in the development of an architecture for generating action for an interactive agent.

1 Introduction

This paper discusses the relevance of work from the human sciences (i.e. fields such as anthropology, sociology and social psychology) to the development of systems or agents that are concerned with observing or with engaging in interaction. Our primary interest is in the latter of these two types of artificial agent: however, we believe that the work is also relevant to identification of interaction and we discuss this briefly in section 2. Section 3 considers the question of generation of appropriate behaviors for an interactive agent, drawing heavily on concepts from the field of conversation analysis (CA). Finally, we briefly consider the issue of assessing and measuring interaction in section 4 before concluding.

Before proceeding, it is necessary to clarify what we mean by interaction. We see interaction as a reciprocal activity in which the actions of each agent influence the actions of the other agents engaged in the same interaction, resulting in a mutually constructed pattern of complimentary behavior.[1] Key in this

---

[1] For those readers who are familiar with the framework of autopoiesis our definition of interaction might be reminiscent of the notion of 'structural coupling' which de-

I. Wachsmuth and T. Sowa (Eds.): GW 2001, LNAI 2298, pp. 254–268, 2002.
© Springer-Verlag Berlin Heidelberg 2002

definition is the co-constructed nature of the interaction: this aspect of interaction is emphasized in Clark's notion of joint action [3] and in di Paolo's work with interacting virtual agents [4]. At the present time we are most interested in visual interaction, by which we mean an interaction in which action is detected primarily in the visual channel. In other words, we are interested in interaction in which the actions of participants are detectable by sight, as in the case of movement, gesture and the like.[2]

We also wish to note that we are discussing systems that deal with the output from an underlying machine vision system: in principle, the ideas discussed here should be compatible with any machine vision system capable of identifying actions of interest and labeling its identifications with a probability of correctness. Finally, we should mention that we are primarily interested in interactive *robots*: thus we often refer to robots rather than artificial agents more generally in the following discussion. We nonetheless believe that this discussion is relevant to any artificial agent that is intended to interact with humans, including virtual agents e.g. embodied conversational agents [5,6].

2   Identification of Global Structures

In this section we consider structures of interaction that operate at a global level[3] i.e. structures that are applied to relatively long sequences of interaction rather than operating at a more local level.[4]

An example of a global structure would be a greeting as described by [7]: here we have the idea of a greeting as composed of a number of phases, from distant salutation through approach to a final close salutation. Each phase in this structure has a set of typical associated actions. We briefly discuss here means of identifying such a sequence if it is observed as part of a vision sequence — the interested reader is referred to [2] for a more detailed discussion of the same ideas. We note that the ideas in this section draw particularly on the work of Kendon [8,9] which, in turn, reflects the familiar AI/cognitive science notion of scripts [10]. We also note that [11] describes similar work to Kendon in this respect.

For our purposes we consider a global structure as being composed of a number of phases, as noted above, each with a set of associated actions. Each action has a meaning for each phase in which it occurs: thus if an action occurs

---

scribes the relationship between a living organism and its environment [1]. Indeed, we explore these similarities in more detail in [2].

[2] Note that we are not claiming that visual resources are involved in the *production* of such actions, which will usually take the form of movement and gesture. This is merely a term for describing the channel through which the actions are detected.

[3] Note that the ideas in this section are mainly *not* drawn from conversation analysis, which generally sees conversation as locally managed.

[4] In [2] we discuss large and small *units* of interaction rather than long and short *sequences* of interaction. We now prefer the term 'sequences' as it implies a less rigid unit and thus is a better description of the case in human interaction.

in more than one phase it can have a different meaning in each phase (e.g. a 'wave hello' vs. a 'wave goodbye'). We view the problem of identifying an action sequence of a given interaction as one of assigning probabilities of being an instance of a given interaction type to each observed potential action sequence. We assume that the basic machine vision system itself is able to assign some kind of base probability to each action it observes and we take this probability as our starting point in selecting an action sequence. We can then consider heuristics based on the structure of human interaction to weight these probabilities.[5]

Continuous Phase Progression Heuristic. It seems that a sequence that proceeds through each and every phase from beginning to end should be weighted more heavily than one that skips some phases of the preferred sequence or ends without reaching the end of the preferred sequence, although both of these events happen in the course of normal interaction [7,12]. Some 'deviations' from the preferred sequence may be more common than others and this should be reflected in weights assigned according to this heuristic. We note that variations in the preferred sequence in normal interaction will in themselves be actions in the interaction and thus of interest — however, in the present case we are interested in classifying an observed sequence of behaviour as belonging or not to one member of a set of classes of types of interaction (e.g. greeting, argument, making a purchase). We are therefore working at a very coarse granularity and not attempting to determine the full implications of everything that we see.

Globally Improbable Phase Transition Heuristic. A further heuristic is simple weight of numbers. If we observe three actions in phase x of an interaction, then an action in phase $x + 1$, then a further four actions in phase x, we should probably conclude that the transition to phase $x + 1$ never happened and that the observation of the action was spurious.

Adjacency Pair Heuristic. In cases where a given action creates the expectation of a given response (e.g. Kendon [7] notes that a head nod is often responded to with another head nod) then interpretations of the data that show this pair should be weighted more strongly than alternative interpretations.

Contiguous Action Heuristic. In a turn-taking interaction it seems reasonable to expect minimal overlapping and leaving of gaps between consecutive actions most of the time [13,12,14].[6] In such interactions we can assign greater

---

[5] We note that these heuristics are not necessarily to be used in a real-time system. Some are simply inappropriate for this purpose: for example, the boundary signaling heuristic will involve reinterpretation of assumptions regarding prior actions and hence potentially will require reanalysis of the whole sequence of actions-so-far several times over in the course of a single sequence of interaction.

[6] Hutchby and Wooffitt [14] also points out that in certain types of interaction there may be more gap and overlap. It is also worth noting that gaps can be acts in an

probability to interpretations that minimize overlapping and gaps between the actions of interactants.[7]

Boundary Signaling Heuristic. The actions of the first phase of an interaction define its beginning: similarly, the actions of the last phase of an interaction define its end. This allows us to reduce the set of action sequences of interest to those that begin with a first phase action of some interaction and to weight action sequences that terminate with an action from the end phase of a matching interaction more heavily than those that do not.

3   Local Structures

We now move from structures that organize longer sequences of interaction to those that organize shorter sequences. These local structures[8] often work at the level of action and response[9] and so seem especially suited to incorporation into systems that generate interactive action as they can provide a means for selecting an action based only on the prior action and, possibly, on the likely nature of the next action. This minimizes the past memory or future planning required by the agent, allowing considerably computational savings. We examine structures observed by conversation analysts in their studies of human interaction and consider the application of similar concepts to a robotic agent. To provide some context we consider the case where these structures are applied to the agent in the 'dancing with strangers' experiment [15], but note that these interactional structures could be applied in architectures other than the very simple one considered here. More generally, CA has already been applied in the development of new interfaces by researchers in human-computer interaction [16].

3.1   The Dancing Robot

Dautenhahn [15] describes an experiment in which a robot coordinates its movements to a human's, modifying its movement behaviour in response to reinforce-

---

interaction in themselves [13,12] but, as noted above, we are dealing with a coarse grained classification rather than a detailed analysis at this stage.

[7] Note that rules of conversation, a special case of auditory interaction, will not necessarily correspond to the case of visual interaction. In this case, the visual channel seems much more open to overlapping actions than the auditory channel - an individual who speaks while someone else is speaking creates interference in the channel, but it is certainly possible to move and gesture simultaneously with another individual without necessarily causing interference. Thus the value of this heuristic is uncertain at this time.

[8] We note that what we describe as 'local' and 'global' structures are both seen as locally managed in the CA view. Here we are using the term 'local' to describe an approach that deals with short sequences of interaction (as opposed to a 'global' approach involving longer sequences of interaction), rather than in its CA sense.

[9] Note, though, that the response is itself an action in the interaction i.e. a response is itself an action requiring a response in turn.

ment from the human's hand movements which are classified into six categories: left, right, up, down, clockwise and anti-clockwise. Our discussion here is confined to the 'autonomous-select' condition of the experiment: in this case the robot cycles through a few basic behaviors with the human's hand movements selecting one or a few of these behaviors by reinforcement.[10] The experiment employs an association matrix relating inputs and outputs. A weight in the matrix is activated when the two agents perform the matching behaviors i.e. the weight that exists for the pair c_1/r_1 where c_1 is the human's movement input and r_1 the robot's movement output is activated when the human performs movement c_1 and the robot performs movement r_1. A weight is increased if it is activated in two consecutive time steps and decreased when it is not activated. This condition is called temporal coordination between the movements of human and robot. In this way, using a simple reinforcement mechanism, the robot will 'learn' to perform particular movements in response to particular human movements and a simple interaction can develop between robot and human. It is easy to imagine extensions of this experiment in which the movements of interest are not necessarily the six categories of hand movement but instead are any set of arbitrarily defined actions.

3.2   The Interactively Active Robot

Interaction is often viewed as actively constructed by the interactants: this is also the case in our own definition in section 1. In the basic 'dancing' case as described above, however, the interaction is entirely led by the human interactant: the robot's only contributions are simple responses. We could get closer to 'real' interaction by having the robot more actively constructing the interaction. To achieve this the robot must have goals in the interaction. Ordinarily a human might approach an interaction with various goals in mind. The robot's goals will obviously be much simpler than those in the human case: specifically, they can be the production of some specific sequence of actions on the part of both interactants. In order to be able to achieve its goals the robot needs to understand the likely effects of its own actions on the human: to this end it needs to build up a set of mappings from its own behavior to the human's. This seems simply achieved by a similar system to that which constructs the mappings from the human's behavior to the robot's. To begin, we take the set R to be the set of all of the robot's possible actions and the set C to be all of the human's actions that the robot can recognize and thus respond to and we take a network in which all human actions have a weight connecting them to all robot actions and vice versa (i.e. action c_1 is connected to every action in the set R, as is c_2, c_3, c_4 etc. In turn, each action in the set R is connected to each action in the set C by a different weight). Note that weights are not symmetric i.e. the weight mapping r_1 to c_1 (w_{r_1/c_1}) is not the same as the weight mapping c_1 to r_1 (w_{c_1/r_1}). Now, instead of having all non-activated weights decay in each time step we have the weight w_{r_1/c_x} decrease in cases where action r_1 is followed

---

[10] The reader is referred to the original paper for more information.

instead by some other action c_y: the weight w_{r_1/c_y} would also increase, of course. In this way the weights can adapt to observed behavior. We can consider any weight above some arbitrary threshold t_w to be a valid action. Thus, we could have structures such as the following: $r_1/c_1|c_2|c_3$ where r_1 is the first part of the pair and c_1, c_2 and c_3 are the set of expected actions of the human following action r_1: of these three alternatives, one action should be produced. We then have 'action chains' composed of action/response pairs as the basic structure of our interaction where the first action of a sequence defines a set of valid responses, each of which in turn has a further set of valid responses and so on. For the robot to generate a target sequence it needs to both create a situation where the correct mappings exist for the sequence to have a high likelihood of being produced and to successfully produce the first action of the sequence. However it should not take control of the interaction to such an extent that, for example, when it has the goal sequence $r_1/c_1, r_2/c_2, r_3/c_3, r_2/c_4$, it produces r_1 constantly until c_1 is produced in response and so on until it eventually succeeds in forcing the interaction sequence that it was seeking (indeed, it is not necessary that it ever reaches the target sequence, but it should attempt to 'guide' the human in this direction: the final pattern of behavior will thus be an amalgam of the goals of both interactants). We can achieve a degree of control by modifying the weights mapping human action to robot response: in this way the robot acquires the ability to guide the interaction in a given direction without taking all control away from the human. Thus if we have the structure $c_1/r_1|r_2|r_3|r_4$ the weight mapping c_1 to r_2 could be multiplied by some arbitrary factor α, where $1 < \alpha$. Similarly, the other weights in this structure could be multiplied by some arbitrary factor β, where $0 \leq \beta < 1$. Exact values for α and β would have to be determined empirically, but in general the higher the value of α and the lower the value of β the more control the robot has and the less control the human has. Thus the robot's likelihood of producing a part of its goal sequence is increased without it necessarily slavishly repeating the same response on each occasion that a relevant human action is produced. We also need, of course, to consider the reverse case: the human's action produced in response to the robot's action. We obviously cannot achieve this through simple adjustment of weights, as the weights mapping robot action to human response obviously have no effect on the human's actions. One solution is to catch the case where the human coincidentally produces the correct response and attempt to reinforce this behavior, e.g. by producing some rewarding action (e.g. pretty light displays, 'happy' sounds etc.). We note that further alternative goals are possible, such as maintaining a given distance from one another (as in [4]), maintaining a given heading with respect to one another, moving to a certain area in the space within which the agents are interacting, etc. It would also be possible to introduce something similar to the homeostatic control used by the robotic head Kismet [17,18,19]. The precise nature of the goals is not important — what matters is that the agents be able to mutually influence one another in order to achieve these goals.

3.3   Conversation Analysis and Computational Architectures

We wish to apply conversation analytic concepts in the construction of a computational system for an interactive robot. In doing this it is necessary to consider the appropriate application of these concepts in designing artificial interactive systems. It is particularly important to understand the nature of conversation analytic 'rules'. These are not rules in the usual sense as they do not define steps that are followed by people to produce conversation but instead describe the manner in which it is produced. People can be said to 'orient to' these rules, rather than to follow them: the rules will not necessarily be followed all the time, but violations will usually be noted and inferences drawn from them.[11] Thus not following a rule is an interactive act in itself. We take the following example of the effect of violation of a CA rule (here, a perceived failure to produce an expected second part of an adjacency pair):

(Two colleagues pass in the corridor)
1   A: Hello.
2   B: ((almost inaudible)) Hi
3   (Pause: B continues walking)
4   A: ((shouts)) HEllo!

(Example from [14] p. 42). The notation conventions used in this paper are described in appendix A.)

In this example A appears not to hear B's simultaneous and quiet greeting and this violates B's expectation that the greeting will be responded to. In this case the interpretation that may be drawn from the assumed failure to complete the adjacency pair is that person A is being snubbed, or alternatively that he has not been heard. In either case, as far as A is concerned the normal expected sequence has not been followed and thus it is appropriate for A to engage in repair. Thus we can see that the rule that a greeting is followed by a greeting is violable in that A and B have interacted even though B does not realize that his greeting is being responded to, but that this violation (deliberate or otherwise) itself constitutes an action in the interaction.[12] While this example is conversational, we can consider a similar case in a 'dancing' interaction in which one robot performs an action with a given expected response which the other fails to provide (in this case, the other might fail to move at all, or else fail to change its existing movement).

We note that although the methodology of CA seems appropriate for research into visual interactive behavior and has indeed been used in this manner [20,21],

---

[11] In (more accurate) CA terms, the rules are indexical practices locally oriented to and locally produced: talk is context sensitive in that it orients to cultural notions of how a conversation should proceed but also context building in that the act of selecting particular ways of structuring talk affirms, renews or subverts the culturally given notion of structure.

[12] This example also demonstrates that a non-action (perceived or actual) can constitute an action in the interaction

many of CA's rules are designed to deal with conversational interaction and may not be directly applicable to visual interaction. Nonetheless, we believe that principles and observations derived from this field can be applied in our case. We are particularly interested in the concept of joint action [3], the idea of turn taking as a local management system [13] and the notion of repair [14,3,13]. We focus particularly on repair in this paper as human interaction is very flexible and not strictly bound by inviolable rules, which presents obvious problems for a computational controller. We see repair[13], combined with a notion of the rules that the controller uses as violable, as creating the potential for a controller that is able to use such rules without being constrained to always follow them. Before continuing with our consideration of repair, however, we would like to note advantages stemming from seeing interaction in terms of joint action and of the local management view of turn-taking. Firstly, the concept of joint action [3], (also exhibited in [4]) gives us a clearer idea of what interaction actually is and a means of determining if some form of interactive behavior has actually been achieved e.g. by contrasting the behavior of two interacting agents with the case where one agent is attempting to interact but the other is simply 'playing back' a recording of the actions of an agent in an earlier interaction [4]. Secondly, the idea of turn-taking as a local management system is also promising. The evidence that CA offers that this is how human conversational turn-taking works suggests that such a system is certainly a workable way of managing interaction (or at least that it would provide a part of a complete system for such management). This would simplify the process of managing an interaction considerably by not requiring any reference to external rules of larger scope, such as some of the heuristics introduced in section 2.[14] Local management of transitions between interactants and problems that occur reduces the need to maintain memory of preceding actions and the need to plan ahead beyond the next turn — in a global approach such planning could extend from the present point to the expected end of the interaction. These points provide obvious computational advantages.

3.4   Robotic Repair

In CA, repair is a means of correcting a misunderstanding or a mistake in an interaction, or of correcting a deviation from the normal rules of interaction. Conversation analysts generally consider four kinds of repair[15] [14,12]. These

---

[13] Frohlich and Luff [22] provides an example of the use of repair in an existing computational system

[14] External context is important in CA. However, the 'context-free' aspect of certain CA concepts, such as the conversation turn-taking system described in [13] is the part that we focus on in this paper as it seems to offer the greatest computational advantages.

[15] Clark [3] describes an additional three categories of repair: preventatives, which prevent a problem from occurring; warnings, which warn of a future unavoidable problem and repairs, which refers to repair of the type considered here in which action is taken after a problem in order to resolve it. While these ideas are interesting and relevant they will not be considered here due to space limitations.

four kinds of repair are generally labelled as follows, where 'self' is the interactant whose action is being repaired and 'other' is the other interactant:

1. Self-initiated self-repair: problem detected and repaired by self
2. Other-initiated self-repair: problem pointed out by other but repaired by self
3. Self-initiated other-repair: problem pointed out by self, repaired by other (e.g. prompting for help with a memory lapse)
4. Other-initiated other-repair: problem both pointed out and repaired by other

The ability to engage in repair is essential in interaction: errors and misunderstandings are likely to arise and must be corrected if the interaction is to be successful. In applying concepts from CA in a visual interaction we might expect to face some difficulties, as CA usually deals with sequences of talk-in-interaction rather than sequences of movement. However, we find that it is possible to consider equivalent situations in a purely visual interaction. We will consider each of the above classes of repair in turn, first in cases where the human is 'self', then in cases where the robot is 'self'.

Human Error: Self-Initiated Self-Repair. Let us consider the case where the human performs action c_1 but meant to perform c_2. He might well rapidly replace c_1 with c_2, something that he could accomplish in a number of ways. He could transform c_1 to c_2 by changing action part-way through (this is likely to confuse virtually any present machine vision system); he could abort c_1 part-completed and perform c_2 instead or he could complete c_1, abort and perform c_2, immediately or after an arbitrary amount of time. In all of these cases we may expect negation behaviour of some kind, such as head-shaking or saying 'no', although such behaviour is not guaranteed to occur.[16] In each of these cases, if c_1 has been identified then the robot may already be engaging in some manner of response: this is most likely, of course, in the cases where c_1 is completed. We could build a short delay into the robot, giving it time to detect a second action following negation behaviour. If negation behaviour is detected then the robot can terminate any response that it is currently engaged in and wait for the next non-negating action. However 'negation behaviour' is a very abstract concept and could include a range of behaviors, including vocalizations, head shaking, hand waving, a pause in action and so on. It seems unlikely that a robot could be programmed to recognize visual 'negation behaviour' generally, although it could be programmed to recognize specific instances of such behaviour if particular kinds of negation behaviour are found to occur frequently.

Human Error: Other-Initiated Self-Repair. One form of repair is repetition of an action that has not produced an expected response, as in the following example from [14](p. 42):

---

[16] Other negation behavior is possible, as long as the robot is able to perceive and correctly interpret it. One might also think of a set of predefined and easy to recognize words/gestures, a basic 'vocabulary', possibly domain-specific, that can be used in human-robot interaction.

1   Child: Have to cut these Mummy. (1.3) Won't we
2   Mummy.
3   (1.5)
4   Child: Won't we.
5   Mother: Yes.

Here the child responds to the mother's failure to answer her question (a violation of adjacency pair structure) by repeating the question until getting an answer.[17] This seems a sufficiently simple behavior for the robot to be able to engage in it. Thus if the human performs an action that violates the robot's expectations (remember that inaction can be considered a form of action), repetition of the elicitor provides a means for attempting to produce the expected behavior instead.

Human Error: Self-Initiated Other-Repair. It certainly seems possible that the human would be unsure about the next move in a movement sequence, either through uncertainty about some (possibly implicitly) assumed set of rules or through a memory lapse. There are two ways that this might be visually communicated: a cessation of activity, or the generation of some action but in an 'uncertain', perhaps hesitant manner. Humans are good at detecting uncertainty in others and it seems that a variety of visual cues might be involved in this, perhaps including facial expression and general bodily tension and hesitancy. However, getting a robot to detect uncertainty seems extremely difficult at best. If the robot could detect the human's uncertainty then it could prompt the human in some way e.g. by speaking or by repeating the previous action in hopes of triggering a memory through association.

Human Error: Other-Initiated Other-Repair. In the conversational case this would take the form of the human saying something and the robot producing a corrected version of the human's word or sentence. In the visual case we could argue that the human performing an action other than what the robot expected, with the robot then demonstrating the action the human should have performed, constitutes a roughly equivalent structure. This still leaves us, though, with the problem of how the human is to know that the robot's action is a correction and not simply a response. The robot would have to provide an explicit signal along the lines of the negation behaviour described earlier before demonstrating or describing the correct action.

Robot Error: Self-Initiated Self-Repair. We consider the case where a robot is responding to an action that it has assumed to be c_1 that, as the human continues moving, turns out to be c_2 as an example of the robot catching its own error and thus initiating its own repair. In this case the appropriate behaviour would seem to be to stop, perhaps to engage in some kind of negation behaviour,

---

[17] We note that a similar strategy is employed by an autistic child in [23].

and then to perform the correct action. Of course, if the human continues to act through all this then the robot should start acting again from the latest action: presumably the human has ignored or missed the repair behaviour (or the original error) if he continues to act through it.

Robot Error: Other-Initiated Self-Repair. For this to work the robot needs to be able to detect the human's initiation, which again leaves us with the difficult problem of detecting and correctly interpreting uncertainty and surprise reactions. In this case, the robot could shift to the next most probable interpretation for the current action, or maintain the present interpretation but shift to a less likely expectation. It should engage in negation behaviour first to make it clear that it is abandoning the previous action and switching to a different one.

Robot Error: Self-Initiated Other-Repair. This seems very useful, as it would cover cases where the robot does not know what action to take. If the robot can signal its uncertainty then the human can show it what to do, for instance demonstrating that the robot should move forward by approaching the robot, or that it should move away by retreating away from the robot.

Robot Error: Other-Initiated Other-Repair. Provided that the robot can detect the human's initiation, this would function very similarly to the self-initiated other-repair case. The robot would have to stop as soon as it detected the human's initiation and observe the demonstrated action, then repeat that action itself. Detection of initiation, as always, is the significant problem here.

3.5   Summing Up

We note that a common problem in the above section is the issue of identifying cases where a problem has occurred — it seems that we can get around this problem in many cases by prespecifying a simple 'negation behaviour'. It seems that by assuming the use of explicit signalling (negation) behaviour by both parties to indicate that repair is being engaged in, repair of a limited kind does become an option even in our simple visual interaction. However, the detection of such negation behaviour seems problematic and it is unclear to what extent negation behaviour is commonly used in human interaction: we do not know if we can expect such behavior or not. Repair is potentially very useful as it gives both parties in the interaction a greater ability to make their expectations clear and to influence the behaviour of the other interactant. In our particular application domain, where we study the interaction of children with autism with a mobile robot [24,25,23], repair seems especially important for the children, some of whom already attempt to give the robot verbal instructions as to how it should behave which, of course, it ignores. For these children the ability to correct the robot and to get it to behave in a way that they consider appropriate could be very valuable.

4   Measurement

Our goal in all of this, of course, is to create an agent capable of having a 'successful' interaction with a human, based on criteria used for the measurement of human–human interaction rather than on external optimization criteria. It is difficult to define exactly what constitutes a successful interaction, which partly follows from the difficulty of measuring interaction. There are a number of measures we can consider, though. First, of course, there is the possibility of using CA to analyze our interactions [23], giving us a rigorous qualitative assessment of the success of the interaction. Equally, we can assess the satisfaction of the human with the interaction through interviews and questionnaires. We can also suggest more quantitative measures, such as duration of interaction or time human spends gazing at the robot [23]. We also note the existence of the tool THEME, a statistical program which seeks patterns in movement and which has been used for the study of interactive behavior[26,27,28]. [4] also provides means of examining the interactive behaviour of virtual agents, albeit in a quite different context. In each case we can use a robot controlled by a simpler, non-interaction-aware controller (e.g. following heat sources, as in the Aurora project [24,25,23]) as a control.

5   Conclusion and Other Work

We have considered application of a number of ideas derived from the human sciences in the development of artificial interactive agents. Our initial, brief consideration of heuristics applicable to identification of longer sequences of interaction provides us with the beginning, at least, of a means of constructing a system capable of identifying observed interactive behaviors, albeit in a coarse way. Our more detailed consideration of extension of a simple interactive agent architecture to encompass something closer to 'true' interaction and to incorporate some CA-inspired structures suggests a means of building an agent capable of generating interactive behaviors in a very simple interaction. Hopefully we have demonstrated the usefulness of considering the design of interactive agents in the light of knowledge from the human sciences. We also do not wish to give the impression that we are only interested in the dancing interaction: we are merely using this as an example. Future work will create agents with an architecture based on section 3 of this paper and study their interactions with both humans and other interactive agents. At the moment the agents in question are Khepera robots [29] operating in the Webots simulation environment [30]: the controllers involved will also be ported to real Kheperas.

It seems to us that the sequences of interaction described by CA may usefully form the basis for computational modeling and robot controllers if context is appreciated and the diversity of possible initiations and responses considered, with the action produced always being a matter of local activity and contingencies. At this point we are only dealing with a very limited part of the full complexity of human interaction as described by CA, but we believe that our approach represents a reasonable starting point.

Acknowledgements

This work is partially supported by EPSRC grant GR/M62648. We would like to thank Paul Dickerson and John Rae for discussions and explanations about conversation analysis and for their involvement in the Aurora project.

References

1. Maturana, H. R., Varela, F. J.: Autopoiesis and Cognition. D. Reidel, Dordrecht, Holland (1980) 255
2. Dautenhahn, K., Ogden, B., Quick, T.: From embodied to socially embedded agents: Implications for interaction among robots. Cognitive Systems Research Journal (to appear) Special Issue on Situated and Embodied Cognition. 255
3. Clark, H. H.: Managing problems in speaking. Speech Communication **15** (1994) 243–250 255, 261
4. di Paolo, E.: Behavioral coordination, structural congruence and entrainment in a simulation of acoustically coupled agents. Adaptive Behavior **8** (2000) 27–48 255, 259, 261, 265
5. Yoon, S. Y., Burke, R. C., Blumberg, B. M., Schneider, G.: Interactive training for synthetic characters. In: Proceedings of the Seventeenth National Conference on Artificial Intelligence and Twelfth Conference on Innovative Applications of Artificial Intelligence, Austin, Texas, USA, AAAI Press / The MIT Press (2000) 249–254 255
6. Cassell, J., Sullivan, J., Prevost, S., Churchill, E., eds.: Embodied Conversational Agents. MIT Press (2000) 255
7. Kendon, A.: A description of some human greetings. [8] 255, 256
8. Kendon, A.: Conducting Interaction: Patterns of Behavior in Focused Encounters. Cambridge University Press, Cambridge, UK (1990) 255, 266
9. Kendon, A.: Features of the structural analysis of human communicational behavior. In von Raffler-Engel, W., ed.: Aspects of Nonverbal Communication. Swets and Zeitlinger, Lisse, Netherlands (1980) 255
10. Schank, R. C., R, A.: Scripts, plans, goals and understanding. Lawrence Erlbaum Associates Inc, Hillsdale, NJ (1977) 255
11. Collett, P.: Mossi salutations. Semiotica **45** (1983) 191–248 255
12. Levinson, S. C.: Pragmatics. Cambridge University Press, Cambridge, UK (1989) 256, 257, 261
13. Sacks, H., Schegloff, E. A., Jefferson, G.: A simplest systematics for for the organization of turn-taking for conversation. Language **50** (1974) 696–735 256, 257, 261
14. Hutchby, I., Wooffitt, R.: Conversation Analysis. Polity Press, Cambridge, UK (1998) 256, 260, 261, 262, 267
15. Dautenhahn, K.: Embodiment and interaction in socially intelligent life-like agents. In Nehaniv, C., ed.: Computation for Metaphors, Analogy and Agents. Springer-Verlag, Berlin, Germany (1999) 257
16. Luff, P., Gilbert, N., Frohlich, D., eds.: Computers and Conversation. Academic Press Ltd, London, UK (1990) 257, 267
17. (Ferrell), C. B.: Early experiments using motivations to regulate human-robot interaction. In: Proceedings of 1998 AAAI Fall Symposium: Emotional and Intelligent, The Tangled Knot of Cognition, Orlando, FL, USA. (1998) 259

18. Breazeal, C.: Sociable machines: Expressive social exchange between humans and robots. PhD thesis, Department of Electrical Engineering and Computer Science, MIT (2000) 259

19. Breazeal, C., Scassellati, B.: Infant-like social interactions between a robot and a human caretaker. Adaptive Behavior **8** (2000) 49–74 259

20. Bryan, K., McIntosh, J., Brown, D.: Extending conversation analysis to non-verbal communication. Aphasiology **12** (1998) 178–188 260

21. Robinson, J. D.: Getting down to business: Talk, gaze and body orientation during openings of doctor-patient consultations. Human Communication Research **25** (1998) 97–123 260

22. Frohlich, D., Luff, P.: Applying the technology of conversation to the technology for conversation. [16] chapter 9 261

23. Dautenhahn, K., Werry, I., Rae, J., Dickerson, P., Stribling, P., Ogden, B.: Robotic playmates: Analysing interactive competencies of children with autism playing with a mobile robot. In Dautenhahn, K., Bond, A., Cañamero, L., Edmonds, B., eds.: Socially Intelligent Agents - Creating Relationships with Computers and Robots. Kluwer Academic Publishers (to appear) 263, 264, 265

24. Werry, I., Dautenhahn, K.: Applying mobile robot technology to the rehabilitation of autistic children. In: Proceedings of the Symposium on Intelligent Robotic Systems (SIRS'99), Coimbra, Portugal (1999) 265–272 264, 265

25. AURORA: http://www.aurora-project.com. Last referenced 21st January 2002 (2002) 264, 265

26. Magnusson, M. S.: Hidden real-time patterns in intra- and inter-individual behavior: Description and detection. European Journal of Psychological Assessment **12** (1996) 112–123 265

27. Tardif, C., Plumet, M. H., Beaudichon, J., Waller, D., Bouvard, M., Leboyer, M.: Micro-analysis of social interactions between autistic children and normal adults in semi-structured play situations. International Journal of Behavioral Development **18** (1995) 727–747 265

28. Grammer, K., Kruck, K. B., Magnusson, M. S.: The courtship dance: Patterns of nonverbal synchronization in opposite-sex encounters. Journal of Nonverbal Behavior **22** (1998) 3–29 265

29. K-Team: http://www.k-team.com. Last referenced 21st January 2002 (2002) 265

30. Cyberbotics: http://www.cyberbotics.com. Last referenced 21st January 2002 (2002) 265

A Some CA Notation

The following is adapted from [14].

CAPITALS	Speech is noticeably louder than that surrounding it.
!	Exclamation marks are used to indicate an animated or emphatic tone.
.	A stopping fall in tone - not necessarily the end of a sentence.
?	A rising inflection - not necessarily a question.
<u>Under</u>	Underlined fragments indicate speaker emphasis.
(n)	The number in brackets indicates a time gap in tenths of a second.
(( ))	A nonverbal activity e.g. ((banging sound)), or a transcriber's comment.
[	Indicates the end of a spate of overlapping talk.

Are Praxical Gestures Semiotised in Service Encounters?

Isabelle Dumas

Research Group in Communicative Interactions – UMR 5612, University Lumière
Lyon 2, 5 avenue Pierre Mendès-France, 69676 Bron Cedex, France
Isabelle.Dumas@univ-lyon2.fr

Abstract. Empirically based on the study of praxical gestures in service
encounters, this paper questions the possibility of semiotisation for such
gestures. Praxical gestures are supposed to be extra-communicative and
thus "unsemiotised". In reality, although they are not coded in a
systematic way, their meaning is fundamentally connected to their
context of realisation. An analysis of the contexts in which praxical
gestures appear in service interactions demonstrates that they become
"semiotised" when they are put into context and that they have a full
part in the script of service interactions.

1   Introduction

One of the characteristics of service encounters is that they are anchored in a *script*,
that is "a predetermined stereotyped sequence of actions that defines a well-known
situation" [5], p. 41. More characteristic of the particular type of interactive situation
studied is the fact that we have a combination of verbal and non-verbal activities. In
service interactions, some communicative gestures are present, such as smiles or
greetings for example, but these gestures are not the object of the present study. The
types of non-verbal behaviors which are relevant to this type of script are those which
have not been described much in the descriptive tradition of non-verbal
communication specialists because they are judged as being *extra-communicative* [1],
p. 2037. In the literature, communicative gestures are studied but rarely if ever, extra-
communicative gestures. These have nevertheless been identified and sometimes
called *praxical gestures* [2], p. 47, or *physical doings* [3], p. 38. Goffman provides an
example of these gestures in service interactions in the following description: "a
customer who comes before a checkout clerk and places goods on the counter has
made what can be glossed as a first checkout move, for this positioning itself elicits a
second phase of action, the server's obligation to weigh, ring up and bag" (*ibid.*).

The problem raised by these praxical gestures is that they do not seem to be
semiotised because they are not coded in a systematic way, as some other gestures
may be. If this is the case, they are not communicative.

Up to now, the author of this paper has worked essentially on speech acts and
verbal activities in general but she has a growing interest in the integration of the non-
verbal component in her descriptive model. She is interested in praxical gestures in as
far as they enable the researcher to examine globally an interaction by shedding light
on its function.

I. Wachsmuth and T. Sowa (Eds.): GW 2001, LNAI 2298, pp. 268-271, 2002.
© Springer-Verlag Berlin Heidelberg 2002

2   Data

Working on a comparative study of service interactions, the author was led to work from case studies, that is from empirical data. Several corpora were collected but the present study will only centre on two of these. Each corpus is composed of about forty recordings of a particular type of service encounter. Each recording corresponds to an interaction and begins and ends with a greeting. The corpora were recorded with a hidden microphone and/or camera. The shopkeepers knew they were being recorded while the customers did not. The first corpus was collected in a stationer's, the second in a tobacconist's.

3   Are Praxical Gestures Semiotised in Service Encounters?

Most gesture specialists agree that there are *communicative* gestures as opposed to other *extra-communicative* gestures. However, they fail to reach a consensus on where exactly to place the cut-off point between these two types of gesture. They consider that there is a continuum with communicative gestures at one end and extra-communicative ones at the other. Along this continuum, various types of gestures are placed which are more or less loaded in communicative value with various degrees of semiotisation. Problems arise however when one tries to define the criteria necessary to categorise a gesture as communicative.

In what measure can praxical gestures be said to have meaning? Grosjean and Kerbrat-Orecchioni [4] partly answer this question when they compare speech acts and non-verbal activities. The following example of a praxical gesture illustrates this. A lights a cigarette and B immediately opens the window. The praxical end of B's gesture takes on the symbolic value of a reproach, i.e. B opens the window to show A that he is bothered by the cigarette smoke. Here, the praxical gesture clearly works as an implicit speech act.

Following this line of reasoning, it is easy to think of other examples such as the distinction between closing a door and slamming a door. In this case, the speech act provides some information about the state of mind of the individual closing the door. Yet there are other ways of giving meaning to the action, the main one being the connection with the context. Speech acts always make sense because they are coded in language. However, praxical gestures are not coded in a systematic way: their meaning is fundamentally connected to the context in which they are realised. It is when they are put into context that praxical gestures become semiotised in service interactions. Their meaning is fundamentally connected to their context of realisation.

Thus praxical gestures are functional. They are simultaneously oriented towards the intentional transformation of the physical environment and towards the regulation of symbolic practices [6], p. 367.

4   Contexts in which Praxical Gestures Appear

Since the semiotisation of praxical gestures is established by their context, it is advisable to examine the contexts in which they appear. Two types of praxical gestures have thus been identified: gestures in reaction to a verbal appeal and gestures in reaction to a non-verbal appeal.

4.1 Praxical Gestures Semiotised in Reaction to a Verbal Appeal

This category of praxical gesture is the easiest to distinguish since an utterance appeals to a non-verbal reaction. In the corpora collected, these utterances generally correspond to the speech act of request. The request for goods can take two different forms and be uttered with either an indirect or an elliptical formulation. The indirect formulation of the request appears fairly frequently in the corpora and takes several forms that will not be dealt with here in detail. The indirect formulations represent more than 90% of the customer's requests.

Example at the stationer's: *I'd like a cheap blue pen*

The elliptical formulation of the request appears fairly infrequently and represents less than 10% of the requests.

Example at the tobacconist's: *A big box of matches*

While it is important to be aware of the difference in the formulation of the request, it has no particular effect on the semiotisation process of the praxical gesture to follow.

The problem with these formulations of the request is establishing whether they are conventional indirect requests or not. Among several criteria, a speech act can be said to be conventionally indirect when the reaction it is supposed to provoke takes place immediately. After a true conventional indirect request, a non-verbal reaction is expected to follow, i.e. the service of the customer. Of course, all the requests in the corpora are conventional indirect requests implying the immediate service of the customer. Besides there are several degrees of conventionalisation that will not be dealt with in detail here. What should be noted is that the immediacy of the non-verbal reaction is all the more striking as the request is more routinised because the shopkeeper reacts non-verbally to a request, not to a question or a request for a piece of information. The following example of a pure conventional indirect request at the tobacconist's shows that the shopkeeper serves the customer immediately.

> **Customer**: *I'd like a packet of Kim orange please*
> *The tobacconist turns round to take a packet of tobacco and puts it on the counter.*

On the other hand, when the request is lower in the degree of conventionalisation, the shopkeeper generally answers the customer verbally before giving him the goods, as can be seen in this example at the stationer's.

> **Customer**: *Have you got any cartridges for this fountain pen?*
> **Shopkeeper**: *Yes, I'll just get you one*
> *The shopkeeper hands the customer a cartridge.*

The interconnection between the speech act and the praxical gesture that follows makes it become meaningful, functional and semiotised, whether the service of the customer is delayed or not.

4.2 Praxical Gestures Semiotised in Reaction to a Non-verbal Appeal

The second category of praxical gestures is found when a gesture from one participant implies a gestural reaction on the part of the other. In the corpora this gestural reaction to a non-verbal appeal can be found in two types of situation.

On Sundays when you buy the local newspaper in France, a T.V. guide is given to you. So the customer enters the stationer's shop, takes the newspaper, and puts the newspaper on the counter. This gesture functions as a non-verbal request for the supplement and the stationer puts it on top of the newspaper.

At the tobacconist's you can buy a French national lottery ticket, which is a scratch card, for ten francs each. You win if you find the same sum of money three times. The most frequent return is ten francs, which is the price of your ticket. Often customers do not ask for the repayment of their winnings. They prefer to exchange them for another ticket. The regular customers do not ask for this exchange. They put their ticket on the counter and this gesture substitutes the request non-verbally.

Since the non-verbal appeal functions as a non-verbal request, the result is exactly the same as with the verbal request: the praxical gesture following the non-verbal appeal becomes both functional and semiotised.

5   Conclusions

The description of the contexts in which praxical gestures appear in service interactions establishes their semiotic character even more firmly. Praxical gestures can be considered as semiotised when they are put into context for at least two reasons. First of all, they accompany the verbal statements and are not redundant. Moreover the fact that they can be inserted into the script of the interaction is the biggest proof of their communicative, functional and semiotised character since it is impossible to relate the progress of a service encounter without referring to these praxical gestures. They are involved in the stream of action in the first degree.

References

1. Cosnier, J.: Communication non Verbale et Langage. In: Psychologie Médicale, vol. 9, n°11 (1977) 2033-2049
2. Cosnier, J. & Vaysse, J.: La fonction référentielle de la kinésique. In: Prothée, Théories et Pratiques Sémiotiques, vol. 20, n°2 (1992) 40-47
3. Goffman, E.: Forms of Talk. University of Pennsylvania Press, Philadelphia, Second printing (1983)
4. Grosjean, M. & Kerbrat-Orecchioni, C.: Acte Verbal et Acte Non Verbal: ou Comment le Sens Vient aux Actes. In: Actes du colloque "Les relations intersémiotiques", Lyon (2002, forthcoming)
5. Schank, R.C. & Abelson, R.P.: Scripts, Plans, Goals and Understanding: an Inquiry into Human Knowledge Structure. Lawrence Erlbaum Associates, Hillsdate (N.J.) (1977)
6. Streeck, J.: How to Do Things with Things. In: Human studies, n°19. Kluwer Academic Press (1996) 365-384

Visually Mediated Interaction Using Learnt Gestures and Camera Control

A. Jonathan Howell and Hilary Buxton

School of Cognitive and Computing Sciences, University of Sussex
Falmer, Brighton BN1 9QH, UK

Abstract. In this paper we introduce connectionist techniques for visually mediated interaction to be used, for example, in video-conferencing applications. First, we briefly present background work on recognition of identity, expression and pose using Radial Basis Function (RBF) networks. Flexible, example-based, learning methods allow a set of specialised networks to be trained. Second, we address the problem of gesture-based communication and attentional focus using Time-Delay versions of the networks. Colour/motion cues are used to direct face detection and the capture of 'attentional frames' surrounding the upper torso and head of the subjects, which focus the processing for visually mediated interaction. Third, we present methods for the gesture recognition and behaviour (user-camera) coordination in the system. In this work, we are taking an appearance-based approach and use the specific phases of communicative gestures to control the camera systems in an integrated system.

Keywords. Gesture Recognition; Computer Vision; Visually Mediated Interaction; Camera Control; Face Recognition; Time-Delay Neural Networks

1   Introduction

Visually mediated interaction (VMI) is a process of facilitating interaction between people, either remotely or locally, using visual cues which are often similar to those used in conventional, everyday interaction with other people. The aim is to enhance interaction, overcoming limitations due, for example, to distance or disability. This involves many visual competences such as recognising facial expression, gaze, gesture and body posture which are all used in human communication and interaction. Gestures are often spontaneous but can also be intentional, where we can distinguish between verbal (sign languages) and nonverbal (pointing, emphasis, illustration) usage. In our work here we are mainly concerned with intentional, nonverbal gestures which are relevant for communication in VMI. Also, we use gaze which can provide an important cue for discourse/interaction management. In particular, gaze direction is often associated with diectic pointing to indicate objects or people of interest (focus of attention) as part of the behavioural interaction.

I. Wachsmuth and T. Sowa (Eds.): GW 2001, LNAI 2298, pp. 272–284, 2002.
© Springer-Verlag Berlin Heidelberg 2002

In general, however, the robust tracking of non-rigid objects such as human faces and bodies involved in the machine analysis of this kind of interactive activity is nontrivial due to rapid motion, occlusion and ambiguities in segmentation and model matching. Research funded by British Telecom (BT) on *Smart Rooms* [18] and the ALIVE project [15] at MIT Media Lab have shown progress in the modelling and interpretation of human body activity. This used the *Pfinder* (Person Finder) system [23], which can provide real-time human body analysis. Gesture recognition then involves techniques such as Hidden Markov models (HMMs) which can be parameterised to provide information such as direction of pointing [22] or human interaction analysis using the Conditional Expectation Maximisation (CEM) approach [14]. Recently, we have been developing computationally simple view-based approaches to action recognition under the ISCANIT project. These new systems start to address the task of intentional tracking and behavioural modelling to drive visual interaction.

Our original work centred around example-based learning techniques for a pose-varying face recognition task [11]. The particular task considered was the recognition in real-time of a known group of people within indoor environments. It could not be assumed that there would be clear frontal views of faces at all times and so a key capability was to identify faces over a range of head poses [9]. An important factor in this approach was the flexibility of the example-based *Radial Basis Function* (RBF) network learning approach, which allowed us to reformulate the training in terms of the specific classes of data we wished to distinguish. For example, we could extract identity, head pose and expression information separately from the same face training data to train a computationally cheap RBF classifier for each separate recognition task [2,7]. Further studies using image sequences [8] and recognition of simple temporal behaviours like head rotation using Time-Delay RBFs [6] suggested the computational feasibility of fast training and on-line recognition using these techniques.

2   Gesture Recognition

Our initial research to support *Visually Mediated Interaction* (VMI) concerned the development of person-specific and generic gesture models for the control of active cameras. A Time-Delay variant of the Radial Basis Function (TDRBF) network was used to recognise simple pointing and waving hand gestures in image sequences [10]. The ISCANIT Phase I gesture database was developed as a source of suitable image sequences for these experiments. Characteristic visual evidence can be automatically selected during the adaptive learning phase, depending on the task demands.

A set of *interaction-relevant* gestures were modelled and exploited for reactive on-line visual control. These can then be interpreted as user intentions for live control of an active camera with adaptive view direction and attentional focus. In particular, pointing (for direction) and waving (for attention) are important for deliberative control and the reactive camera movements may be able to provide the necessary visual context for applications such as group video-conferencing as well as automated studio direction.

Previous approaches to recognising human gestures from real-time video as a nonverbal modality for human-computer interaction have involved computing low-level features from motion to form *temporal trajectories* that can be tracked by Hidden Markov Models or Dynamic Time Warping. However, for this work we explored the potential of using simple image-based differences from video sequences in conjunction with the RBF network learning paradigm to account for variability in the appearance of a set of predefined gestures. The computational simplicity and robust generalisation of our alternative RBF approach provided fast training and on-line performance, highlighting its suitability as a source of interactive responses required by applications with active camera control.

The new experimental results were able to show that high levels of performance for this type of *deliberative gesture recognition* can be obtained using these techniques both for particular individuals and across a set of individuals. Characteristic visual evidence was effectively selected and can be used, if required, even to recognise individuals from their gestures [12]. Previous TDRBF network experiments had learnt certain simple behaviours based on y-axis head rotation [9], distinguishing between left-to-right and right-to-left movements and static head pose. Such tasks are simplified by their constant motion, so that arbitrary short segments (2/3 frames) of the whole sequence can be used to identify the overall direction of head turning. Due to the complex motion involved in the ISCANIT Phase I gestures, characteristic parts of the complete action needed to be contained in the time window presented to the network in order that it can be recognised. The initial requirement to present the entire gesture sequence for recognition meant that event signalling could only be done retrospectively.

Subsequent work has further refined the training for the gesture information to reduce the amount of data required [13]. Taking advantage of the *tri-phasic nature* of the waving and pointing gestures, each gesture can be split into a pre-, mid- and post-gesture sequence. Each of these can be trained as its own gesture class. This gives some predictive power from the pre-gesture and focusses on the characteristic movement of the mid-gesture. It also gives some temporal invariance by allowing the two phases of the gesture to be different in length to each other (the original studies imposed a fixed relationship between the overall gesture length and its three phases). Rapid signalling of the gesture event can be obtained in this way, as it can be performed without waiting for the post-gesture phase, see Fig. 1.

2.1   The Phase I Gesture Database

The ISCANIT Phase I gesture database [10,16] was created to provide a source of single-person gesture data. It concentrated on two specific behaviours which could be used to move the camera or adapt its field of view: *pointing*, which is interpreted as a request to pass camera attention, and is implemented by zooming out and panning in the pointing direction, and *waving*, which is interpreted as a request for camera attention, and implemented by panning towards the waver and zooming in. We have two types of each behaviour, giving four gestures in all, shown in Table 1.

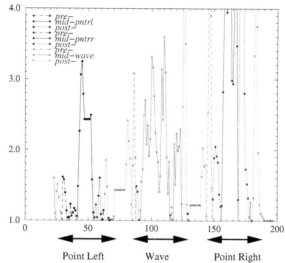

Fig. 1. Output for the multi-phase TDRBF gesture network with a test sequence with different background, lighting and person to that encountered during training. Each line represents a pre-, mid- or post-gesture class, and shows the confidence level of output when its class is the maximum, and is zero at all other times

Table 1. Definitions for the ISCANIT Phase I gesture database

Gesture	Body Movement	Behaviour
pntrl	point right hand to left	pointing left
pntrr	point right hand to right	pointing right
wavea	wave right hand above head	urgent wave
waveb	wave right hand below head	non-urgent wave

Four examples of each gesture from three people were collected, 48 image sequences in all. Each sequence contains 59 378×288 8-bit monochrome images (collected at 12 frames/sec for roughly 5 seconds), a total of 2832 images.

2.2 Summary

– Simple preprocessing techniques such as frame differencing and thresholding can be effective in extracting useful motion information and segmenting gestures in time.
– Different types of TDRBF network can be trained to distinguish gestures over specific time windows, for instance person-specific gesture models (trained and tested on one person) and generic gesture models (trained on one person, tested on other people).

- The TDRBF network can distinguish between arbitrary gestures, with a high level of performance.
- Some characteristics of an individual's expression of gestures may be sufficiently distinctive to identify that person.
- The TDRBF network can learn such data both as complete gesture sequences [10] and as specific gesture phases within a tri-phasic structure [13].
- Splitting multi-phasic gestures into separate phase classes not only gives more precise timing of gesture events, but also allows the gesture recognition network to provide prediction hypotheses by identifying pre-gesture classes.

In summary, the Time-Delay RBF networks showed themselves to perform well in our gesture recognition task, creating both person-specific and generic gesture models. This is a promising result for the RBF techniques, considering the high degree of potential variability (present even in our constrained database) arising from different interpretations of our deliberative gestures by each individual. Note that this is in addition to variability in position, lighting etc. that had to be overcome in earlier face and simple behaviour recognition work.

3   Interpretation of Group Behaviour

While full computer understanding of dynamic visual scenes containing several people may be currently unattainable, we have investigated a computationally efficient approach to determine areas of interest in such scenes. Specifically, we have devised a method for modelling and interpretation of single- and multi-person human behaviour in real time to control video cameras for visually mediated interaction [21]. Such machine understanding of human motion and behaviour is currently a key research area in computer vision, and has many real-world applications. *Visually Mediated Interaction* (VMI) is particularly important to applications in video telecommunications. VMI requires intelligent interpretation of a dynamic visual scene to determine areas of interest for effective communication to remote users.

Our general approach to modelling behaviour is *appearance-based* in order to provide real-time behaviour interpretation and prediction [10,16,20]. In addition, we only use views from a single pan-tilt-zoom camera with no special markers to be worn by the users. We are not attempting to model the full working of the human body. Rather our aim is to exploit approximate but computationally efficient techniques. Thus, our models are able to support partial view-invariance, and are sufficient to recognise people's gestures in dynamic scenes. Such task-specific representations need to be used to avoid unnecessary computational cost in dynamic scene interpretation [1,20].

For our purposes, *human behaviour* can be considered to be any temporal sequence of body movements or configurations, such as a change in head pose, walking or waving. However, the human body is a complex, non-rigid articulated system capable of almost infinite spatial and dynamic variations. When attempting to model human behaviour, one must select the set of behaviours to be modelled for the application at hand. For VMI tasks, our system needs to identify regions of interest in a visual scene for communication to a remote user. Examining the case in which the scene contains people involved in a video conference, the participant(s) currently involved in communication will usually constitute the appropriate focus of attention. Therefore, visual cues that indicate a switch in the chief communicator, or 'turn-taking', are most important. Gaze is a significant cue for determining this focus of communication, and can be approximated by head pose. *Implicit behaviour* can be defined as any body movement sequence that is performed subconsciously by the participant, and here, it is head pose that is the primary source of implicit behaviour.

However, head pose information may be insufficient to determine a participant's focus of attention from a single 2D view, due to loss of much of the 3D information. Then, it is necessary to have the user communicate explicitly with our VMI system through a set of pre-defined behaviours. *Explicit behaviour* can be defined as a sequence of body movements that are performed consciously by a participant in order to highlight regions of interest in the scene. We used a set of pointing and waving gestures as explicit behaviours for control of the current focus of attention. As we have seen, such gestures can be reliably detected and classified in real-time [10,16].

Our approach to modelling group interaction involves defining the *behaviour vector* of a participant to be the concatenation of measured implicit and explicit behaviours (head pose angles and gesture model likelihoods). From this a *group vector* can be defined as a concatenation of the behaviour vectors for all people present in the scene at a given time instant, and *group behaviour* is just a temporal sequence of these group vectors. Given the group behaviour, a high-level interpretation model can determine the current area of focus. In our scenarios, the region of interest is always a person so we track the head of each individual. The output need only give an indication of which people are currently attended in the high-level system and is called the *camera position vector*. This has a boolean value (0 or 1) for each person in the scene indicating whether that person is currently attended. This information can then be used to control the movable camera, based on the position of the people in the scene.

Given a particular group behaviour, we constructed a *scene vector*, which contains the previous camera position vector information as feedback. This allowed the current focus of attention to be maintained, even when no gestures or head turning occurred. To learn the transformation from scene vector to camera position vector, we developed an effective Time-Delay RBF Network, trained on half of our sequence database and tested for generalisation on the other half [21]

The bottom section of Fig. 2 shows the training signal, or target camera vectors, traced above the actual output camera vectors obtained during tests with the trained RBF network. It can be seen that the network follows the general interpretation of group behaviour, although the transition points from one focus of attention to another do not always exactly coincide. However, these transition points are highly subjective and very difficult to determine with manual coding,

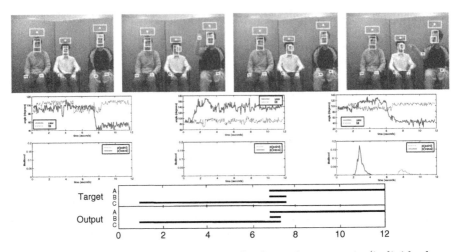

Fig. 2. Group behaviour recognition results for **point** scenario (individuals are labelled A, B and C from left to right): Frames from sequence (top), plots showing pose angles and gesture likelihoods (middle), and target/output camera position vectors (bottom)

so this result is not surprising and the results give switches of attention that are acceptable at the perceptual level.

3.1   The Phase II Group Interaction Database

The ISCANIT Phase II database [21] contains examples of group interaction in a static scene. This database contains 15 sequences, each between 240 and 536 frames in length, a total of 5485 320×240 24-bit colour images. We constrain the complexity of the data by restricting behaviour to certain fixed scenarios, shown in Table 2, and by always having three participants, who remain sitting for the complete sequence. Each scenario is a group behaviour in which the participants perform gestures and change their head pose in a fixed pre-defined order. The exact timing of the events varies between different instances of the same scenario, but the focus of attention switches from one region to the next in the same order.

3.2   Summary

- A framework has been devised for tracking people and recognising their group behaviours in VMI contexts. This requires high-level information about group and individual interaction in a 'scene vector' to learn a 'camera control vector', specified by a temporal model.
- The scene vector provides ongoing probabilities of the dynamic head-pose and gesture phases for interacting participants and the camera control vector provides reactive direction and zoom.
- Pre-defined gestures and head pose of several individuals in the scene can be simultaneously recognised for interpretation of the scene.

Table 2. Scenario descriptions for the ISCANIT Phase II group interaction database, involving three participants A, B and C

Scenario	Description
wave	Person C waves and speaks, A waves and speaks, B waves and speaks
wave-look	Person C waves and speaks, A waves and speaks, B waves and speaks. Each time someone is speaking the other two participants look at him
question	Person C waves and speaks, A and B look at C, A interjects with a question, C looks at A to answer, then looks back at camera
point	Person C waves and speaks, A and B look at C, C points to A, C and B look at A, A looks at camera and speaks
interrupt	Person C waves and speaks, A and B look at C, a person enters from the left, A, B and C watch as the person leaves, C looks at the camera and continues speaking, A and B look at C All participants look at the camera unless stated otherwise.

- A scene vector-to-camera control transformation can be performed via a TDRBF network, using example-based learning.

We have been able to show how multi-person activity scenarios can be learned from training examples and interpolated to obtain the same interpretation for different instances of the same scenario. However, for the approach to scale up to more general applications, it must be able to cope with a whole range of scenarios. The approach implicitly requires such a system to extrapolate to novel situations in the same way that we do. Unfortunately, there is no reason to believe that current computer architectures are capable of such reasoning and our temporal models fall far short of full intentional semantics. Therefore, a significant issue in future work will be the feasibility of learning generalised temporal structures and default behaviours from sparse data.

4   Towards an Integrated System

In this section we present our work towards a complete connectionist system for understanding the visual aspects of human interaction which could be used, for example, in video-conferencing applications. First, we present methods for face detection and capture of attentional frames to focus the processing for visually mediated interaction. This frame can be used for recognising the various gesture phases that can then be used to control the camera systems in the integrated system, as discussed in previous sections.

4.1   Capturing the Attentional Frame

Our techniques here used colour/motion cues from the image sequence to identify and track the head. Once we know the position and size of the head, we can define an *attentional frame* around the person. The attentional frame is a 2-D area around the focal user that contains all the body movement information relevant to our application, which is all movement of the head and right arm. To allow people to move closer or further away from the camera, this information is normalised for size (relative to head size) around an arbitrary standard position from the camera.

Our main priority is to find *real-time solutions* for our application. Therefore, we used two computationally cheap pixel-wise processing techniques on our image: thresholded frame differencing, giving motion information, and Gaussian mixture models [17], giving skin colour information. These were combined to give a binary map of moving skin pixels within the image, and we used local histogram maxima to identify potential 'blob' regions. A box which was large enough to contain the head at all distances in our target range was then fitted over the centroid of each of these regions.

A robust approach to head tracking using colour/motion blobs is what we call *temporal matching*: the tracker only considers blobs from the current frame which have been matched to nearby blobs from previous frames. This excludes any anomalous blobs that appear for one frame only in an image sequence, and promotes those that exhibit the greatest temporal coherence. Having found the position and size of the head, we can extract the attentional frame from around the person.

4.2   Pose-Invariant Face Detection

The previous section described how we isolated small areas of moving skin-tones from the overall image. This reduces computation and network size, by allowing the face detector to work only within a small subset of the full spectrum of possible objects typically encountered in an office environment. Specifically, we can consider the restricted form of face detection where we need to distinguish a face only from other moving skin-tone blobs (typically hands).

In order to perform effective face recognition, we need to identify the position of the central face area (eyes, nose, mouth), rather than the entire skin area on the head (which also includes forehead, neck, ears, etc). Our face detection task, therefore, is to distinguish centred faces from both non-centred faces and other moving skin-tone blobs. This is in contrast to conventional face detectors, eg. [19], which have to find (frontal) faces in any patch of image. In such a case, there is a finite number of faces and essentially an infinite range of non-face images that can be encountered, and therefore a huge number of training images are required to reliably teach the system.

In our highly constrained task, we need to train a RBF network with a much smaller range of examples than would be needed for a full face detector: in this case, centred faces (of varying pose) for the face class, and non-centred faces for the non-face class. Our system also provides a level of confidence based on the difference between the two output values (face/non-face) from the network [7], which allows discarding of low-confidence results where data is noisy or ambiguous.

Our training examples need to take variable head-pose into account, so the central face region of a person can be recognised at all normal physiological pose positions. Facial information is only visible on a human head from (roughly) the front $\pm 120°$ of x- and y-axis movement, and z-axis movement is physiologically constrained to around $\pm 20°$ (when standing or sitting) [6]. The face region is centralised on the nose, rather than the face, for all profiles, as this allows non-occluded face information to remain roughly in the same position. This has previously been shown to more useful for pose-varying face recognition [6]. We can then easily determine a coarse estimate of head-pose, such as left, frontal or right, from the output grid. This qualitative level of head-pose was found to be very useful for group interaction analysis [21].

4.3   The Integrated System

A complete video-conferencing active camera control system requires high-level interpretation of group and individual interaction [21]. As we have seen, our approach is to provide sufficient information, contained in a 'scene vector' [21], for this interpretation to take place and for the system to provide camera control information, the 'camera control vector'. The scene vector typically provides head-pose and gesture probabilities for the participants in the field of view. If individuated control of the system is required, then we need to identify who these people are (from a small known group). Two extra stages, therefore, are needed: gesture and (pose invariant) identity recognition. Sections 2 and 3 discuss practical techniques for tackling these tasks in real-time, using the RBF and TDRBF networks [8,10].

To complete our integrated system, we need to pass this gesture and head-pose information, with identity if appropriate, to a higher-level interpretation network [21], as discussed in Section 4. In addition, we have to adapt our system to cope with multiple people in the scene, which increases the complexity of the low-level processing stage. There will be more head blobs to find, but by assigning attentional frames to each person, and analysing each of these separately, it is hoped that problems due to occlusion from other members of the group will be kept to a minimum. This will allow a full implementation of the multi-user system with generalised attentional switching.

4.4   Summary

- We can use colour/motion cues to effectively segment and track human heads in image sequences.
- An attentional frame can be extracted relative to the head position and size to allow the real-time recognition of hand gestures through time.
- By extracting colour/motion regions from the overall image, the face detection task is greatly simplified.
- A face detection network can be used to give a qualitative estimate of head-pose for predictive control using implicit behaviour.

– Splitting multi-phasic gestures into separate phase classes not only gives more precise timing of gesture events, but also allows the gesture recognition network to provide prediction hypotheses for explicit behaviour control.

Although it has been possible to fully integrate real-time recognition, tracking and on-line intentional control for single users, there are still some outstanding problems for multiple interacting users. A major issue with the machine learning approach to multi-participant behaviour interpretation is the feasibility of collecting sufficient data. The multiplicity of possible events increases exponentially with the addition of extra participants. Therefore, it is difficult to know which scenarios to collect beforehand in order to evenly populate the space of possible scenarios with the training set. The use of high-level models such as Bayesian belief networks might provide a combination of hand-coded *a priori* information with machine learning to ease training set requirements.

5   Conclusions and Further Research

It is clear that there are many potential advantages of visually mediated interaction with computers over traditional keyboard/mouse interfaces. For example, removing system-dependant IT training and allowing the user a more intuitive form of system direction. However, we have also seen that there are still many challenges for integrating multi-user interaction analysis and control due to the ambiguities and combinatorial explosion of possible behavioural interactions. We have demonstrated how our connectionist techniques can support real-time interaction by detecting faces and capturing 'attentional frames' to focus processing. To go further we will have to build our VMI systems around the task demands which include both the limitations of our techniques and potentially conflicting intentions from users. Connectionist techniques are generally well suited to this kind of situation as they can learn adaptive mappings and have inherent constraint satisfaction.

Further research is taking two main paths: 1) the development of gesture-based control of animated software agents in the EU ESE PUPPET project; and 2) the development of context-based control in more complex scenarios in a proposed EPSRC project. The first extends the use of symbolic (action selection) and mimetic (dynamic control) functions in gesture-based interfaces where pointing could indicate the current avatar and movement patterns could control animation parameters. The second involves recognition of complex behaviours that consist of a sequence of events that evolve over time [3,4,5]. As yet there has been little work that combines automated learning of behaviours in different contexts. In other words, it is usually only simple, generic models of behaviour that have been learnt rather than learning when and how to apply more complex models in a context sensitive manner.

Acknowledgements

The authors gratefully acknowledge the invaluable discussion, help and facilities provided by Shaogang Gong, Jamie Sherrah and Stephen McKenna during the development and construction of the gesture database and in collaborative work with the group interaction experiments.

References

1. H. Buxton and S. Gong. Visual surveillance in a dynamic and uncertain world. *Artificial Intelligence*, 78:431–459, 1995. 276
2. S. Duvdevani-Bar, S. Edelman, A. J. Howell, and H. Buxton. A similarity-based method for the generalization of face recognition over pose and expression. In *Proc. IEEE Int. Conf. Face & Gesture Recognition*, pp. 118–123, Nara, Japan, 1998. 273
3. R. J. Howarth and H. Buxton. Visual surveillance monitoring and watching. In *Proc. European Conference on Computer Vision*, Cambridge, UK, 1996. 282
4. R. J. Howarth and H. Buxton. Attentional control for visual surveillance. In *Proc. International Conference on Computer Vision*, Bombay, India, 1998. 282
5. R. J. Howarth and H. Buxton. Conceptual descriptions from monitoring and watching image sequences. *Image & Vision Computing*, 18:105–135, 2000. 282
6. A. J. Howell. *Automatic face recognition using radial basis function networks*. PhD thesis, University of Sussex, 1997. 273, 281
7. A. J. Howell. Face recognition using RBF networks. In R. J. Howlett and L. C. Jain, editors, *Radial Basis Function Networks 2: New Advances in Design*, pp. 103–142. Physica-Verlag, 2001. 273, 280
8. A. J. Howell and H. Buxton. Towards unconstrained face recognition from image sequences. In *Proc. International Conference on Automatic Face & Gesture Recognition*, pp. 224–229, Killington, VT, 1996. 273, 281
9. A. J. Howell and H. Buxton. Recognising simple behaviours using time-delay RBF networks. *Neural Processing Letters*, 5:97–104, 1997. 273, 274
10. A. J. Howell and H. Buxton. Learning gestures for visually mediated interaction. In *Proc. British Machine Vision Conference*, pp. 508–517, Southampton, UK, 1998. 273, 274, 276, 277, 281
11. A. J. Howell and H. Buxton. Learning identity with radial basis function networks. *Neurocomputing*, 20:15–34, 1998. 273
12. A. J. Howell and H. Buxton. Gesture recognition for visually mediated interaction. In *Proc. Int. Gesture Workshop, GW'99*, pp. 141–152, Gif-sur-Yvette, France, 1999. 274
13. A. J. Howell and H. Buxton. RBF network methods for face detection and attentional frames. *Neural Processing Letters*, 15, 2002 (In Press). 274, 276
14. A. Jebara and A. Pentland. Action reaction learning: Automatic visual analysis and synthesis of interactive behaviour. In *Proc. International Conference on Vision Systems (ICVS'99)*, Las Palmas de Gran Canaria, Spain, 1999. 273
15. P. Maes, T. Darrell, B. Blumberg, and A. Pentland. The ALIVE system: Wireless, full-body interaction with autonomous agents. *ACM Multimedia Systems*, 1996. 273
16. S. J. McKenna and S. Gong. Gesture recognition for visually mediated interaction using probabilistic event trajectories. In *Proc. British Machine Vision Conference*, pp. 498–507, Southampton, UK, 1998. 274, 276, 277

17. S. J. McKenna, S. Gong, and Y. Raja. Face recognition in dynamic scenes. In *Proc. British Machine Vision Conference*, pp. 140–151, Colchester, UK, 1997. 280
18. A. Pentland. Smart rooms. *Scientific American*, 274(4):68–76, 1996. 273
19. H. A. Rowley, S. Baluja, and T. Kanade. Human face detection in visual scenes. In *Advances in Neural Information Processing Systems*, volume 8, pp. 875–881, Cambridge, MA, 1996. 280
20. J. Sherrah and S. Gong. Fusion of 2D face alignment and 3D head pose estimation for robust and real-time performance. In *Proc. IEEE Int. Workshop Recognition, Analysis, and Tracking of Faces & Gestures in Real-Time Systems*, pp. 24–31, Corfu, Greece, 1999. 276
21. J. Sherrah, S. Gong, A. J. Howell, and H. Buxton. Interpretation of group behaviour in visually mediated interaction. In *Proc. Int. Conf. Pattern Recognition*, pp. 266–269, Barcelona, Spain, 2000. 276, 277, 278, 281
22. A. D. Wilson and A. F. Bobick. Recognition and interpretation of parametric gesture. In *Proc. Int. Conf. Computer Vision*, pp. 329–336, Bombay, India, 1998. 273
23. C. R. Wren, A. Azarbayejani, T. Darrell, and A. P. Pentland. Pfinder: Real-time tracking of the human body. *IEEE Trans. Pattern Analysis & Machine Intelligence*, 19:780–785, 1997. 273

Gestural Control
of Sound Synthesis and Processing Algorithms

Daniel Arfib and Loïc Kessous

CNRS-LMA
31 chemin Joseph Aiguier, 13402 Marseille Cedex20 France
arfib@lma.cnrs-mrs.fr

Abstract. Computer programs such as MUSIC V or CSOUND lead to a huge number of sound examples, either in the synthesis or in the processing domain. The translation of such algorithms to real-time programs such as MAX-MSP allows these digitally created sounds to be used effectively in performance. This includes interpretation, expressivity, or even improvisation and creativity. This particular bias of our project (from sound to gesture) brings about new questions such as the choice of strategies for gesture control and feedback, as well as the mapping of peripherals data to synthesis and processing data. The learning process is required for these new controls and the issue of virtuosity versus simplicity is an everyday challenge…

1   Introduction

We will describe here the first phase of a project named "le geste créatif en Informatique musicale" (creative gesture in computer music) where we link sound synthesis and processing programs to gesture devices. Starting from a data base of non-real time algorithms written in MusicV or Csound, we choose some of these "frozen music" algorithms and translate them into real-time programs where some additional anchors allow a gestural control by musical or gestural devices

This research has already been used for the gestualisation of an entire piece of music, le Souffle du Doux, mostly linking it to a "radio-drum" device and a keyboard [1]. We are now choosing other data sources to give a broad view of digital music programming styles, and trying other peripherals, either obvious or non-obvious.

This experiment raises different questions which can be practical – you need to be very pragmatic to go on stage- but also psychological: what is most important: sound or gesture, what is most efficient, to provide an easy gesture or an expert one? Most of these questions deal with the mapping system that is used to link the gesture to the sound algorithms, and the choice of the strategies that are used.

I. Wachsmuth and T. Sowa (Eds.): GW 2001, LNAI 2298, pp. 285-295, 2002.
© Springer-Verlag Berlin Heidelberg 2002

2   Real-Time and Non-real-Time Programs

Traditionally, computer music has always made a big differentiation between non real time programs, where a sound sequence is calculated without any interaction between performer and systems that are directly linked to a performance device. This difference was bound to a philosophical division (composing and performing are two different arts) but also and mostly to a technical point of view: machines that allowed real time interaction (such as synthesizers) were not the same ones that were computing musical scores. Nowadays this division is slowly dissolving as we will now show for some Macintosh programs that we use.

2.1   The MusicV and CSound Area

Music V and Csound are two programs which define a language that composers can use to describe not only events, but also the construction of the sound itself. They are driven by alphanumerical lines. Music V was the first one to appear and be thoroughly documented [12] and has given rise to many examples or computer pieces. This is the language we have used for years in our laboratory at CNRS. More recently, it appears that it is not maintained any longer. Csound is an equivalent program which comes in several flavours. An impressive work has been done to provide the community with an extremely well-documented and maintained stable version [6]. These programs rely on a block structure where "unit generators" are linked together to constitute an "instrument" that is activated by events and follows driving curves.

Example of a Csound Program

```
;-----------------------------; PARAMETER LIST
;p4 : AMPLITUDE
;p5 : PITCH
;p6 : OSCILLATOR FUNCTION
;p7 : WAVESHAPING FUNCTION
;p8 : ENVELOPE (DISTORTION INDEX) FUNCTION
;p8 : ENVELOPE (post correxction) FUNCTION

;-----------------------------; INITIALIZATION BLOCK
ifr     =         cpspch(p5)      ; PITCH TO FREQ
ioffset =         .5              ; OFFSET
;-----------------------------------------
kenv    oscilli 0,1,p3,p8              ; ENVELOPE (DISTORTION
INDEX)
kamp    oscilli 0,1,p3,p9          ;POST cCORRECTION
ain1    oscili  ioffset,ifr,p6    ; FIRST OSCILLATOR
awsh1   tablei  kenv*ain1,p7,1,ioffset; WAVESHAPING OF
OSCILLATOR
asig    =         p4*kamp*awsh1
        out       asig; OUTPUT
        endin
```

These programs were not meant to run in real time, but rather to remove the limitations intrinsic to real-time. The notion of gesture is directly linked to the triggering of these events and the shape of the driving curves. Though a real time version of Csound exists that can be driven by MIDI, these programs are really dedicated to a compositional job more than a performing one. Some front ends allow for some kind of graphical input. However, from my point of view, it is easier to start with a program which is dedicated still the beginning to real-time issues.

2.2   The MAX-MSP Area

Max-Msp is a Macintosh program that uses a graphical interface to describe objects which are very similar to the MusicV "unit generators". It is a real time program and is intended to be a performance tool. It is directly linkable to MIDI devices and has drivers for peripherals such as graphic tablets and for some joysticks and driving wheels.

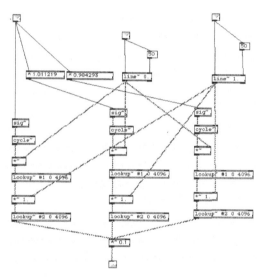

Fig. 1. A Max-MSP Patch uses the same kind of unit generators as the Csound program, but it uses a graphical interface and it runs in real time

One may wonder if there is a link between the real-time and non real time approaches. A previous study [1] consisted in finding a methodology to translate Music V programs to Max-Msp patches. If the construction of the instrument itself is quite straightforward, the constitution of the driving curves is not totally evident. While Music V and Csound use dozens of curve generators, Max-Msp mostly loads precalculated tables or uses tables described graphically using breakpoints. Also the NOT activation, which means the ensemble of parameters that define the structure of an entire note, should use a list structure and be dispatched -and eventually mapped to synthesis parameters- inside the instrument.

Though this operation is not at all trivial and is still hand made, we have proved that it is possible to translate a MusicV or Csound instrument to a Max-Msp program and this is for us an essential key for the gestural control.

3   Gesture to Sound and Sound to Gesture

The traditional approach is to start from peripherals that people buy or build and ask themselves: what can I do with it? Afterwards what can bind to this device, be it a sound, an image, a program. Our project is starting from the opposite question: when you have a synthetic sound of which you know the algorithm [16], what kind of gesture can you superimpose to make it work?

A good way to deal with this question is to draw a dual path which symbolizes the gesture to sound and sound to gesture connection [14,17].

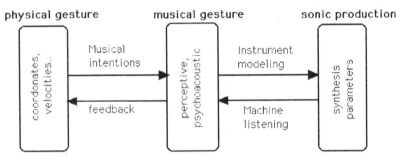

Fig. 2. The link between physical gestures and sonic production can be seen as a dual process of action and feedback, where an intermediate level represents a perceptual space of listening parameters, or conversely a psychoacoustic space of action factors

In Fig2, we can see that starting from the right (sonic production) we want to influence the synthesis parameters, which means to give expressivity or improvise, by way of physical gestures. The key is to build an intermediate space that can be linked to both worlds: starting from the sound, it represents the perceptual facts as can be recovered from perceptual criteria and put into psychological spaces through data reduction methods. A strategy is then defined by linking the intermediate space to both gestural and synthesis parameters.

4   Decision and Modulation Gestures

Here we deal with a simplified taxonomy of gestures that start from sound allures, a more complex one can be found in [22]. Sounds have the particularity that they are really bound to time. Their perception relies on the arrow of time and they cannot be seen as "objects" disconnected from this dimension. As a matter of example in percussive sounds the first 50 ms immediately trigger the perception of their percussive nature and are a particular clue for the recognition of the timbre of the instrument.

Decision gestures are so named because they decide of a particular set of values at a given time. Striking MIDI keyboards is something of that kind. On a Max drum (Fig.1), the crossing of a certain plane triggers the production of an event, where the data are the X and Y crossing position, and the velocity is calculated by the time used to cross the distance between two planes. Decision gestures influence the sound by an abrupt change of parameters or by the initialization of a process. A trivial example is the triggering of an event, but even this note start-up is not so simple to define. For example the crossing of a plane gives two coordinates which must then be mapped over a number of synthesis parameters that are typically more than two.

Fig. 3. A radio-drum consists of two sticks containing radio emitters and a plate upon which antennas are placed. This device allows a gestural capture of the position but also calculates decision coordinates by the help of the crossing of a virtual horizontal plane. In this figure the left hand makes a decision gesture while the right one modulates using a circular gesture

Conversely sounds also have a steady state which lasts for some time where it is essential to keep them alive. Any natural sound has an evolution which is not only a plain reproduction of a steady state spectrum. As an example the vibrato of a voice is a way of producing a lasting sound without being dull. As Mathews states and demonstrates [13], the perfect freezing of a scanned synthesis process immediately gives way to a synthetic feeling whereas its natural evolution produces a feeling of life.

Modulation gestures are gestures that can drive the sound over a long period of time. They usually rely on the detection of movement in space, and can use 3D positioning, rotation or inertial devices, or more simply linear and rotating potentiometers. Changing one device for another means two steps. The first step is the programming of the device itself to obtain a set of values describing the physical parameters. As a matter of example, a USB game driving wheel gives the following parameters: the value of three axes (the wheel and two pedals) plus a certain number of small switches. The second step consists of the mapping of these variables to the patch.

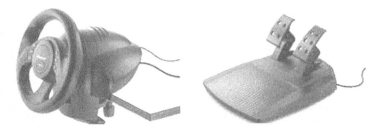

Fig. 4. A conventional racing wheel and its associated pedals is a perfect source for an easy modulation gesture

5   Some Strategies

We will now show an example of modulation gestures linked to a virtual instrument which uses the combination of both waveshaping [2] and filtering (formant) algorithms.

5.1   A Simple Patch

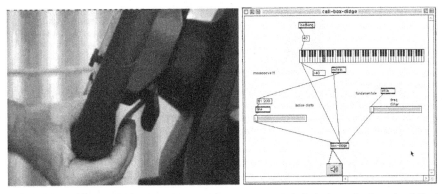

Fig. 5. A Max-Msp patch using three controls: here the pitch is linked to a keyboard, while the distortion index and the center frequency of the filter are linked to the acceleration pedal and the driving wheel

In this case, the waveshaping instrument is the sum of three distortion units, each of them being biased in frequency by a very small amount to produce a choir effect. The shape of the distortion is particular in the sense that it is a periodical function. The sound it produces is reminiscent of the FM synthesis described by Chowning [8]. The filtering uses a resonant filter whose controls are the filtering central frequency and the bandwidth. This instrument is really dedicated to modulation gestures, because the sound itself gives the impression of clay that we want to mold.

5.2   The Voicer, a Musical Instrument

This musical instrument allows the articulation of French vowels defined by three formants, as well as the control of the pitch in a way that allows intonation and nuances. A graphic tablet and a joystick are used as controllers for this instrument.

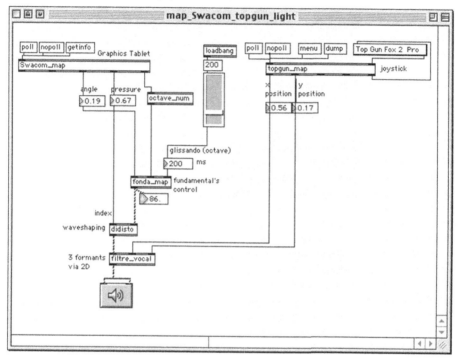

Fig. 6. The voicer Max-Msp patch where the pitch control is linked to a graphic tablet and the articulation of vowels to a joystick via an interpolation plane

The Max-MSP patch uses an all pole filter structure that is superimposed on a sound source. The pitch variation is linked to the coordinates of the tip of a pen on a graphic tablet in way that is not linear: the angle in a polar representation provides the pitch and vibrato is possible in some zones of the circle. Lateral buttons make an octave up or down jump. The coordinates of the joystick axes are linked to an "interpolation plane" which converts them to synthesis parameters. On an interpolation plane, each referenced object corresponds to a set of synthesis parameters and the position of one point on the plane interpolates between the values of the parameters of these objects following "attraction curves" [19]. So two coordinates for the joystick position allow control of n parameters (in our case 6 which define the cascaded filter) and, this way, a navigation in the interpolation space is possible.

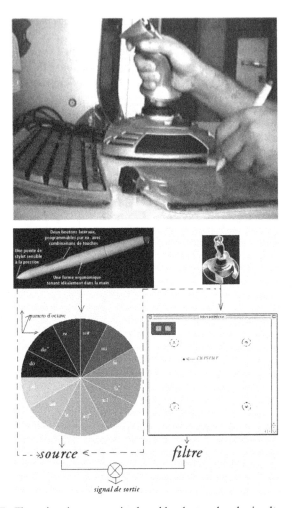

Fig. 7. The voicer instrument is played by the two hands simultaneously

One plays this musical instrument, named the voicer, with the two hands controlling the graphic tablet pen and the joystick. It is very expressive, because of its intuitive control of the pitch, with glissando, vibrato and subtle variations, and the articulation of the vowels ordered in a practical way on the XY plane [20]. The control of temporal envelopes linked to the pressure of the tip of the pen allows the playing of linked or detached notes. One can follow scores, or improvise, and this has been used in concert for a musical piece named "Spectral shuttle" where it makes a dialog with a saxophone processed in real-time.

6   Future: Mapping, Feedback, Learning and Virtuosity

Mapping, feedback, learning and virtuosity are important considerations in our realization. Connecting a device to sound creation involves a reflection on these domains. Therefore I describe here the framework that is behind our research.

Visual feedback is an important step in the driving of digital audio effects [3], and it has been shown that the visual interface is a definitive help for the creation of an artistic transformation of sounds. This holds for synthesis too.

Mapping is an essential area for a proper control. As stated in chapter 3 the building of a "psychoacoustic" level requires evaluation of some perceptive criteria. This is a combination of curve extraction and data reduction. Sound analysis has already provided methods to project data into a 3D or nD perceptive space[7,9,11,18,21] and linking physical parameters coming from the device to these parameters is a mapping which corresponds to some form of pedagogy: musical intentions must be transcribed to perceptive data.

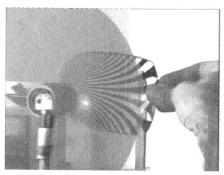

Fig. 8. A gesture used on a photosonic instrument. The light is displaced by the left hand while the filter is controlled – horizontally and vertically- by the right hand

The existence of such mappings has a very strong influence on the learning curve of the instrument: there must be some coherence between the gesture itself and the sonic result. But also, as stated by some authors [10], simplicity is not the ultimate goal: virtuosity is common in performances from difficult instruments, and it looks like the effort required to learn an instrument is also a key to the possibility to play it well. As a matter of example the gestures that one uses to control a photosonic instrument [5] are very precise (rings over the photosonic disk are 3 mm wide). However this instrument is very playable and its complete transcription into a real-time numeric simulation is part of our project.

Nevertheless we have only dealt here with an experimental approach, which is self-consistent: by building instruments that are playable, one tends to create situations where presence or immersion feeling exist. This way, one improves things in a practical and a theoretical frame. These "intuitive mappings" can also be the object of a further study in the sense that they give all the elements for the discovery of proper relationship between the instrumentist and his instrument.

7   Conclusion

This first phase of our project has already shown that it is possible to use the large knowledge base of non real time music programs such as Music V and Csound to build real time systems combining Max-Msp applications and gestural devices. The diverse strategies allow some classifications of gesture and they also emphasize the role of mapping in the sound to gesture and gesture to sound relationship. Future directions will concentrate on a more precise knowledge of the learning process and on the virtuosity these links permit.

Acknowledgments

This research is sponsored by the National Center for Scientific Research (CNRS) and the Conseil Général des Bouches du Rhône (France) under the project name "le geste créatif en Informatique musicale".

References

1. Daniel Arfib & Loic Kessous: "from Music V to creative gesture in computer music", proceedings of the VIIth SBC conference, Curitiba, June 2000, available in CD format
2. Daniel Arfib: "Digital synthesis of complex spectra by means of multiplication of non-linear distorted sine waves", Journal of the Audio Engineering Society 27: 757-768.
3. Daniel.Arfib: "Visual representations for digital audio effects and their control", proceedings of DAFx99, Trondheim, decembre 1999, pp 63-66
4. Daniel.Arfib, Jacques.Dudon : "A digital version of the photosonic instrument", proceedings ICMC99, Pekin, novembre 1999 , pp 288-290
5. Daniel Arfib & Jacques Dudon: "Photosonic disk: interactions between graphic research and gestural controls", in CD-ROM "Trends in Gestural control of music", editeurs M. Wanderley & M. Battier , publication Ircam, 2000
6. Richard Boulanger: The Csound Book, Perspectives in Software Synthesis, Sound Design, Signal Processing and Programming , MIT press, 2000
7. Gérard Charbonneau: "Timbre and the Perceptual Effects of Three Types of Data Reduction". Computer Music Journal 5(2) p.10-19, 1981.
8. John Chowning: "The Synthesis of Complex Audio spectra by Means of Frequency Modulation. Journal of the audio Engineering Society 21:
9. 561-534. Reprinted in C. Roads and J. Strawn editions 1985 "Foundations of Computer Music". Cambridge, Massachusetts: MIT Press.
10. John Grey: "An Exploration of Musical Timbre". PhD dissertation, department of Psychology, Stanford University, 1975.
    and: John Grey: "Multidimensional Perceptual Scaling of Musical Timbres" J.A.S.A vol.61 n°5 May 1977 p 1270-1277

11. Andy Hunt, Marcelo M. Wanderley, Ross Kirk: "Towards a Model for Instrumental Mapping in Expert Musical Interaction", proceedings ICMC 2000, Berlin
12. Stephen McAdams, Suzanne Winsberg, Sophie Donnadieu, Geert De Soete, Jochen Krimphoff; "Perceptual scaling of synthesized musical timbres: common dimensions, specificities, and latent subject classes", Psychological Research, 58, 177-192 (1995) available on
    <http://mediatheque.ircam.fr/articles/textes/McAdams95a/>
13. Max Mathews: "The Technology of Computer Music" (1969). MIT Press, Cambridge, MA.
14. William Verplank, Max V. Mathews, Robert Shaw: "Scanned Synthesis, proceedings of the Icmc 2000, Berlin
15. Eric Metois: "Musical Gestures and Audio Effects Processing", DAFx98 conference, barcelona, available on <http://www.iua.upf.es/dafx98/papers/>
16. Joe Paradiso: "American ionnovation in electronic musical instruments", available on <http://www.newmusicbox.org/third-person/index_oct99.html>
17. Jean-Claude Risset et Wessel: "Exploration of Timbre by Analysis and Synthesis", The Psychology of Music, Deutsch eds. Orlando Academics, 1982.
18. Sylviane Sapir: "Interactive digital audio environments: gesture as a musical parameter", proceedings of DAFx00, available on
    <http://profs.sci.univr.it/~dafx/papers.html>
19. Wessel David L.: "Timbre Space as a Musical Control Structure", Rapport Ircam 12/78, 1978, available on
    <http://mediatheque.ircam.fr/articles/textes/Wessel78a/>
20. Vect pour Max Macintosh, vectFat 1.1 (par Adrien Lefevre), available on
    <http://www.adlef.com/soft/vect/vectdoc.html>
21. A. Slawson "Vowel quality and musical timbre as function of spectrum envelope and fundamental frequency" 1968,Journal of Acoustic Society of America, n°43, p 87-101.
22. Thierry Rochebois, Thèse, available on <http://www.ief.u-psud.fr/~thierry/these/>
23. Claude Cadoz & Marcelo M. Wanderley: Gesture-Music, in M. Wanderley and M. Battier (eds): Trends in Gestural Control of Music- Ircam - Centre Pompidou, 2000.

Juggling Gestures Analysis for Music Control

Aymeric Willier and Catherine Marque

UTC UMR 6600
BP 20529, 60205 Compiègne Cedex France
aymeric.willier@utc.fr

Abstract. The aim of this work is to provide jugglers with gestural control of music. This is based on the willing to control music by recycling mastered gestures generated by another art. Therefore we propose the use of a gestural acquisition system based on the processing of the electromyographic signal. The recordings are done during a three-ball cascade, of electromyogram from chosen muscles, which play a specific role in the juggling gesture. Processing of those signals is proposed in order to control musical events by means of parameters related to juggling gesture.

1 Introduction

The aim of our work is to provide jugglers with gestural control of music. Many juggling shows actually propose to join juggling and music [1][2][3]. As juggling expresses itself through gestures and the motion of objects, it mainly solicits our visual sense. Joined to music it also attempts to touch our ears, but this goal is achieved mainly through cooperative work with musicians. An interesting way to increase the relationship between juggling and music would be to obtain a direct and expressive control of music through the juggling gestures: thus we suggest that the gestures made by the jugglers to move the objects, could also control music. We therefore propose to use a gesture capture system, to extract meaningful information about juggling and juggling gestures, and to condition them for the real-time control of music. In order to design such a system, we have to take into account the constraints imposed by juggling such as: the need to be portable, not obstructive...We thus plan to use a gesture capture system based on the processing of Electromyographic signal.

1.1 Gestural Control of Music

As the possibilities offered by computer generated music are numerous and constantly increasing in power, the question of their control offers a productive and debated subject of research [4][5]. Many hardware devices have been proposed with very personal approaches depending on their builder's needs. Institutions [6][7][8] as well as corporations [9] are also interested in the research on gestural control of sound synthesis and in the design of new controllers.

I. Wachsmuth and T. Sowa (Eds.): GW 2001, LNAI 2298, pp. 296-306, 2002.
© Springer-Verlag Berlin Heidelberg 2002

We can distinguish between two main approaches in the design of controllers: instrument like controller or alternative controller. The first ones are controllers designed on the traditional musical instrument paradigm, such as keyboards, guitar, drums or wind instruments. Those controllers are the most common ones. Their main advantage is that they do not request the production of a special and dedicated effort in their learning. The seconds ones, alternative controllers, may be parted into two categories according to the two following approaches: with the willing of developing a new musical instrument (Biomuse [10], the hands [11][12]) or with the willing of recycling a mastered gesture provided by another artistic activity. In this later case, dance is a great source of experimentation: for example, the digital dance system designed at DIEM [13], the Bodysynth [14], or many experiments mentioned on the Dance and Technology zone Web site [15].

Our personal approach to the question of controllers, according to our willing to give the control of music to jugglers, enters this latter category. Those kinds of controllers also give an indirect answer to J-C Rise's remark about the musical gestures, stating that it is an expert gesture, not a spontaneous one, hardly learned [16]: recycling mastered gestures of another artistic form to produce music would not require an additional effort from the performers, to the one already made in learning this gesture for their initial artistic purpose.

1.2   Gestural Acquisition

Gestural acquisition can be done using many kinds of motion capture system, and the technologies used for motion capture are numerous: acoustic, magnetic, optical, mechanical, video... In the particular case of the design of a new controller for music, much of those technologies are also encountered: D. Roger has proposed an interesting review of them [17]. In order to understand better the limitations and possibilities offered by each of them, a classification has been proposed by A. Mulder [18]. Three main categories are distinguished under the following names: Inside-In, Inside-Out, and Outside-In. Systems from the first category are mainly used to track motion of some body parts, and the sensors suited for the phenomenon that they track, are located on the body. The two second ones are mainly used to track motion of the body in a defined space. These systems, from the Inside-Out category, track, on the body, a phenomenon created outside of the body, but affected by the motion of the body. For Outside-In systems the configuration is reversed. Our choice to provide freedom of displacement, and to focus on the gestures that occur during juggling, and not on the ball motion, explains our interest for the Inside-In category.

Apart from the technical point of view, M. Wanderley suggested an interesting distinction between direct and indirect gesture acquisition in the field of musical application [19]. He has noticed that most of the gesture acquisition systems are direct, in that sense that they track the physical realization of the gesture in terms of distance, position, or angles...Fewer ones may be described as indirect ones, as they track gesture in their effect, such as systems extracting information on gesture made to produce sound on an acoustic instrument, through the analysis of this sound. Hybrid approaches coupling direct and indirect gesture acquisition are also noticed, for example for the hyper instruments designed by T. Machover [20]. But another kind of indirect gesture acquisition system consists in the capture of gestures through the physiological signals that are at their origin. Our work will focus on such another indirect gesture capture system, extracting information about the gesture through the analysis of the electrical signal linked to the muscles' activity. This signal is called Electromyogram (EMG). Its main advantage for this work is to be suitable for non-invasive, non-obstructive, portable, and light recording system.

1.3   EMG-Based Gesture Capture Systems

The use of EMG has been proposed and explored for the control of music by B. Knapp [10], H.S. Lusted [21] and B. Putnam [22] in the beginning of the nineties as they built the Biomuse. This controller is actually still used by A. Tanaka [23], who has defined gestures that give him a fine control of sound synthesis and structural parameters of music using the Max software [24]. T. Marrin has also used EMG in the context of musical gestures, in her study of conducting gestures [25]. She has used it in combination with motion tracking system and others physiological sensors, to attempt to analyze and interpret the expression in conducting gestures [26]. Our approach differs from the previously mentioned ones, in that sense that we will use the EMG as the base of a gesture capture for existing mastered gestures. It differs from the biomuse, which proposes to define gestures giving a fine musical control, and it differs from the conductor's jacket, because our approach focuses on gestures which are not produced for musical purpose.

2   Methods and Results

Our approach is the following: first we determinate, by observation of the pertinent gestures, the main muscles involved in their realization. Then, we observe the EMG of those muscles, thus validating our muscles choice. Next, we analyze the temporal correspondences between juggling phases and muscles activities. And finally, we propose a processing of chosen EMG, which provides us with signals suitable for the control of music. We chose to work on juggling because of personal interest for this fast evolving art of gestural expression, but also because it permits us to limit as a first approach our observation to upper limbs (even if juggling is not limited to upper limb motions), and because juggling offers some definite actions [27], the combination of which is the base of a gestural expression.

2.1   Gesture Description and Muscle Choice

As a first approach to our problem, we will focus on a basic juggling pattern called cascade. This pattern is performed with an odd number of balls. The three-ball cascade is usually the first pattern learned. A cascade consists in alternative throws of both hands; during the three-ball pattern, a hand throws a ball when the last ball thrown by the other hand reaches its zenith.

Focusing on what happens to a ball, we can describe the pattern as alternative flights and displacements by the hand, separated by catches and throws. Focusing on what happens to a hand, it is seen as alternative displacements with the ball, or without it, also separated by throws and catches (Figure 1).

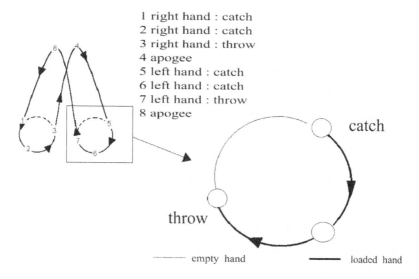

1 right hand : catch
2 right hand : catch
3 right hand : throw
4 apogee
5 left hand : catch
6 left hand : catch
7 left hand : throw
8 apogee

catch

throw

——— empty hand   ——— loaded hand

Fig. 1. The three-ball Cascade

Key moments of the pattern can be those catches and throws and conjoint displacements of balls and hands. It may be interesting to compare first, the temporal activities of muscles working during the juggling and the timing of catches and throws. The first step has been to select the muscles to observe. We have to choose muscles that are playing a specific role in juggling and that can also be recorded with surface electromyography. Observing the motion of the arm during juggling, we assist to a circumduction of the hand. It can be described as a combination of external and internal rotation of the arm, combined with flexion and extension of the elbow. Muscles that act for flexion and extension of the wrist are also solicited.

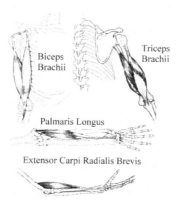

Biceps Brachii

Triceps Brachii

Palmaris Longus

Extensor Carpi Radialis Brevis

Fig. 2 Muscles observed with SE

The muscles that may be observed with surface electromyography (SEMG) are the rotators of the arm, internal and external as Pectoralis Major (PM) and Teres Minor

(TM) the flexors and extensors of the elbow, Biceps (BB) and Triceps (TB) brachial, and the flexors and extensors of the wrist: Palmaris Longus (PL) and Extensor Carpi Radialis Brevis (RB) (Figure 2).

2.2   EMG

2.2.1   Recording

For the recording of SEMG we have used a four channels HF electromyographic signal acquisition system by John+Reilhofer. We have used four pairs of Ag/AgCl electrodes that are located in parallel with the muscle fiber direction, with a constant interelectrode distance of 2 cm. Prior to the electrode attachment, the skin is prepared, using a standard method to lower the skin resistance below 10 KOhms. A ground electrode is located away from the considered muscles, over the shoulder joint. The use of a HF transmission system, between the conditioner/transmitter box carried by the juggler and the receptor, gives the juggler a total freedom of motion during the recording. With the same intention, a particular care is taken in the attachment of electrodes and leads. The conditioner/transmitter box contains a filter for the four EMG (1000 Hz low pass). The four outputs of the receiver are recorded on a magnetic tape recorder. The data are then 500 Hz low pass filtered and digitized at a sampling frequency of 1000 Hz, and then digitally processed on a computer using Matlab software.

2.2.2   Description of the Task

Jugglers were asked to perform the following task:

- Three-ball cascade at their usual speed for this pattern, which is considered as the ease speed of execution.
- Three-ball cascade at a higher speed, the highest speed that they can maintain during more than 30 seconds.
- Three-ball cascade at a lower speed, the lowest that they can maintain assuming they keep the three-ball rhythm.

As an optional task, we have also recorded five ball cascades and other patterns such as juggling two or one ball in the rhythm of three. Other patterns, including bouncing juggling have been recorded (we won't discuss them in this paper).

We have made recordings with seven professional jugglers. Some recordings have also been made with amateur jugglers whose juggling experience range from one to twenty years, with a weekly training ranging from less than one hour to more than two daily hours. This will permit us to study the effect of training on organization of muscular activity for a defined task.

2.2.3   Recordings

A first observation of SEMG recorded during juggling has permitted us to notice that Biceps and Triceps Barchii are almost always active (Figure 3). This may be explained by the supination of the forearm oscillating around a maintained half-flexed position. The supination is related to an activation of the BB, and the half-flexed position is the result of a coactivation of BB and TB. The oscillating motion around this position is created by variations in the activity of both muscles, which does not appear significant in the SEMG recordings. Those oscillations are in fact slow, and could be assumed by the Brachialis, deep muscles that is not reachable with SEMG. As they come faster, variation in the BB activity seems to be more evident.

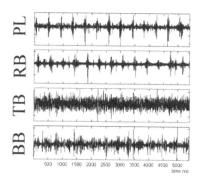

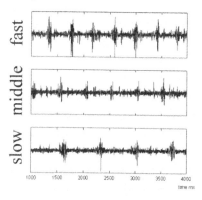

Fig. 3 SEMG during the three-ball cascade for PL, RB, BB and TB

Fig. 4 Effect of the variation of speed on the SEMG recorded on PL

The SEMG observed for the flexor and extensor of the wrist and for internal and external rotators of the arm have exhibited a more specific activity, which occurs at the same rhythm as juggling. As the juggling rhythm is one of the parameter that we want to extract from the SEMG, recording of SEMG those for muscles may have been of interest. Finally flexor and extensor of the wrist were first selected, expecting that their activities be linked with those two specific moments of the cascade pattern: throw and catch.

This relation between cyclical burst of activity and juggling rhythm is confirmed by the observation of the evolution of SEMG recorded on PL with the variation of juggling rhythm proposed on Figure 4. The first musically interesting information would be the extraction of rhythm of juggling, so that music and juggling may be synchronized.

2.2.4   Effect of Training

An other evident observation on this recording of the SEMG of PL for a professional trained Juggler compared to an amateur one (Figure 5) is that the organization of muscular activity is more precise for the professional trained juggler. Bursts of activity are shorter and more localized in time. Between two bursts, there is almost no activity corresponding to rest phases. This confirms that the acquisition of an expertise in the realization of the gesture is linked to an optimization of the activation of muscles. SEMG used as a tool for the observation of temporal activity of muscle reveals this phenomenon with emphasis. This observation has led us to use only trained people for the processing of EMG related to chosen expert gestures.

Untrained

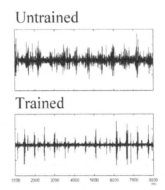

Trained

Fig. 5 SEMG of PL recorded on untrained and trained juggler for the three-ball cascade

2.3   Juggling Phases Identification

The second step of this study has been the identification of the two key moments of the three-ball cascade: throws and catches of the balls. This has been made by the comparison of the SEMG, with the signals obtained with two other sensors: accelerometer and Hall effect sensor.

2.3.1   Sensor

We have first used an accelerometer attached to the backside of one hand. This one dimension accelerometer provides us with the acceleration of the hand perpendicular to the palm. It permits us to detect the arrival of the ball in the hand, as a discontinuity in acceleration. The signal provided by this sensor does not allow the precise detection of the throw: acceleration of the hand being always present before the throw; we expected that the ball would leave the hand at the maximum of acceleration, but we were not able to prove it.

Therefore, we also have also used a Hall-effect sensor, attached to the palm of the hand. One of the balls used for this experiment has been filled with a magnet, so that its arrival in the hand is detected by the Hall effect sensor through the variation of the magnetic field created by this ball. The departure of this ball is also detected at the end of the variation of the magnetic field. According to the fact that magnet orientation is not always the same when the ball is caught, the sign and amplitude of the variation is never predictable. Thus, the signal obtained with this sensor is then rectified, and also low pass filtered.

2.3.2   Results

Only one recording using both sensors has been done with simultaneous recording of SEMG of LP and RB. Most of the recordings made with the other jugglers have been done with only accelerometer and SEMG, as the use of both sensors at the same time may disturb the usual motion of juggling.

Figure 6 shows the recording of signals obtained with simultaneous recording of the accelerometer and the Hall effect sensor. The accelerometer signal permits us to locate the ball catch, which is related to the discontinuities, as the one quoted on the figure. On the Hall effect sensor signal, we can see the perturbation of the magnetic field that occurs when a ball is in the hand. As only one of the three balls has been filled with a magnet, this perturbation occurs only each three cycles evidenced on the accelerometer signal. The concordance between the discontinuity of the acceleration and the appearance of the magnetic perturbation permits us to locate precisely the catch. The end of the magnetic perturbation allows the time localization of the moment when the ball leaves the hand, which is located short before the maximum of the accelerometer signal.

Hall effect sensor

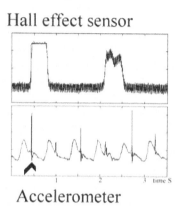

Accelerometer

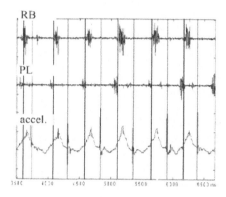

RB

PL

accel.

Fig. 6 Signal obtained with the Hall effect sensor (rectified signal) and the accelerometer for the three-ball juggling. Units on the vertical axis are arbitrary units

Fig. 7 SEMG of RB (upper trace) and PL (lower trace) recorded together with the accelerometer for the three-ball juggling

Catch and throw of balls being located at the discontinuity and before the maximum of the accelerometer signal, it is now possible to locate phases of juggling on simultaneous recording of SEMG and accelerometer signals: on Figure 7, empty hand phases are indicated in white and ball in hand phases are indicated in gray. It is then possible to locate activities of muscles according to throws and catches.

Concerning the PL, its activity is located before the throw. According to the electromechanical delay, which is the time separating EMG appearance and effective production of force by the muscle, this activity of this flexor of the wrist at this time could assume for the up holding of the hand prepared for the throw.

Concerning the RB, its main burst is located at the throw time. We can suggest an explanation: after the ball leaves the hand, it may be necessary to use the wrist extensor to avoid a flexion of the wrist. The hand has to be kept aligned with the forearm to prepare for the following catch. The RB burst of activity should then be relating to the preparatory phase keeping the wrist aligned.

2.4   Processing

The last part of our study has been to extract information from the SEMG, which may be used for music control. With this aim we have first computed the IEMG. The

IEMG provides an estimation of the energy of the signal that may be used easily for continuous control. The row data are rectified and the obtained data are smoothed on 50 ms. As represented on the Figure 8, for RD and PL off both hands, the IEMG offers a smooth signal suggesting its ability to be used as a continuous controller for sound synthesis. A. Tanaka has suggested that it is really suitable for the control of timbral evolution [24]. IEMG will therefor be encoded as continuous control MIDI signal.

The IEMG showing cyclical peaks, as the SEMG shows cyclical bursts of activity, we have also developed peak detection on the IEMG. Our first try using a fixed threshold was not efficient, mainly because the amplitude of the peaks was changing a lot over time. We then used a dynamically adapting threshold: this threshold is calculated for each sample as a fixed percentage of the maximum value observed for the last N values of the IEMG. N is also dynamically corrected according to the number of samples separating the last two detected peaks. This method has showed a good ability to detect peaks in the IEMG during the juggling cascade, even at varying speed of execution.

Looking at the result of the simultaneous peak detection on RB and PL for both arms (Figure 9), the time organization of muscular activity during the cascade is even more evident. For each arm, a burst of activity on the PL is followed by a burst of activity on the RB. According to the previously made observation that the RB burst of activity occurs at the throw time, peak detection on the SEMG of this muscle appears to be suitable to synchronize musical events and throw. Peak detection is therefor encoded as Note on/off MIDI signal.

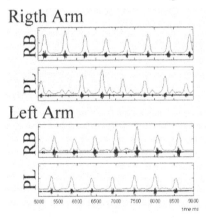

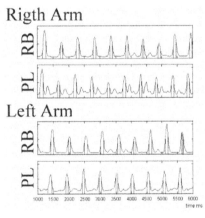

Fig. 8 SEMG and related computed IEMG obtained on the three-ball cascade for PL and RB of the right arm (two upper traces) and left arm (two lower traces)

Fig. 9 IEMG and Peak detection computed for RB and PL of right arm (two upper traces) and left arm (two lower traces) during the three-ball cascade

3   Perspective and Conclusion

The first part of our work has permitted us to evidence the muscle used in juggling. Indeed, very few works has been concerned with the biomechanical aspect of juggling gestures. We have now evidenced the different muscles involved in these gestures, for the global movements (displacement of arms and hands) as well as the ones involved in catch and throw. The actual processing provides us with a musical control in terms of juggling rhythm, related to the peak detection, and of continuous control related to IEMG. IEMG is actually computed in real time, and encoded as continuous controller for the MIDI protocol. Our peak detection algorithm will be implemented in real time soon. We will then encoded it as on/off Midi event with the aim to trigger events at the precise time of throw. The localization of occurrences of two consecutive throws permits us to estimate the period of hand motion for the three-ball cascade.

The work still to be done is parted in two. The first part is related to the recording of the muscles involved in the arm rotation. We hope to extract from their SEMG information that is more related to the throw dynamics. The second part is related to the SEMG processing. We will try to find an accurate way to detect catch time through the analysis of SEMG. Thus, knowing about throw and catch time will help us to determinate the ball loaded period, or even the flight time for a ball. Indeed, C. Shannon has proposed a direct relation between flight time of a ball, holding time, and unloaded hand time, for a given number of juggled objects [28].

Indeed as base of their expression, jugglers use variations in rhythm of pattern, which is perceptible in the hand motion period, and they also can change the way they perform the pattern, so that, for a fixed period of hand motion, flight time can vary in proportion with holding time [29]. This observation suggests us to use hand motion period and time of fly as two parameters for the control of music through the juggling. We also plan to improve our peak detection for other juggling patterns, mainly arrhythmic ones, proposed by jugglers in the optional part of the recordings. As we would also like to explore the dynamic aspect of juggling gestures, we will explore the spectral evolution of EMG over time, using time frequency analysis tools. Indeed, time frequency content of SEMG seems to be related to the dynamic of movement [30]. The analysis of specific frequency bands (to be determined) would then allow a real time control of music parameters, in relation with gesture dynamic.

References

Websites

1. Juggling Juke box by James Jay http://www.jamesjay.com/juggling/jukebox/hightech/
2. Chant de balles, Vincent De Lavenere http://www.chantdeballes.com/contact.html
3. Kabbal http://www.lefourneau.com/creations/00/kabbal/kabbal.htm
6. STEIM http://www.lefourneau.com/creations/00/kabbal/kabbal.htm
7. IRCAM http://www.ircam.fr/equipes/analyse-synthese/wanderle/Gestes/Externe/index.html
8. ACROE http://www-acroe.imag.fr/
9. Yamaha Miburi http://www.yamaha.co.jp/news/96041001.html
14. Bodysynth http://www.synthzone.com/bsynth.html
15. Dancetech http://art.net/~dtz/
17. Roger, D. Motion Capture in Music: Research, http://farben.latrobe.edu.au/motion/. (1998).
21. Lusted, H. S. and B. Knapp. Controlling computers with neural signals. Scientific American. (1996). http://www.sciam.com/1096issue/1096lusted.html

Papers

4. Trends in gestural control of music. CD-Rom edited by M.Wanderley and M.Battier. Paris, IRCAM-Centre Pompidou.
5. Wanderley, M. and P. Depalle. Contrôle Gestuel de la Synthèse sonore. Interfaces Homme-Machine et création musicale. Hermes. Paris. (1999).
10. Knapp, R. B. and H. S. Lusted "A Bioelectric Controller for Computer Music Applications." Computer Music Journal 14(1) (1990). 42-47.
11. Waiswisz, M. "The hands, a set of remote midi-controllers." Proc. Int. Music Conf. (ICMC 1985)(1985). 313-18.
12. Paradiso, J. Electronic music: new ways to play. IEEE Spectrum Magazin. (1997).
13. Siegel, W. "The Challenges of Interactive Dance - an overview and case study." Computer Music Journal 22(4). (1998).
    http://www.daimi.au.dk/~diem/dance.html
16. J. C. Risset, round table "les nouveaux gestes de la musiques" mentioned by M.Wanderley in [19]
18. Mulder, A. Human movement tracking technology, Tech. Rep., Simon Fraser University, school of kinesiology. (1994)
19. Wanderley, M. M. (1997). Les nouveaux gestes de la musique. Tech. Rep., Paris, IRCAM, Equipe Analyse/Synthèse. (1994).
20. Machover, T. Hyperinstruments a progress report 1987-1991, Tech. Rep., Massachusetts Institut of Technology. (1992).
22. Putnam, B. The use of the electromyogram for the control of musical performance, Ph.D. Dissertation, Stanford University. (1993).
23. Bongers, B. An interview with Sensorband. Computer Music Journal 22(1). (1998). 13-24.
24. Tanaka, A. Musical performance practice based on sensor-based instruments. Trends in gestural Control of Music. CD-Rom. Paris, IRCAM-Centre Pompidou. (2000).
25. Marrin, T. and R. Picard. A methodology for mapping gestures to music using physiological signals. Presented at the International Computer Music Conference, Ann Arbor, Michigan. (1998).
26. Marrin, T. and R. Picard. The Conductor's jacket : a device for recording expressive musical gestures. International Computer Music Conference, Ann Arbor, Michigan. (1998). 215-219
27. Durand, F. and T. Pavelack. Le psychojonglage, le livre de la jongle². Toulouse. (1999)
28. Shannon, Lecture at Carnegie Mellon university, July 1983, reported in [30]
29. Beek, P.J. Juggling Dynamic, Ph.D. Dissertation, Free University Amsterdam (1989)
30. Karlson , S. J. YU and M. Akay . Time-Frequency analysis of myoelectric signals during dynamic contractions: A comparative study. IEEE Transaction on Biomedical Engineering. 47(2).(2000) 228-237

Hand Postures for Sonification Control

Thomas Hermann, Claudia Nölker, and Helge Ritter

Neuroinformatics Group, Faculty of Technology, Bielefeld University
D-33501 Bielefeld, Germany
{thermann,claudia,helge}@techfak.uni-bielefeld.de

Abstract. *Sonification* is a rather new technique in human-computer interaction which addresses auditory perception. In contrast to speech interfaces, sonification uses non-verbal sounds to present information. The most common sonification technique is parameter mapping where for each data point a sonic event is generated whose acoustic attributes are determined from data values by a mapping function. For acoustic data exploration, this mapping must be adjusted or manipulated by the user. We propose the use of *hand postures* as a particularly natural and intuitive means of parameter manipulation for this data exploration task. As a demonstration prototype we developed a hand posture recognition system for gestural controlling of sound. The presented implementation applies artificial neural networks for the identification of continuous hand postures from camera images and uses a real-time sound synthesis engine. In this paper, we present our system and first applications of the gestural control of sounds. Techniques to apply gestures to control sonification are proposed and sound examples are given.

1 Introduction

Central goal in data mining is the discovery of structures and regularities in high-dimensional and often very large datasets [2]. One strategy to reach this goal is *exploratory data analysis*, which aims at providing human researchers with suitable interfaces that enable them to explore complex datasets and facilitate the detection of patterns in the data. Besides a rapid growth of scientific visualization, which aims at supplying data presentation for the visual sense, *sonification*, the rendering of auditory presentations of data, has attracted increasing interest in recent years [6].

The different nature of sound as compared to visual displays poses special demands on a human-computer interface that shall support interactive sonification. While many visualizations can already work with static images, sound is an intrinsically dynamic medium so that *only* dynamic displays do make sense. Therefore any useful interface has to allow real-time control. This is particularly important when acoustic analogues to visual pointing are required, using techniques such as interactively positioned acoustic markers or rhythmical patterns. Since sonification, especially with the technique of parameter mapping demands the adjustment of many control parameters in order to optimize the sonification, it becomes tantamount to provide easy-to-use, intuitive methods for the

I. Wachsmuth and T. Sowa (Eds.): GW 2001, LNAI 2298, pp. 307–316, 2002.
© Springer-Verlag Berlin Heidelberg 2002

real-time control of multidimensional parameter sets. In most computer interfaces such a parameter adjustment is done by moving sliders or changing other elements in a graphical user interface with the mouse or keyboard. However, the parameters can only be adjusted one after another, which requires time, is often inconvenient and can be limiting if dynamical control of high-dimensional parameter vectors is required. Compared to contemporary graphical widgets, the *human hand* is a much more flexible tool for specifying several parameter settings simultaneously and in a dynamic fashion [5]. Our hand has about 20 degrees of freedom and we possess high skills in adapting our hand movements to perform certain tasks (e.g. playing violin, manipulating clay, etc.). Developing a gestural interface for simultaneous multidimensional control tasks e.g. for music composition and sound design is an idea that has first been exploited in the Theremin but then left unstudied for a long time and been revived only rather recently. Mulder [7] coined the name "sound sculpting" for the editing of sound and employed two data gloves to change the attributes (shape, position, orientation) of the virtual object by hand gestures. In this paper, we aim at developing a purely vision-based hand posture driven interface to manipulate sonifications in order to facilitate interactive exploratory data analysis. Section 2 gives an overview of our system, in section 3 we demonstrate that real-time control of multiple parameters can be done with the system. For this purpose, an interface for the gestural control of sound synthesis parameters is presented. Section 4 discusses different techniques to apply hand postures to control sonifications. The paper ends with a conclusion.

2   System Overview

The developed system consists of three main components: the hand posture recognition, the sonification modules and a database which supplies the data to be analyzed. As a platform both for the implementation of the hand posture recognition and the sonification routines, the graphical programming environment NEO/NST is used [9]. It allows to interface to a MySQL database, which holds the data to be used for sonification.

2.1   Hand Posture Recognition

The system for the recognition of continuous hand postures is based on camera images of a hand in front of a black wall. A hierarchical approach with artificial neural networks is used to find the fingertips in images of the hand. The images are processed at a frame rate of 11 Hz on a 200 MHz Pentium PC if only a single camera is used [10]. In the mono system, where only one of the two cameras is used, only the x and y coordinates are obtained. Using two cameras, the z-coordinate can also be extracted from the images.

Figure 1 shows the setup our stereo vision system that we have developed for that purpose. One processing option is to reconstruct the posture of the human hand by using a geometric hand model [11]. In this hand model, each hand

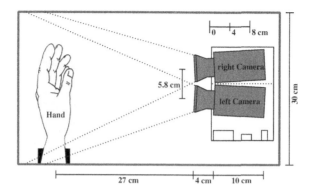

Fig. 1. This figure shows the hand-box interface, a contact-free visual recognition system for 3d hand postures

posture is parameterized by 20 angles (for each finger 3 flexion angles for the finger segments and one angle to describe abduction/adduction). A Parameterized Self-Organizing Map (PSOM)[14] is used to perform the inverse kinematics, i. e. to determine the joint angles from the fingertip positions detected in the stereo vision front-end. However, the 20 degrees of freedom (DOF) of the model are not fully independent. The effective number of DOFs is only 15 since the last two joints of each finger are driven by the same tendon and their angles are roughly proportional [11]. A simpler and more direct possibility is to use the identified fingertip positions directly as parameters.

2.2   Sound Rendering

For sound synthesis we use the program *Csound* [1]. Csound is a software sound synthesis system which allows to define instruments in an orchestra description and to control these instruments by so-called score events. Since score events can be processed by Csound in real-time mode and thus faster than our current frame rate of 11 Hz, it is well suited to our application. The system works on a Linux operating system. An overview of the system components and the data flow is given in Figure 2. Interactive data visualization is not yet integrated but planned as a future extension.

3   Real-Time Control of Soundscapes Using Hand Postures

3.1   Hand Postures as a "Virtual Instrument"

To demonstrate that the system allows usable real-time manipulation of sound, we first implemented a simple application which does not yet invoke external data, but just controls a "soundscape" by hand postures. For the control, we

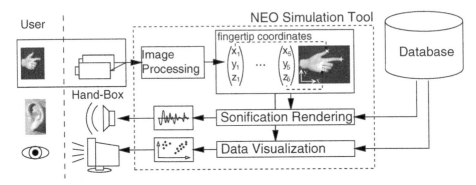

Fig. 2. Sketch of the System. The system is controlled by mouse actions and keyboard events. Hand postures of one hand are used to control parameters of the sound synthesis in real-time

use the (x, y) coordinates of the fingertips from the mono camera system. The coordinate frame is shown in Figure 2. For sound synthesis, we used 4 oscillators whose amplitudes are modulated with a low frequency. The sound signal thus is given by

$$s(t) = g \cdot \left(\sum_{i=1}^{4} s_i(2\pi f_c^i t) \cdot \left(1 - \sin(2\pi f_m^i t)\right) \right) \tag{1}$$

where $s_i(\cdot)$ are periodic functions with period 2π, f_c^i is the fundamental frequency and f_m^i is the modulation frequency of the ith tone. This sound model has 9 control parameters: the gain g and the frequencies f_c^i, f_m^i, $i = 1, \ldots 4$, which are to be controlled by the hand postures. This is done by assigning the gain g to the thumb and each of the four oscillators to one of the other fingers. The gain is controlled from the thumb's y-coordinate as this is easier to control than the x-coordinate. For the fingers we chose the x-coordinate to control the modulation frequencies between 0.4 Hz (for flexed fingers) and 2 Hz (for straight fingers). Manipulating the fundamental frequencies in the same manner from the fingertips' y-coordinates results in mostly disharmonic sounds. Therefore we assigned a frequency to each finger and used the y-coordinate of each fingertip to switch between two tones. The tones were chosen so that all combinations are musically pleasant:

finger	pitch for $y < 0$	pitch for $y > 0$
index finger	C6	→ D6
middle finger	E5	→ F5
ring finger	G4	no change
little finger	C4	no change

The finger-specific sound signals and the mapping from hand posture parameters to sound parameters is illustrated in Figure 3. For the mapping from a fingertip

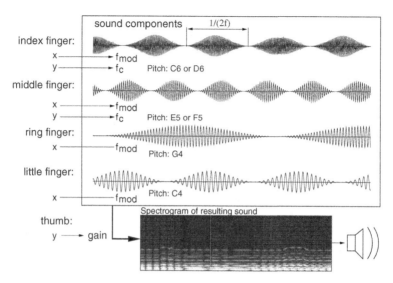

Fig. 3. The plot shows the amplitude modulated sine signals used for the virtual instrument. The superposition leads to a complex sound signal whose spectrogram is shown below the signal plots

coordinate x to control a value $v(x)$ we used a linear mapping given by

$$v(x) = v_{min} + (v_{max} - v_{min})\frac{x - x_{min}}{x_{max} - x_{min}}. \qquad (2)$$

Another possible control mode is differential operation: the fingertip coordinates are taken as increments resp. decrements for a control value. While differential control may be suited for some tasks, it seemed to be more intuitive here to use as a direct mapping control.

This very simple interface allows the real-time synthesis of "soundscapes" which have a relaxing effect and can be thought of as playing a simple instrument just by moving the hand. With this kind of interface, dynamic ambient sounds can be created by sound artists. A sound example is found on our website [8].

3.2 Formant Space Control

As a second example, we allow the fingers to control different formants in an articulatory speech synthesis model for vowels by using *granular synthesis* [?]. In human voice, each vowel has a unique signature of formants, which can also be measured from recorded speech data. In this application, we provide the user with an interface to simultaneously control several formants and to explore the vowel sound space and transitions within this space. Experiences have shown that the flexion of fingers is easier to control than the abduction/adduction. For this reason we use only the fingertip's x-coordinates to drive synthesis parameters in this example. Each formant has three control parameters: the formant center frequency, the bandwidth and the gain. Synthesis is performed by applying a resonance filter on an excitation signal for each formant and mixing of the obtained signals which can be expressed as

$$s(t) = g_1 H(f_1^c, \Delta f_1) s_0(t) + g_2 H(f_2^c, \Delta f_2) s_0(t) \quad , \tag{3}$$

where $s_0(t)$ is a source signal consisting of a series of grain pulses and H describes a one-pole resonance filter with given center frequency f^c and bandwidth Δf. The control parameters of the synthesis model are (g_1, f_c^1, g_2, f_c^2) . For the bandwidth Δf_i we selected constant values to reduce the complexity of the control task. For the mapping from fingers to parameters we assigned one formant to two neighboring fingers: fore finger and middle finger control the 1st formant, ring finger and pinkie control the 2nd formant. The thumb is used to control the pulse frequency of $s_0(t)$ which determines the fundamental pitch of the voice. Figure 4 illustrates the mapping and a spectrogram of an example sound. With this gestural interface, complex vowel transitions can be created and controlled after a short time of practice. A sound example can be found on the website [8].

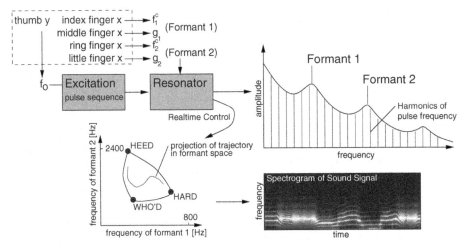

Fig. 4. This figure shows an idealized spectrum envelope for a vowel sound rendered by granular synthesis. The spectral envelope peaks are called formants and their center frequency and gain is driven by gestural control. The formant space plot of the first two formant frequencies (left) shows that vowels between "heed", "who'd" and "hard" can be synthesized by controlling two formants

4 Hand Postures for Data Mining Tasks

4.1 Using Hand Postures to Control Parameter Mapping Sonifications

The most common technique for rendering auditory data presentations is Parameter Mapping [6,12]. Parameter Mapping composes a sonification for a dataset by superimposing sound events (tones) whose acoustic attributes (e.g. onset, pitch, timbre, duration, amplitude, brightness) are driven by variable values of the dataset as illustrated in Figure 5. A typical mapping from data values to attribute values is the linear mapping as described in equation (2). Whereas the variable ranges are determined by the given dataset, the attribute ranges and the mapping have to be chosen be the user. We propose to apply gestural control to manipulate the attribute ranges as follows: four attributes are assigned to the four fingers. Only the x-coordinates of the fingertips are used to manipulate the ranges. The hand posture either controls the center values v_i or range r_i of the attributes. Toggling between these modes is done by pressing a key on the keyboard with the other hand. Holding a mute key on the keyboard allows to move the fingers without an effect on the attribute range. Assume that the parameters r_i are chosen to be controlled. Releasing the mute key, the actual fingertip coordinates are stored as reference coordinates $\hat{x}_j$, $j = 1, \ldots, 4$ and the actual range settings are stored as well as $\hat{r}_j$, $j = 1, \ldots, 4$. By changing the hand posture, new attribute ranges r_j are computed by

$$r_j = \hat{r}_j \exp\left(\frac{x - \hat{x}_j}{\sigma}\right) \quad \forall \; j = 1, \ldots, 4 \;, \tag{4}$$

where σ controls the sensitivity of the gestural interface. During such parameter tuning, the parameter mapping sonification is repeated in a cycle.

So far, the thumb is completely unused. It can be either used to drive a 5th attribute. However, flexion of the thumb mainly affects the y-coordinate so that we propose to use this instead of the x-coordinate for parameter control. Another

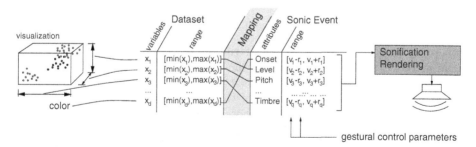

Fig. 5. Illustration of Parameter Mapping Sonification. Variable ranges are determined by the dataset at hand. Attribute ranges and the mapping are controlled by the user's hand postures

possibility is to use the flexion of the thumb as a trigger signal, e.g. to start/stop sonification or to permute the mapping. A mapping permutation here means to associate other dataset variables with the sound attributes under control.

4.2   Using Hand Postures to Amplify Contrast in Sonifications

The previous approach demonstrated gestural control for interactive optimization of a parameter mapping. The following interface modifies this strategy so that the user may find a mapping from data to sound such that the perceptual differences (or contrast) between the sounds are maximized for two datasets under consideration. The practical use for such an interface may be motivated with an example: assume a medical dataset with measurements of cell descriptions from breast cancer tumors is given. Consider that two groups of data records are available, one that contains measurements of benign tumors, the other containing the same variables for malignant tumors. Finding a mapping such that data samples from the two datasets sound different would allow a doctor to classify a new measurement by comparing its sound with the sound of the available data.

Finding a mapping from high-dimensional data to a perceptual space which optimally differentiates some property of interest is a rather old idea [3]. In projection pursuit, data is projected onto a linear subspace whose axes are changed to maximize deviation from a Gaussian distribution in the subspace [13]. While in visualization the criterion can be formalized by a structure valuation index, auditory perception is much less well understood. For that reason, we propose to allow a human listener to evaluate and modify the mapping.

Practically, we propose to play an equally sized subset of data points from each group in turn. The sonification for the groups should not last longer than about 1 sec, so that the effect of the user's control can be perceived in short time.

The user has to choose q acoustic attributes (e.g. pitch, amplitude, duration, brightness) and to specify their ranges r_j $j = 1, \ldots, q$. As an initial mapping, the principal axis of the joined datasets (including all records from both datasets) are taken as row vectors of a transformation matrix $\mathbf{A}$. The acoustic attributes for the parameter mapping of a data point x are then computed from

$$a = a_0 + \mathbf{SRA}x \qquad (5)$$

where a_0 holds center values for the attributes and $\mathbf{R}$ is a q-dimensional orthogonal matrix. The scaling is done with the diagonal matrix $\mathbf{S}$ by setting $S_{ii} = r_i$. As starting values, a unit matrix $\mathbf{R}_0 = \delta_{ij}$ is taken.

R is now updated at each cycle of the sonification by rotating it within a 2d-subspace. Assume that $q = 3$ is chosen. Then rotations around the x_0-axis with ϕ_0, around the x_1 axis with ϕ_1 and around the x_2-axis with ϕ_2 are possible. The rotation angles ϕ_0, ϕ_1, ϕ_2 are now controlled by the fingertip's x-coordinates of three fingers in the same manner as the value ranges are controlled in the previous section. As an alternative control interface, the orientation of the hand itself could be used to control the projection axis of up to three attribute axis within a selected 3d subspace of the dataset. Such a one-to-one control would be very intuitive. With our current system, however, it is more difficult to realize as the system is trained with data for a standard orientation of the palm.

The implementation of this application is still proceeding and the reader is referred to the web site [8] where we report our experiences with this gestural interfaces and sound examples.

5 Conclusion

Hand postures provide an attractive interface which shows great promise for the control of dynamic data displays. In this paper, an interface for the use of hand postures to control auditory displays has been presented. This research is in a rather early stage. However, we think, that it shows a very valuable extention of human-computer interfaces for multi-parameter control tasks. This paper describes our system and presents first examples of real-time gestural control of sound. Our experiences with the gestural control of synthesis parameters within the virtual instrument and the formant space exploration have been that intentional manipulation of the sound by hand postures is possible and after some training the user's performance improves. For application of our system for data mining tasks, two approaches have been presented which aim at controlling Parameter Mapping Sonifications. The implementation and testing of these techniques with our system are the next steps within this work.

References

1. R. Boulanger. *The Csound Book*. MIT Press, 2000. 309
2. U. M. Fayyad et al., editor. *Advances in Knowledge Discovery and Data Mining*. MIT Press, 1996. 307
3. Friedman, J. H. and Tukey, J. W. A projection pursuit algorithm for exploratory data analysis. *IEEE Transactions on Computers*, 23:881–890, 1974. 314
4. T. Hermann and H. Ritter. Listen to your Data: Model-Based Sonification for Data Analysis. In M. R. Syed, editor, *Advances in intelligent computing and mulimedia systems*. Int. Inst. for Advanced Studies in System Research and Cybernetics, 1999.
5. I. A. Kapandji. *The physiology of the joints: Upper limbs*. Churchill Livingstone, 1982. 308
6. G. Kramer, editor. *Auditory Display - Sonification, Audification, and Auditory Interfaces*. Addison-Wesley, 1994. 307, 313
7. A. G. E. Mulder and S. S. Fels. Sound Sculpting: Manipulating sound through virtual sculpting. In *Proceedings of the 1998 Western Computer Graphics Symposium*, pages 15–23, 1998.
   http://www.cs.sfu.ca/~amulder/personal/vmi/publist.html. 308
8. C. Nölker and T. Hermann. GREFIT: Visual recognition of hand postures. http://www.techfak.uni-bielefeld.de/~claudia/vishand.html, 2001. 311, 312, 315

9. H. Ritter, "The graphical simulation toolkit NEO/Nst,"
   http://www.techfak.uni-bielefeld.de/ags/ni/projects/
   simulation_and_visual/neo/neo_e.html. 308

10. C. Nölker and H. Ritter. GREFIT: Visual recognition of hand postures. In
    A. Braffort, R. Gherbi, S. Gibet, J. Richardson, and D. Teil, editors, *Gesture-Based Communication in Human-Computer Interaction: Proc. International Gesture Workshop, GW '99, France*, pages 61–72. Springer Verlag, LNAI 1739,
    http://www.techfak.uni-bielefeld.de/techfak/ags/ni/publicfr_99d.htm,
    1999. 308

11. C. Nölker and H. Ritter. Parametrized SOMs for hand posture reconstruction. In
    S.-I. Amari, C. L. Giles, M. Gori, and V. Piuri, editors, *Proceedings IJCNN'2000. Neural Computing: New Challenges and Perspectives for the New Millennium*,
    pages IV-139–144, 2000. 308, 309

12. C. Scaletti, "Sound synthesis algorithms for auditory data representations," in
    Auditory Display, G. Kramer, Ed. 1994, Addison-Wesley. 313

13. B. D. Ripley, Pattern Pecognition and Neural Networks, Cambridge University
    Press, 1996. 314

14. J. Walter, C. Nölker and H. Ritter. The PSOM Algorithm and Applications.
    Proc. Symposium Neural Computation 2000, pp. 758–764, Berlin, 2000 309

Comparison of Feedforward (TDRBF) and Generative (TDRGBN) Network for Gesture Based Control

Helen Vassilakis, A. Jonathan Howell, and Hilary Buxton

School of Cognitive and Computing Sciences, University of Sussex
Brighton, BN1 9QH, UK.
{helenv,jonh,hilaryb}@cogs.susx.ac.uk

Abstract. In Visually Mediated Interaction (VMI) there is a range of tasks that need to be supported (face and gesture recognition, camera controlled by gestures, visual interaction etc). These tasks vary in complexity. Generative and self-organising models may offer strong advantages over feedforward ones in cases where a higher degree of generalization is needed. They have the ability to model the density function that generates the data, and this gives the potential of "understanding" the gesture independent from the individual differences on the performance of a gesture. This paper presents a comparison between a feedforward network (RBFN) and a generative one (RGBN) both extended in a time-delay version.

1   Introduction

Recently, we have been developing computationally simple view-based approaches to learning for recognising identity, pose and gestures, and learning to map these to users intentions. We have also developed new behavioural models based upon head and body movements [7,9]. These models not only enable analysis of behaviour, but also allow us to perform intentional tracking through improved prediction of expected motion. Thus, we have been able to integrate an interactive, connectionist system incorporating face detection and tracking with active camera feedback control, recognition of face [4,6], pose [7], expression [2] and gesture [5,7], plus active intentional control of attention [8,9]. The users have minimal awareness of the system, which can identify them if appropriate, then estimate and predict essential body parameters such as head pose, sitting, talking, pointing and waving gestures. An essential task in this system is mapping a vector of head pose and gestures from a group of interacting users onto the camera control by predicting who is the current focus of attention. This gesture-based control is a task where we might expect unsupervised learning in a generative network to confer advantages.

There are now many view-based models of perceptual tasks in the literature, split into two major groups: *pure feedforward network models* and *network models with feedback connections*. The term feedforward describes the networks where the information signals flow towards one direction. Feedforward models include the well-

I. Wachsmuth and T. Sowa (Eds.): GW 2001, LNAI 2298, pp. 317-321, 2002.

known Radial Basis Function (RBF) network (see [1]) that has been used for many visual tasks, as discussed above. In feedback systems information flows both directions. Feedback models include architectures using a *generative* approach, according which the network tries to model the density function that is assumed to have generated the data. In generative feedback models the system is essentially using top-down knowledge to create a neural representation and measuring discrepancies with bottom-up representation from the visual input to adjust the output to most likely interpretation. The Rectified Gaussian Belief Network (RGBN) (see [3]) is an example of this class that we have tried for gesture-based control. The human visual system is capable of providing this kind of feedback control in 1-200ms, and we would like similar performance in our artificial systems to avoid disruptive delays. This puts strong constraints on the type of feedback processing we can afford, as there is no time for costly online iterative convergence. However, it is worth considering if pre-learnt generative networks can improve performance over more standard feedforward networks as above.

2   Network Models

The network models used in this paper (RBFN and RGBN) are applied to static visual tasks, such as image recognition. While RBF models are commonly used, the potential of the RGBN has not been explored in real applications. It is possible that both models can be extended to time-delay versions, to satisfy the needs of gesture recognition problems. In this paper we compare the two on a gesture recognition problem. This section describes RBFN and RGBN in their static form and explains how they can be extended for temporal tasks use, such as gesture recognition.

2.1   RBF Network

RBFN uses basis functions to form a mapping from a d-dimensional input space to a one-dimensional target space (see Fig 1a). The output of a unit in such a network is given by the expression:

$$y_k(x) = \sum_{j=1}^{M} w_{kj}\, \varphi_j(x) + w_{ko}, \tag{1}$$

where M is the number of basis functions, w_{kj} the weights, w_{ko} the bias and φ_j the basis functions. An example of such a basis function is the Gaussian:

$$\varphi_j(x) = \exp\left(-\frac{\|x - \mu_j\|}{2\sigma_j^2}\right), \tag{2}$$

x is the input vector and μ_j the vector determining the centre of the basis function.

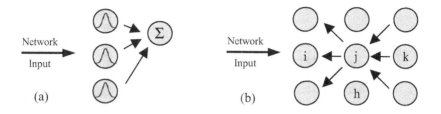

Fig. 1. RBFN (a) and RGBN (b)

2.2   Rectified Gaussian Belief Network

RGBN (see Fig 1b) consists of units with values that are either zero or positive real. This is the only non-linearity in the system. The values of the units follow the Gaussian distribution with a standard deviation σ_j and a mean $\hat{y}_j$ where:

$$\hat{y}_j = g_{oj} + \sum_k \tilde{y}_k g_{kj} . \tag{3}$$

g_{oj} is the generative bias, $\tilde{y}_k$ the rectified values of the units k in the next layer and g_{kj} the weights. The energy of a unit is defined by the following equation:

$$E(y_j) = \frac{(y_j - \hat{y}_j)^2}{2\sigma_j} + \frac{\sum_i \left(y_i - \sum_h \tilde{y}_h g_{hi} \right)^2}{2\sigma_i} \tag{4}$$

where y_i are the states of the units at the previous layer and h an index over the units in the same layer as j, including j. In order to calculate the posterior probability distribution of the unit states it is necessary to perform Gibbs sampling, which is the main disadvantage of the model. Given these values we can use the online delta rule to adjust the weights.

2.3   Time Delay Version

Time-delay versions of the neural network models above were adapted to the problem of gesture based control. In order to deal with this task, we did a number of experiments under the purpose of identifying a head movement (right or left). To reduce the size of input data, a vectors v describing the gestures of the individuals were extracted from image frames (see Fig 2). These vectors were both simulated and extracted using Gabor filter analysis. It has been shown that 5 image frames are sufficient to identify a head moving gesture, i.e. 5 head pose positions with angles: (-34 ,-17 ,0 , 17,34) [7]. To train the network, a larger vector V that consists of 3 vectors v is used, while the other two vectors were used for testing. This vector becomes the input of the neural network and results are compared for feedforward and generative models.

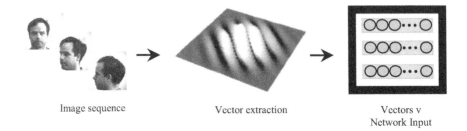

Image sequence   Vector extraction   Vectors v
                                                   Network Input

Fig. 2. Time delay version of RBFN and RGBN

3   Conclusions

In this paper we have addressed the problem of applying temporal data to naturally static network models. We have used the same technique to both feedforward and generative feedback models and showed that it is possible for both networks to deal with gesture recognition problems such as recognizing head movements.

Previous experiments [4,5,6,7,8] have shown that RBFNs are particularly efficient in simple tasks. They are simple in principle, can be easily extended to a time delay version and exhibit excellent performance in terms of both speed and error rates (c. 1%). This makes them highly suitable for real-time applications. However they are specific to the task and they assume *a priori* knowledge of the data. In contrast, generative models such as RGBN offer the advantage of self-organization; valuable in cases with little or no *a priori* knowledge over the data. RGBN can also be successfully expanded in a time delay version. Though robust to noise, their performance is not as good as that of RBFN. Experiments described in section 2.3 using RGBN demonstrated an average error of 20%, quite high for real applications. Additionally this kind of network is slow due to Gibbs sampling, which is used to calculate the *a posteriori* distribution of the network nodes. This fact results to simulation times that are orders of magnitude slower than the RBFN. Therefore applying it to real-time applications could pose real difficulties. The main advantage of self-organizing in combination with the generative nature is still valuable, and if performance issues could be solved, a feedback neural network would offer advantages that could lead into more intelligent models, able to understand and reproduce complicated data such as gestures.

References

1. D. J. Beymer and T. Poggio. Image representations for visual learning. Science, 272:1905-1909, 1996.
2. S. Duvdevani-Bar, S. Edelman, A. J. Howell, and H. Buxton. A similarity-based method for the generalization of face recognition over pose and expression. In IEEE Conference on Automatic Face & Gesture Recognition, Nara, Japan, 1998.

3. G. E. Hinton and Z. Ghahramani. Generative models for discovering sparse distributed representations. Phil. Transactions of Royal Society London, Series B, 352:1177-1190, 1997.

4. A. J. Howell and H. Buxton. Face recognition using radial basis function neural networks. In R. B. Fisher and E. Trucco, editors, British Machine Vision Conference, Edinburgh, 1996.

5. A. J. Howell and H. Buxton. Learning gestures for visually mediated interaction. In P. H. Lewis and M. S. Nixon, editors, British Machine Vision Conference, Southampton, 1998.

6. A. J. Howell and H. Buxton. Learning identity with radial basis function networks. Neurocomputing, 20:15-34, 1998.

7. A. J. Howell and H. Buxton. Towards visually mediated interaction using appearance-based models. In ECCV Workshop on Perception of Human Action, Freiburg, Germany, 1998.

8. A. J. Howell and H. Buxton. Face detection and attentional frames for visually mediated interaction. In IEEE Workshop on Human Motion, Austin, Texas, 2000.

9. J. Sherrah, S. Gong, A. J. Howell, and H. Buxton. Interpretation of group behaviour in visually mediated interaction. In International Conference on Pattern Recognition, ICPR'2000, Barcelona, Spain, 2000.

Author Index

Lecture Notes in Artificial Intelligence (LNAI)

Lecture Notes in Computer Science